*continued on back*

Comparative Statistical Inference

Comparative
Statistical
Inference

Second Edition

VIC BARNETT
Professor of Probability and Statistics
University of Sheffield, Sheffield, UK

1807 175 YEARS OF PUBLISHING 1982

JOHN WILEY & SONS
Chichester · New York · Brisbane · Toronto · Singapore

Library of Congress Cataloging in Publication Data;

Barnett, Vic
  Comparative statistical inference.
  (Wiley series in probability and mathematical statistics. Applied probability and statistics section)
Biliography: p.
  Includes index.
  1. Mathematical statistics. 2. Statistical decision. I. Title. II. Series.
QA276. B2848   1982   519.5′4   81–14806
ISBN 0 471 10076 5   AACR2

British Library Cataloguing in Publication Data:

Barnett, Vic
  Comparative statistical inference.—2nd ed.—
  (Wiley series in probabilty and mathematical statistics: applied probability and statistics section)
  1. Mathematical statistics
  I. Title
  519.5′4   QA276

  ISBN 0 471 10076 5

Typeset by Macmillan India Ltd., Bangalore
and printed by Page Bros. (Norwich) Ltd.

To Audrey, Katy and Emma

Preface

This century has seen a rapid development of a great variety of different approaches to statistical inference and decision-making. These may be divided broadly into three categories: the estimation and hypothesis testing theory of Fisher, Neyman, Pearson and others; Bayesian inferential procedures; and the decision theory approach originated by Wald.

Each approach is well documented, individually. Textbooks are available at various levels of mathematical sophistication or practical application, concerned in the main with one particular approach, but with often only a passing regard for the basic philosophical or conceptual aspects of that approach. From the mathematical and methodological viewpoint the different approaches are comprehensively and ably described. The vast amount of material in the professional journals augments this and also presents a detailed discussion of fundamental attitudes to the subject. But inevitably this discussion is expressed in a sophisticated form with few concessions to the uninitiated, is directed towards a professional audience aware of the basic ideas and acquainted with the relevant terminology, and again is often oriented to one particular approach. As such, the professional literature cannot (nor is it intended to) meet the needs of the student or practising statistician who may wish to study, at a fairly elementary level, the basic conceptual and interpretative distinctions between the different approaches, how they interrelate, what assumptions they are based on, and the practical implications of such distinctions. There appears to be no elementary treatment which surveys and contrasts the different approaches to statistical inference from this conceptual or philosophical viewpoint. This book on comparative statistical inference has been written in an attempt to fill this gap.

The aim of the book is modest; by providing a general cross-sectional view of the subject, it attempts to dispel some of the 'air of mystery' which must appear to the inexperienced statistician to surround the study of basic concepts in inference. In recognizing the inevitable arbitrary and personal elements which must be reflected in any attempt to construct a 'rational' theory for the way individuals react, or should react, in the face of uncertainty, he may be better able to understand why factional groupings have developed, why their members attach 'labels' to themselves and others, and why discussion so easily reaches a somewhat 'emotional' level. By stressing the interrelationships as well as the conceptual conflicts it is hoped that the different approaches may be viewed as a composite theory of inference, the different methods having separate relevance in different situations, depending on local circumstances. The book achieves its object substantially if it does no more than persuade the reader that he is not required to 'stand up and be counted' with those who have committed themselves to one particular attitude towards inference to the exclusion of all others.

The idea of the book originated from my experience, over several years, of running a lecture course on comparative statistical inference. The course was attended by final-year undergraduate, and first-year postgraduate, students in Mathematical Statistics at the University of Birmingham; it was introduced to augment their knowledge of the mathematics and techniques of different approaches to statistical inference and decision theory by presenting them with the spectrum of philosophical attitudes inherent in a comparison of the different approaches. Other universities offer similar courses and this book should provide useful collateral reading for such courses, as well as being a general treatment of comparative statistical inference for a wider audience.

This book is not intended as a comprehensive discussion of the mathematics or methodology of any particular approach, nor as an authoritative history of the development of statistical consciousness. Some historical comment is included to add interest, and the mathematics and methodology are developed to the stage where cogent comparison is possible. This comment and development, however, remains subservient to the prime objective of a comparison of different philosophical and conceptual attitudes in statistical inference.

No detailed mathematical proofs are given, and the treatment assumes only a knowledge of elementary calculus and algebra (perhaps to first-year university level) and an acquaintance with the elements of the theory of probability and random variables. Some familiarity with specific methods of inference and decision theory would be an advantage.

The first two chapters of the book are introductory. Preliminary ideas of inference and decision-making are presented in Chapter 1, and applied in Chapter 2 to the informal construction of various inferential techniques in the context of a practical example. Chapter 3 traces the range of different definitions and interpretations of the probability concept which underlie the different approaches to statistical inference and decision-making; Chapter 4 outlines utility theory and its implications for general decision-making. In Chapters 5 to 7 specific approaches are introduced and developed with a general distinction drawn between *classical* inference on the Neyman–Pearson approach, *Bayesian* methods and *Decision Theory*. Particular attention is given to the nature and importance of the basic concepts (probability, utility, likelihood, sufficiency, conjugacy, admissibility, etc.) both within and between the different approaches. The final chapters (8 and 9) present a sketch of some alternative attitudes, and some brief concluding remarks, respectively.

A subject and author index and bibliography are included, and textual references to *books* relate, by author's name and date of publication, to the bibliography. References to *papers* in the professional journals bear the author's name and an index number, e.g. Savage[7], and relate to the list of references at the end of the *current* chapter. It is hoped that readers may be stimulated to delve deeper into the various topics which are presented. To assist them to do so, particularly when the book is used as collateral reading for a lecture course on comparative inference, certain references in the list at the end of each chapter are marked with a dagger, e.g. †3. Cox, D. R. . . . . This indicates material which seems particularly suitable as the basis for extended study or discussion of the subject matter of the current chapter. The marked references have been chosen, in the main, on the basis of providing a broad, non-detailed, extension or reappraisal of relevant material—often surveying, interrelating or comparing the different approaches. The dagger (†) is in no sense intended as a mark of distinction in terms of the merits of different authors' contributions.

It is necessary to explain one or two particular points of notation at the outset. Frequently we will be concerned with data arising from an assumed parametric model. The data are denoted x, the parameter, θ. Usually no indication is given (or is needed within the general treatment) of dimensionality. Thus, x may be a single observation of a univariate random variable, or of a multivariate random variable, or may be a sample of independent (univariate or multivariate) observations. Similarly, θ may have one, or many, components. In the same spirit, X denotes the general random variable of which x is a realization. The sample space and parameter space are denoted by $\mathscr{X}$ and Ω, respectively. The probability mechanism governing the occurrence of x is represented by the function $p_\theta(x)$, with a correspondingly broad interpretation as a probability or probability density, or where the emphasis demands it as a likelihood. To avoid the unnecessary complication of distinguishing between discrete and continuous variables, and to maintain the presence of $p_\theta(x)$ as a central component in mathematical expressions, an individual style is adopted to denote integration or summation. The expressions

$$\int_{\mathscr{X}} h(x, \theta) \quad \text{or} \quad \int_{\Omega} h(x, \theta)$$

are used to represent the appropriate single, or multiple, integrals or sums of some function $h(x, \theta)$ over the range of variation of x or θ, respectively. The subscript $\mathscr{X}$, or Ω, will also be attached to the expectation operator $E(\cdot)$, to indicate the space over which the expectation is calculated. For example,

$$E_{\mathscr{X}}[g(X)] = \int_{\mathscr{X}} g(x) p_\theta(x)$$

is the expected value of the function $g(X)$ of the random variable, X. (The subscripts $\mathscr{X}$, or Ω, will be omitted if the appropriate space is obvious in the context of the discussion.)

The use of the usual integral sign for this purpose may offend the purist. However, it seems more appropriate to take such a liberty for the sake of the intuitive simplicity of the notation, than to introduce some new symbol and require the reader constantly to remind himself of its meaning.

In particular examples where it is important to distinguish the structure of x or θ the more conventional notation for integrals, sums, density functions, and so on, will be explained and used.

It is a pleasure to acknowledge my indebtedness to all those who have contributed, in various ways, to the production of this book. Its existence owes much to the example and stimulus of colleagues. My debt to the vast literature on the subject is self-evident, from the extent to which I have referred to other writers who have so often expressed ideas far better than I can. I am grateful also to my students, whose comments and enquiries have directed my thoughts to what I feel is a reasonable mode of presentation of the material. My thanks are especially due to a few friends who have given their time to read and comment on sections of the book; in particular David Kendall, Toby Lewis, Dennis Lindley and Robin Plackett. Toby Lewis has been a constant source of help and encouragement; I am very grateful to him. Every effort has been made to ensure that factual details are correct and that historical and interpretative attribution is fair and just; also to avoid any implicit bias towards a particular approach to the subject. It is not easy, however, to assess these matters objectively, and any errors, omissions or bias are unintentional and my responsibility alone.

October, 1972   VIC BARNETT

Preface to Second Edition

Much has happened in the field of inference and decision-making over the decade since the first edition of this book was published. The preparation of a second edition presents a valuable opportunity to provide more detailed treatment of some topics, to offer some discussion of new emphases, techniques and even whole approaches to inference, and to reflect changes of basic attitude to the subject.

In classical inference specific attention is given to multi-parameter problems, and to notions of ancillarity and conditional inference. The revitalization of the distinction between hypothesis tests and 'pure significance tests' is discussed and interpreted. The treatment of the role of likelihood is broadened to encompass comment on modified forms of likelihood (marginal, conditional, etc.), and to expand on the significance of the likelihood principle in the various approaches (particularly its relationship to the concept of coherency in Bayesian inference). Greater attention is given to practical ways of representing and assessing subjective probabilities and utilities, and to work on the application of Bayesian methods. The method of Bayesian prediction is outlined. Two new approaches are briefly described: pivotal inference and plausibility inference.

The above topics represent some of the additions in this second edition. The book has been thoroughly revised throughout to reflect changes of substance and attitude in the intervening period. In particular the reference lists at the end of each chapter, and the bibliography, are much more extensive and contain relevant contributions to the literature up to the time of the revision.

It must be stressed, however, that the overall aim of the book is unchanged. It aims to present and develop the various principles and methods of inference and decision-making to such a level that the reader can appreciate the basic characteristics, interrelationships and distinctions of the different approaches. Accordingly, detailed mathematical development or proof and comprehensive coverage of applications are eschewed (in text, and in references), in order not to cloud the objective of presenting in managable proportions a basic understanding of essential principle, philosophy, method, interpretation and interrelationship.

Sheffield, May 1981   VIC BARNETT

Acknowledgements

The author gratefully acknowledges the assistance of the following publishers who have granted permission for the reproduction of extracted material quoted in the text, sources being given in the References at the end of each chapter or in the Bibliography at the end of the book:

Academic Press, Inc., London.
George Allen & Unwin Ltd., London.
The Annals of Eugenics (Cambridge University Press, London).
Biometrika Trust, London.
Cambridge University Press, London.
Chelsea Publishing Co. Inc., New York.
The Clarendon Press, Oxford.
Gauthier-Villars, Paris.
Charles Griffin & Co. Ltd., London and High Wycombe.
Hafner Publishing Company, New York.
Holt, Rinehart and Winston of Canada Ltd., Toronto.
Institute of Actuaries, London.
The Institute of Mathematical Statistics, California (*The Annals of Mathematical Statistics*).
International Statistical Institute, The Hague.
The Macmillan Company, New York.
Massachusetts Institute of Technology, Cambridge, Mass.
Methuen & Co. Ltd., London.
The Regents of the University of California.
The Royal Society, London.
Royal Statistical Society, London.
Statistica Neerlandica, Utrecht.
The University of Adelaide.
University of Western Australia.

Contents

CHAPTER 1

Introduction: Statistical Inference and Decision-making

1.1 WHAT IS STATISTICS?

It is usual for a book to commence by defining and illustrating its subject matter. Books on Statistics are no exception to this rule, and reading at random from the introductory pages of books on this subject suggests two features frequently encountered in such preliminary definitions. They are often brief and superficial with the aim of merely 'setting the ball rolling', and this is expeditiously achieved by relying largely on the reader's personal interpretation and motivation. In addition, they tend to be directed implicitly towards the particular level and emphasis of the treatment being presented. Some examples illustrate this.

The transition from the early use of the word 'statistics' to describe 'figures relating to the state' to its current use as a label for a particular scientific discipline is still apparent in introductory remarks in elementary books. The stress is on the former or latter aspect depending on whether the book is more concerned with the collection and presentation of data ('descriptive statistics') or with statistical methods for analysis and interpretation of the data. In an introductory book on 'applied statistics', Neter, Wasserman and Whitmore (1978) state:

> . . . statistics refers to the methodology for the collection, presentation, and analysis of data, and for the uses of such data. (p. 1.)

When interest focuses on the formal (mathematical) derivation and detail of the statistical methods, definitions become more specific on the nature of the data to be analysed and on the presence of a chance mechanism operating in the situations which yield the data. Kendall and Stuart (1977) remark:

> Statistics is the branch of scientific method which deals with the data obtained by counting or measuring the properties of populations of natural phenomena. In this definition 'natural phenomena' includes all the happenings of the external world, whether human or not. (p. 2.)

1

Cox and Hinkley (1974) introduce at the outset the idea of indeterminateness in saying:

> Statistical methods of analysis are intended to aid the interpretation of data that are subject to appreciable haphazard variability. (p. 1.)

Fraser (1976) is more formal:

> *Statistical theory* . . . builds on the use of probability to describe variation . . . . (p. 2.)

Definitions of this type are general enough to place little constraint on the subsequent development of the statistical methods or theory being presented. However, many examples may be found where the preliminary definition of the subject matter reflects a particular philosophical or conceptual emphasis in the later material. They may imply a concentration on a single interpretation of the probability concept, or a particular attitude to what constitutes relevant information for statistical study and to how it should be processed. The *frequency* concept of probability is explicit in the comments by Hoel (1971):

> . . . statistics is the study of how to deal with data by means of probability models. It grew out of methods for treating data that were obtained by some repetitive operation . . . . the emphasis in this book will be on the repetitive type situation. . . . the statistician looks on probability as an idealization of the proportion of times that a certain result will occur in repeated trials of an experiment . . . . (pp. 1, 2.)

Chernoff and Moses (1959), in an introductory text on *decision theory*, are dissatisfied with a definition which places emphasis on 'data handling':

> Years ago a statistician might have claimed that statistics deals with the processing of data. . . . to-day's statistician will be more likely to say that statistics is concerned with decision making in the face of uncertainty. (p. 1.)

In an elementary treatment of *Bayesian statistical inference*, Lindley (1965) sees statistics as the study of 'how degrees of belief are altered by data'. Savage (1962) stresses a personalistic function of the subject:

> By [statistical] inference I mean how we find things out—whether with a view to using the new knowledge as a basis for explicit action or not—and how it comes to pass that we often acquire practically identical opinions in the light of evidence. (p. 11.)

These few illustrations are not intended to be comprehensive or even representative, nor are they presented in any spirit of criticism. They merely serve to demonstrate two possible purposes behind an introductory definition of statistics. Definitions of the general type, often deliberately cursory or incomplete, serve ably to motivate a mathematical or methodological treatment of the subject at any level and from (almost) any viewpoint. In cases where the definitions are more specific, more personal, they provide an indication of the emphasis and viewpoint adopted in the subsequent development. For our present needs they underline a feature of the study of statistics which is basic to the purpose of this book: that there are a variety of aspects of the subject in which there is room for discussion and individual viewpoints. Different attitudes to

(i) what is meant by *probability*,

(ii) what constitutes *relevant information* for the application of statistical methods,

(iii) whether or not any limitations need to be placed on the areas of human activity amenable to statistical analysis,

will all inevitably colour and influence the construction of a *Theory of Statistics*. The object of this book is to examine the fundamental nature of statistical theory and practice, by a comparative study of the different philosophical, conceptual and attitudinal (sometimes personal) 'approaches' to the subject. To achieve this it will be necessary to consider basic concepts in detail, and to develop the mathematics and methods associated with the different approaches to the stage where detailed comparison is possible.

For these needs, however, we cannot be content with either a superficial or an idiosyncratic definition of statistics. To commence with an 'emotionally charged' definition is to defeat at the outset the aim of presenting a fair cross-section of views and attitudes. In a sense no definition is needed, since the book as a whole may be regarded as an attempt to provide such a definition. But we must start somewhere, and the best course is to construct a preliminary description of the purpose and function of statistical study which is, on the one hand, general enough to accommodate the widely differing philosophical and conceptual views which exist, and at the same time specific enough to mark out the basic components of *any* theory of statistics. This is no easy task; in attempting to be 'impartially specific' there is the risk that we end up by saying nothing of value. But better to say too little at this stage than too much. The gaps can be filled in as the book progresses.

With this attitude we provisionally define statistics as *the study of how information should be employed to reflect on, and give guidance for action in, a practical situation involving uncertainty.*

This definition contains a variety of ingredients which require fuller description. What is meant by 'a practical situation involving uncertainty'? What constitutes 'information'? What is the implied distinction between the 'reflection' and 'action guidance' function of statistics? The following sections consider these points in more detail.

1.2 PROBABILITY MODELS

In amplifying this definition the natural starting point is to consider what is meant by 'a practical situation involving uncertainty'. We have in mind that circumstances exist, or have been contrived through experimentation, in which different possible outcomes may arise which are of concern to us as the observer, or experimenter. The essential feature of the situation is that there is more than one possible outcome and that the actual outcome is unknown to us in advance: it is *indeterminate*. Our interest may be in knowing what that outcome will be, or in deciding on a course of action relevant to, and affected by, that outcome. A doctor prescribes a drug for a patient—will it be successful in curing his patient? Should we decide to spend tomorrow at the beach—when the weather may or may not be fine, and the state of the weather will seriously affect our enjoyment of the exercise?

Any attempt to construct a theory to guide behaviour in such 'situations involving uncertainty' must depend on the construction of a formal (logical or mathematical) **model** of such situations. This requires the formulation of a concept of **probability**, and associated ideas of independence, randomness, etc., as a mechanism for distinguishing between the different outcomes in terms of their degree of uncertainty. We shall see later that, in response to the nature of the situation and depending on individual attitudes, a variety of philosophical interpretations of the probability concept can exist. These interpretations may colour the theory of statistics developed to deal with the situation, and it is therefore useful at this stage to indicate some of the broad distinctions which exist.

A hint of the dilemma is provided in the example of a doctor prescribing a drug. A simple model for this situation is one which specifies two possible outcomes, that the patient is cured or is not cured, and assigns a probability, p, to a cure, $1 - p$ to no-cure. But how are we to interpret the probability that the patient is cured by the drug? We might adopt the attitude that the patient is 'typical' or 'representative' of a larger population of patients with a similar physical condition who have been, or will be, treated with this drug. This leads to a *frequency*-based view of probability, in which p is related to the proportion of cures, or potential cures, in the larger population. Alternatively, we may prefer to regard the patient as an individual, whose personal physiological and psychological make-up will determine the success or failure of the drug in *his* case: he cannot usefully be regarded as a representative member of some larger population. After all, even if we knew that 80 per cent of the larger population are cured we still do not know whether or not this particular patient will be. If we reject his representativeness in a larger population the frequency concept of probability no longer has relevance, and some other interpretation is needed. One possibility is now to regard the probability of a cure as a measure of the doctor's (and patient's) 'degree-of-belief' in the success of the treatment. Here we are likely to need a *subjective* interpretation of the probability concept. In practice the doctor's decision to prescribe the drug is likely to stem from an informal combination of *frequency* and *subjective* considerations, aimed at an assessment, or 'estimate', of the value of *p* as a guide for action. We shall return to this question of alternative interpretations of probability, and their implications for the construction of statistical theories, in Chapter 3, where the contrasting ideas of *'classical'*, *frequency*, *logical* and *subjective* or *personal* concepts of probability are discussed and illustrated.

The model of the practical situation consists, essentially, of a statement of the set of possible outcomes and specification of the probabilistic mechanism governing the pattern of outcomes which might arise. Inevitably, the model is an idealization of the real situation. Its adequacy depends on how valid and appropriate are the (often simple) assumptions on which it is based. The fundamental concern of the statistician is to construct an *adequate* model, either as a description of the real situation (and there his interest rests) or to suggest a reasonable course of action relevant to that situation. The real-world situation becomes replaced by the model and any description, or action, relates to the model as a substitute for the situation. Some simple examples may clarify this:

(i) A radioactive substance emits α-particles. The substance is characterized by the rate at which it emits these particles. A Geiger counter is placed nearby and records the α-particles which bombard it. The usual assumptions of randomness and independence of occurrence of the α-particles lead to a probability model for this situation—in which the number of α-particles recorded in a fixed time interval, $(0, T)$, has a particular probability distribution, namely the Poisson distribution. This model is fully specified by a single *parameter*, its mean, which is proportional to the rate of emission, λ, of the α-particles.

Thus a fairly complex physical situation has been replaced by a simple probability model, in which the extent of our uncertainty is reduced to a single unknown quantity related directly to our prime interest in the physical situation itself. Methods of probability theory make it possible to deduce the pattern of results we would expect to observe *if a fully specified form of the model is appropriate*. Thus we can calculate, say, the probability of recording no α-particles in a five-second period. In reverse, by comparing the actual recordings on the Geiger counter with the deductions from the probability model we can attempt to both *validate the model* and *estimate the unknown parameter* (the mean).

(ii) A leather manufacturer is concerned with the effect of different methods of tanning on the quality of leather being produced for shoe uppers. Four different methods are to be compared, with the aim of determining whether they produce leathers of different qualities and as a guide to which method of tanning to use. The quality of the leather is measured by an appropriate physical characteristic, for example, tensile strength or flexibility (in reality, by many such characteristics). Several pelts are tanned by the different methods and samples of the resulting leathers tested to determine their tensile strengths (or flexibilities, etc.) as an indication of the relative merits of the different tanning processes.

Any statistical analysis of the data again rests on the construction of an appropriate model and on an assumption of its adequacy. Here the assumptions may be that the observed values of tensile strength (etc.), possibly after a suitable transformation, arise independently from normal distributions with constant variance, differing from one method of tanning to another in at most the values of their means. Thus the model embodies a great deal of structure, being unspecified only to the extent of the unknown values of a few parameters. Again the real situation has been replaced by the model, and the ideas of probability theory may be applied to deduce the characteristic behaviour of data arising from the model,

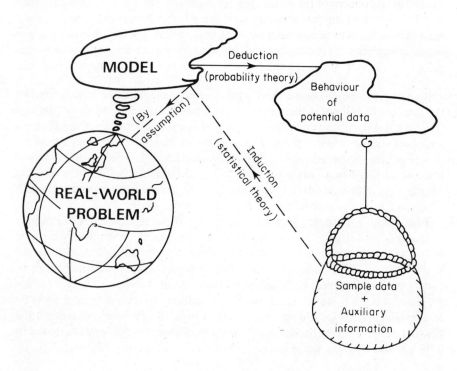

Figure 1.1

and hence by *assumption* from the real situation. The use of a statistical procedure, e.g. analysis of variance, again attempts to *reverse* this process, by using the observed data to reflect back on the model (both for validation and estimation purposes).

What is to be learnt from these examples? Firstly, the relationship between the practical situation, its probability model and the information it generates. Secondly, the roles played by a formal theory of probability and by statistical procedures and methods, in linking these components.

It is the practical situation which is of prime interest, but this is both inaccessible and intangible. An idealization of it is provided by the probability model, with the hope that it constitutes a valid representation. Being logically and mathematically expressed, this model is amenable to formal development. Logical deduction through the ideas of mathematical probability theory leads to a description of the probabilistic properties of data arising from the real situation—*on the assumption that the model is appropriate.* In this way probability acts as the communication channel or 'language' which links the situation (model) with the data: it provides a *deductive* link.

The theory of statistics is designed to reverse this process. It takes the real data *which has arisen* from the practical situation (perhaps fortuitously or through a designed experiment) and uses the data to validate a specific model, to make 'rational guesses' or 'estimates' of the numerical values of relevant parameters, or even to suggest a model in the first place. This reverse, *inductive*, process is possible only because the 'language' of probability theory is available to form the *deductive* link. Its aim is to enable inferences to be drawn about the model from the information provided by the sample data (or, more widely, by incorporating other forms of relevant information, to make more refined inferences) or to construct procedures to aid the taking of decisions relevant to the practical situation. What constitutes 'relevant' information, and the implied differences in the descriptive and decision-making functions of statistical theory, will be taken up in the next two sections.

The different components (practical situation, model, information) and the links between them (deductive or inductive) are represented diagrammatically in Figure 1.1.

1.3   RELEVANT INFORMATION

Returning to the definition given in §1.1, a second component which needs fuller discussion is the 'information' that is to be 'employed'.

In examples (i) and (ii) of the previous section, information took the specific form of 'realizations' of the practical situation: that is, observed outcomes from that situation arising from what are assumed to be independent repetitions of the situation under identical circumstances. This type of information will be termed **sample data**. Some writers would claim that it is only for information of this type, obtained in a repetitive situation (potentially at least, if not practically) that a concept of probability can be adequately defined, and a statistical theory developed. In von Mises' formalization of the frequency concept of probability

(1957; first published 1928, in German) and later in its application to statistics (1964) he says (quoting from his 1964 book, published posthumously):

> Probability . . . theory is the mathematical theory of . . . repetitive events. Certain classes of probability problems which deal with the analysis and interpretation of statistical enquiries are customarily designated as 'theory of statistics' . . .
>
> . . . we limit our scope, roughly speaking, to *a mathematical theory of repetitive events*. (p. 1.)

and later

> . . . if one talks of the probability that the two poems known as the *Iliad* and the *Odyssey* have the same author, no reference to a prolonged sequence of cases is possible and it hardly makes sense to assign a *numerical* value to such a conjecture. (pp. 13, 14.)

Admittedly the restriction of interest to repetitive situations is for the purpose of constructing a theory of probability. But as the 'language' of statistical analysis, there is the implication that only sample data, derived from repetitive situations, may be the subject of such analysis. If the 'language' contains no word for other forms of information how can we 'talk about' these other forms?

A similar attitude is expressed by Bartlett (1962) who, in an essay on the theory of statistics, says that statistics

> *is concerned with things we can count.* In so far as things, persons, are unique or ill defined, statistics are meaningless and statisticians silenced; in so far as things are similar and definite—so many male workers over 25, so many nuts and bolts made during December—they can be counted and new statistical facts are born. (p. 11.)

Again the emphasis is on the repetitive nature of the situation, and information is restricted to sample data.

Bartlett and von Mises are not alone in this attitude that statistical analysis must be based on the use of sample data evaluated through a frequency concept of probability. It is a view widely held and expressed, and acts as the corner-stone of a whole approach to statistics—what we shall call the **classical** approach, stemming from the work of Fisher, Neyman, E. S. Pearson and others. We shall see later how its exponents have considered in some detail alternative views of probability, and of what constitutes relevant information, only to reject them in the aim of producing what they regard as an 'objective' theory of statistics.

But is sample data really the only form of information relevant to a statistical study? Other information certainly exists. The engineer who is considering using a component offered by a particular supplier, for assembly in a new design of aeroplane, can test samples of this component and obtain data to aid his assessment of its suitability. But he may also have information on the reliability of the supplier in terms of a past record of that supplier's ability to provide reliable components in similar circumstances. He may also be able to assess the consequences of using a component which subsequently proves not to meet the required specifications. Both the earlier experience and the potential consequences are *relevant* to the decision on whether or not to use the component offered. Some current approaches to statistics are designed to incorporate such alternative types of information, and any attempt to survey the various approaches to the subject must therefore rest on a wider concept of information than that of just sample data. The broad subdivision of information into *three* categories (earlier experience, sample data, potential consequences), suggested by this example, is a useful one to pursue at this stage. Suppose we consider examples (i) and (ii) of the previous section in more detail.

The leather manufacturer in example (ii) is unlikely to regard his problem as a mathematical one concerning the means of a set of normal distributions. He is presumably interested in using any knowledge gained about the relative merits of the different tanning processes as a guide to future action. It may be that he hopes to adopt one of the processes for commercial production of leather. The mere superiority of one process in terms of the physical properties of the leather is unlikely to be sufficient justification for its adoption. To produce better leather is an advantage to him (psychologically, and in terms of the image of his product); but it may be more expensive to produce and hence need to be priced more highly, or it may involve large capital outlay on new equipment. Higher prices and large capital outlay can be disadvantageous in terms of profitability or liquidity. The decision as to what process to use is thus far more complex than merely a choice of that process which produces the 'best' leather.

The application of statistical methods in this problem illustrates how the object of the study may need to be extended beyond describing (through an appropriate probability model) how different outcomes arise to the wider aim of constructing a policy for action. Here we have an example of a situation where an assessment of the **consequences** of alternative actions is vital. These consequences must be quantified in some appropriate manner and the information they then provide may be critical to the choice of action. This second type of information is thus highly relevant, augments the information provided by the sample data, and demands the construction of statistical techniques through which it may be 'employed'.

Assessment of consequences, and their formal quantification through what is known as the concept of **utility**, must therefore form a part of any comparative study of different approaches to statistics. It is central to a particular approach known as **decision theory**. It is important, however, to recognize that information on consequences may be different in kind to that provided by sample data. Sample data is well defined, objective. Consequences may also lead to objective quantification (the costs of manufacture if a new tanning process is used) but they need not do so. How do we evaluate the effect of the new leather on the company's 'image'? In the same way as situations may demand a subjective interpretation of probability, so an assessment of consequences may involve *subjective* (personal) value judgements. In the sphere of human activity in particular it is often difficult to be objective. Bill cannot decide whether or not to marry Anna-Louise! It is intriguing to ponder what information he might seek to assist in this decision; also it is obvious that his assessment of the consequences will be very personal and difficult to quantify. Let us hope that he does not really need to use the formal machinery of decision theory in this dilemma—but the example is not so exaggerated! Much of the emphasis in utility theory and decision theory is on the construction of a rational model for human behaviour, in the sense of representing how individuals make (or should make) a choice from alternative possible actions in the face of uncertainty.

Even in the study of apparently objective situations we cannot escape the personal element. We saw this in the example of the leather manufacturer. Often a subjective assessment of consequences is forced on us by the sparseness of objective information. The leather manufacturer knows that to market a better quality, but more expensive, leather will price him out of a proportion of the market (he may even know that this currently accounts for 'about 10 per cent' of his volume of sales). He knows also that certain high quality shoe producers may in future be attracted to his products whereas they were not in the past (he might assess this as a 50 per cent chance of 30 per cent extra sales). He possesses, then, some measure of objective information, but to incorporate this factor in a statistical analysis he must *fully* specify the information in numerical terms and this can only be done by incorporating subjective or arbitrary values where his knowledge is incomplete. Alternatively, subjective and personal elements *must* arise when the context of the problem is 'personal' rather than 'global'—(Bill and Anna-Louise, for example). Some writers would argue that this is right and proper: that the individual is making decisions relevant only to *himself*, and that it is therefore *his* (subjective) assessment of consequences that constitutes relevant information. The difficulty of quantification still exists, however!

We find a third type of information potentially arising in the earlier example (i) concerning radioactive decomposition of a substance. The aim here may be merely to characterize the substance in terms of its rate of decay, λ, which is the mean number of α-particles, per second, recorded by the Geiger counter. This can be achieved by representing the situation in terms of the proposed Poisson model and by using the sample data alone to yield an estimate of λ as the only unknown parameter. The reliability or accuracy of the estimate will depend on the extent of the data and the method used to process the data. We can define ways of measuring this accuracy.

Suppose, however, that on chemical or physical grounds we know that the substance has affinities with other substances with known rates of decay. Its own rate of decay should not be too dissimilar to these others. As a result we may 'feel quite confident' that λ is in the range $0.45 < \lambda < 0.55$. This knowledge is a further form of relevant information which we would hope to combine with the sample data to conduct more refined estimation of λ. Such information is derived from *outside* the current situation. It may arise, as in this example, from the general accumulated knowledge of the investigator from other areas of experience: quite often from previous observation of similar situations. Information of this type is termed information *a priori*, or **prior information**. The particular branch of statistics designed to combine *prior information* with *sample data* is known as **Bayesian statistical inference**, but we shall see that prior information can play an important role in *decision theory* and has implications also in the *classical* approach to statistics.

In trying to incorporate prior information in a statistical analysis we again encounter the difficulty of quantifying it. It needs to be expressed in terms of *prior probability distributions*, e.g. in the radioactivity problem we might re-interpret the parameter λ as an observation from some distribution of a random variable Λ (relating perhaps to a 'super-situation' containing the present one). Prior information is now expressed through the distribution of Λ. But we need to be specific about this distribution—and it is hardly specified by feeling 'confident that $0.45 < \lambda < 0.55$'. Again subjective elements arise in expanding our limited knowledge to a complete specification; arbitrary and personal assessments might have to be introduced. For this reason, some statisticians have claimed that prior information (and an assessment of consequences) has no place in a formal theory of statistics—see Pearson's remarks quoted in the next section.

However, an amusing and compelling plea for the use of *subjective* prior information is given in an example by Savage[1].

After reminding us that subjective opinions often rightly influence the way in which a statistical investigation is *approached* (in terms of its design), he claims that subjective principles should also be allowed to influence the *analysis* of an experimental situation. To illustrate this in relation to the use of prior information, Savage presents three different experiments which he says have the 'same formal structure', where the conclusions based on traditional (*classical*) statistical method would be identical, but where the unemployed subjective prior information compels him (and, he is 'morally certain', anyone) to react quite differently in the three situations. The three experiments are as follows.

I. The famous tea-drinking lady described by R. A. Fisher (1966, pp. 11–25) claims to know if the milk, or tea, was poured first into the cup. For ten pairs of cups of tea, each pair having one of each type, she makes the correct diagnosis.

II. A music expert says he can distinguish a page of Haydn score from one of Mozart. He makes a correct assignation for ten pairs of pages.

III. A somewhat inebriated friend, at a party, says that he can predict the outcome when a coin is spun. It is spun ten times, and he is correct each time.

In each case the significance level is the same, 2^{-10}, for a one-tail test of significance. But Savage reacts quite differently in each case. Some 'old wives tales' have some substance; it seems that this *may* be so in the tea-making situation. An expert *should* be able to tell a page of Haydn from a page of Mozart; it is not surprising that he does so, particularly when he is so confident in his ability. Savage says that he does not believe in extra-sensory perception, and is un-impressed 'by this drunk's run of luck'.

The use of prior information in the Bayesian approach also has implications for the interpretation of the probability concept. As a general method of statistical analysis it must certainly be prepared to accommodate a subjective, degree-of-belief, view of probability; at least in the expression of the *results* of the analysis. (See §6.8.) For the prior information either a frequency, or a subjective, concept may be appropriate depending on circumstances. It is hard to see how anything but a degree-of-belief attitude can be applied to the 'confidence that $0.45 < \lambda < 0.55$' in the present example; although similar statements in different contexts may be legitimately interpreted on a frequency basis (see next chapter). Writers differ on the centrality of a subjective view of probability to the structure of Bayesian inference—most would claim it is the only appropriate view; at the other extreme a few would restrict the use of Bayesian methods to situations where prior distributions have a natural frequency interpretation. We shall need to return to this point in more detail, later in this chapter and in subsequent chapters.

We have now distinguished three general forms of information which may, depending on circumstances, be *relevant* to a statistical study. These can be viewed to some extent on a temporal basis—the prior information accumulated from *past* (or external) experience, the sample data arising from the *current* situation, and assessments of consequences referring to (potential) *future* action. The circumstances in which these different forms of information are relevant depend on a variety of factors, as we have seen in the examples above. The ways in which the information is quantified and utilized depend also on many circumstantial factors—in many cases involving subjective evaluation. But even if we judge particular forms of information to be *relevant* to our practical interest (and ignore for the moment the problems of quantification), utilization of the information depends on having available statistical procedures *specifically designed to incorporate the particular forms of information*. This is the 'down-to-earth' level at which we must also contrast and compare the different approaches to statistics. We must examine what different tools and concepts have been developed within the classical, decision-theoretic and Bayesian approaches for the processing of information? The *relevance* of information depends, then, not only on the practical situation but also on whether procedures exist to process it. In introducing decision theory, for example, Raiffa and Schlaifer (1961) remark that its only real novelty lies in the fact that it provides a formal mechanism for handling prior information and consequences. Such information is thus *relevant* to the decision theory approach.

1.4 STATISTICAL INFERENCE AND DECISION-MAKING

The definition of the object of statistics, given in §1.1, implies a distinction between a need to *describe* the practical situation, or a need to *prescribe action* in the context of that situation. Is this a valid, and useful, distinction? The examples of the leather manufacturer and the α-particles suggest that it might be. Let us now look at this question of *function* in more detail.

The interests of the meteorologist and the man-in-the-street in what today's weather will be are quite different. The meteorologist's interest is scientific: he is concerned with providing an informed *description* of the likely situation. The man-in-the-street is involved with using this description to aid him in his *actions*: to decide whether to take an umbrella to work or whether to go trout fishing, for which early morning rain is an advantage, etc.

The distinction between these two modes of interest in a statistical study, the *descriptive* and the *action guidance* functions, arises again and again. We have seen it in the context of examples (i) and (ii) in §1.2. It is a useful distinction to draw for the present purpose of contrasting different approaches to statistics. Any statistical procedure which utilizes information to obtain a *description* of the practical situation (through a probability model) is an *inferential* procedure—the study of such procedures will be termed **statistical inference**. A procedure with the wider aim of suggesting *action* to be taken in the practical situation, by processing information relevant to that situation, is a *decision-making* procedure—the study of such procedures, **statistical decision-making**.

This distinction has often been made in the literature. For instance, Smith[2] says:

> Statisticians often play down something which is obviously true, when it does not quite accord with their line of thought. An example is the statement that there is no difference between inference and decision problems.
>
> A decision problem means the choice between several possible courses of action: this will have observable consequences, which may be used to test its rightness. An inference concerns the degree of belief, which need not have any consequences, though it may. This makes it more difficult to come to agreement on questions of inference than on decisions. For example, the question 'Shall I eat this apple?' is a matter of decision, with possible highly satisfactory or uncomfortable outcomes. 'Is this apple green?' is a question of belief. Of course, the two problems must be closely related, even though they are distinct.

Cox[3] discusses this in greater detail.

A statistical inference carries us from observations to conclusions about the populations sampled. . . . Statistical inferences involve the data, a specification of the set of possible populations sampled and a question concerning the true populations. No consideration of losses [consequences] is usually involved directly [but] . . . may affect the question asked. . . . The theory of statistical decision deals with the action to take on the basis of statistical information. Decisions are based on not only the considerations listed for inferences, but also on an assessment of the losses resulting from wrong decisions, and on prior information, as well as, of course, on a specification of·the set of possible decisions.

Here we see the difference of *function* affecting the forms of information regarded as relevant. (Elsewhere in print Cox acknowledges the occasional relevance of prior information to the *inference* problem as well as to decision-making.)

Lindley (1965*b*, p. xii) also makes this distinction (inference—'how data influences beliefs'; decision-making—'how the beliefs can be used'). He later remarks that the decision-making problem is an extension of the inference problem; but that inference is fundamental to decision-making.

. . . the inference problem is basic to any decision problem, because the latter can only be solved when the knowledge of [the probability model] . . . is completely specified. . . . The person making the inference need not have any particular decision problem in mind. The scientist in his laboratory does not consider the decisions that may subsequently have to be made concerning his discoveries. His task is to describe accurately what is known about the parameters in question. (pp. 66–67).

At a later stage, the more detailed discussions of this issue, such as that by Lindley[4], will be relevant.

So we see decision-making extending the aims of inference: the descriptive function of inference being expanded to suggest rules for behaviour. As a result we may expect decision-making procedures to range more widely than inferential procedures over the types of information they are designed to incorporate and process. Broadly speaking, inference utilizes sample data and, perhaps, other information with a bearing on the description of the practical situation, e.g. prior information. Decision-making augments the inferential knowledge of the situation by also incorporating assessments of consequences.

Consider the case of a car manufacturer who possesses a variety of information on components of his vehicles. He may wish to use this information in various ways. Sample data and prior information may be used to describe the dimensions of a particular component for quoting in a 'reference manual'. But this description needs to be augmented by a knowledge of the effects and implications of using the component if, say, there is some legal obligation to meet prescribed standards or if the component is to complement the overall structure of the car.

Formal decision-making procedures may need to be employed to process a quantified expression of these effects and implications. Such expression may be critical to the appropriate action to take. A machine which produces 'oversize' pistons renders the whole assembly inoperable. The effects are obvious and extreme; the machine *must* be modified. A door handle which is awkward to operate may be detrimental to the sales image of the car. The consequences of retaining or replacing this type of handle are by no means as critical, nor easily quantified. (Advertising stratagems may be all that are required: 'What ingenious door handles! The children will never open them!')

This distinction between the reflection function and the action guidance function must show itself also in the statistical procedures used in the one context or the other. In particular, a decision-making procedure needs formal tools or methods to handle potential consequences, expressed as *utilities* or *losses*.

Most statisticians accept the dual nature of statistics implied by these two functions but we should recognize that not everyone is convinced that such a distinction is useful. Lehmann (1959) sees the distinction as irrelevant. After introducing statistical analysis from the decision-making viewpoint, he remarks:

> However, not all statistical problems are so clear cut. Frequently it is a question of providing a convenient summary of the data or indicating what . . . [can be said about an] unknown parameter or distribution . . . for guidance in various considerations but . . . not . . . [as] the sole basis for any specific decisions. In such cases the emphasis is on the inference rather than the decision aspect of the problem, *although formally it can still be considered a decision problem if the inferential statement itself is interpreted as the decision to be taken.* (pp. 4–5, italics inserted.)

Yet again, Kerridge[5] admits the distinction but sees the responsibility of the statistician restricted to inferential objectives, except in 'emergencies'. He suggests:

> It is not primarily the responsibility of a statistician to make decisions for other people—not in general at any rate. . . . It is for someone else to say what decisions should be made with [inferential] . . . information. In other words, ideally, it is the statisticians job to inform not to decide.

Deming[6] adopts a similar standpoint in describing his views on the 'principles of professional statistical practice'.

The two quotations prompt further comment. They represent the two extremes of attitude to the inference/decision-making distinction, embracing the range of views on this issue expressed in the literature. Furthermore, their contexts lead the authors to express these viewpoints in a rather more extreme way than they are accustomed to express themselves elsewhere on this issue. But taking them at their face value one is tempted to ask of Savage's remarks how in general one is to set up an appropriate numerical expression for, or even a broad description of, the consequential information which is essential to a decision-making reformulation of a pure inference (description) problem. Then again, whilst Kerridge's attitude might seem narrow (even escapist) a case can be made for trying to counter the view sometimes expressed that the statistician should shoulder the whole burden of both interpretation *and decision*, even when he is in a less favourable position to assess consequences than is the person in day-to-day contact with the particular problem. Finally, there can on occasions be an advantage in separating out the message of sample data alone, from the modifications that one must make in the light of additional prior, or consequential, information. This is important, for example, in appreciating the role of *classical* statistics in relation to some other approaches, and it is a theme we will need to return to on several occasions.

1.5   DIFFERENT APPROACHES

The object of this book is to present the different attitudes to statistics, without adopting any as a partisan. Lehmann and Kerridge represent minority attitudes with regard to the question of the two different functions of statistics, and if we are to review the range of different approaches to statistics it is essential to retain the distinction between inference and decision-making. It is apparent that most statisticians regard this distinction as fundamental (Smith[2]), also that a large number are committed to a view of statistics which embraces some decision-making aspect. The vast literature on *decision theory*, at all levels of mathematical sophistication, applicability or philosophical comment, bears witness to this.

The distinction between inference and decision-making, coupled with the earlier subdivision of different types of relevant information, puts us in a position now to make a preliminary classification of the different approaches to statistics. For the moment we distinguish *three* main approaches (summarizing their distinguishing features in anticipation of much fuller discussion later).

(a) **Classical Statistics**, originating with R. A. Fisher, J. Neyman, E. S. Pearson, etc. This includes the techniques of *point* (and *interval*) *estimation*, *tests of significance* and *hypothesis testing*. At first sight it might be judged to be an inferential approach, utilizing sample data as its only source of relevant information—although we will find it necessary to qualify this assessment in certain respects later. The recent re-awakening of interest in the distinction between *significance testing* and *hypothesis testing* is crucial to the issue of the relative inferential/decision-making role of classical statistics. Any particular approach needs to incorporate concepts, tools and interpretations for its 'internal management'. In these respects classical statistics leans on a frequency concept of probability, represents the sample data through what is termed their *likelihood*, and sets up certain criteria based on *sampling distributions* to assess the performance of its techniques. For instance, point estimators may be required to be *unbiased* or *consistent*, hypothesis tests are based on 'tail-area probabilities' and at a more general level data are shown to be best handled in predigested form as *sufficient statistics* where possible. The methods embody *aggregate* measures of their own 'reliability' or 'accuracy'. (See Chapter 5.)

Some comment is needed on the use of the term 'classical' to describe this approach. It is commonly so described, in recognition of its traditional role as a formal principle of statistical method more widely applied and of somewhat earlier origin (in terms of detailed formal development) than either *Bayesian inference* or *decision theory*. But the approach is also labelled in many other ways: *sampling-theory, frequentist, standard, orthodox* and so on. We shall retain the term '*clssical*' throughout the book, but this is not to be confused with the similar label attached to a particular view of the probability concept discussed in Chapter 3. It is interesting to note that Buehler[7] calls the *Bayesian* approach 'classical'; the classical approach 'modern'—an extreme example of the non-standardization of terminology in the area of comparative statistics.

(b) **Bayesian Inference**, again essentially an inferential procedure, but admitting (indeed demanding) the processing of prior information as well as sample data. The prior information is modified by the sample data through the 'repeated use of Bayes' theorem' (Lindley, 1965, p. xi) to yield a combined assessment of the state of knowledge of the practical situation. The *likelihood* again plays a crucial role. Inferential statements are expressed through *posterior probability distributions*, and hence embody their own measure of accuracy. The idea of *sufficiency* is again relevant, but *not so* the method of sampling used to obtain the sample data. The expression of prior information through *conjugate* families of prior distributions, when appropriate to the practical problem, is of mathematical and interpretative convenience. This approach cannot rest on a frequency interpretation of probability alone; a *subjective* interpretation is almost inevitable, and probabilities tend to be regarded as conditional. Central to the basic development of Bayesian methods is the notion of *coherence* (an idea which goes back to Ramsey[8]: see, for example, Lindley (1971b, pp. 3–6)). This is a concept which seeks to express in precise terms what is required of individuals if they are to react 'rationally' or 'consistently' to different prospects in situations of uncertainty. The use of Bayesian methods is not restricted to situations where tangible prior information exists; the formal expression of *prior ignorance* is important and arouses great interest and some controversy. (See Chapter 6.)

(c) **Decision Theory**, stemming from the work of Wald, and first presented in his detailed treatment (1950). As the name suggests, this approach is designed specifically to provide rules for action in situations of uncertainty, i.e. *decision rules*. It inevitably incorporates assessments of the *consequences* of alternative *actions*, expressed through the mathematical theory of *utility* in the form of *losses* or *loss functions*. The value of any decision rule for prescribing action on the basis of sample data (and any prior information) is measured by its *expected loss*, or *risk*. The aim is to choose a decision rule with 'minimum risk'; and the concepts of the *admissibility* of a decision rule and of *complete classes* of decision rules are central to the study of optimality. There is no derived probabilistic assessment of the accuracy of decision rules; their relative merits are measured by their associated risks. This approach may be regarded as the stochastic extension of the deterministic *games theory*, due to von Neumann and Morgenstern (1953; first published 1944): it is concerned with 'games against nature', and a *minimax* principle is one basis for the choice of an 'optimum' decision rule. In as far as prior information is incorporated the methods used are essentially those of Bayesian inference. No particular philosophical view of probability is inevitably implied in decision theory; usually probability statements are frequency-based, although when prior information is processed a subjective attitude is often adopted. Whilst decision theory does not depend on, or demand, the use of Bayesian methods or a subjective interpretation of probability, the study of optimum decision rules is simplified on the Bayesian approach. At the formal level, adoption of prior distributions allows us to delimit (in many situations) the range of admissible decision rules, irrespective of whether or not any particular prior distribution is appropriate to the real problem in hand. At a fundamental level, Ramsey[8] shows that Bayesian decision theory is the inevitably correct way in which decisions *should* be made if we were entirely rational: in that *coherence* implies the existence of prior probabilities, and of utilities, and the need to maximize expected utility. See Lindley (1971a) for a down-to-earth explanation of this *normative* role of Bayesian decision theory. But not everyone is entirely 'rational' in their actions and reactions, and attitudes to decision theory vary from one statistician to another. Some see it as an objective procedure for prescribing action in real and important situations where it is possible to quantify consequences, others as a tentative formal model of the way people behave (or ought to behave) in the day-to-day running of their lives. (See Chapter 7.)

On this simple classification the main characteristics of the three approaches may be summarized in the manner indicated in the following table:

Approach	Function	Probability concept	Relevant information
Classical	Inferential (Predominantly)	Frequency-based	Sample data
Bayesian	Inferential	'Degree-of-belief'; subjective. Possible frequency–interpretable components	Sample data Prior information
Decision theory	Decision-making	Frequency-based	Sample data Consequential losses or utilities
		('Degree-of-belief': subjective, when prior information is incorporated)	(Prior information)

This broad classification of approaches to statistical inference and decision-making oversimplifies the true structure. At this stage the descriptions under (a), (b) and (c) are intended to provide merely an intuitive feeling for distinctions between the approaches, and a first contact with some of the vocabulary of inference and decision-making.

1.6 ARBITRARINESS AND CONTROVERSY

Whilst the above subdivision proves a useful framework for developing particular concepts, criteria and techniques in the later chapters, its oversimplicity can lead to a false sense of security. It may tempt us to think that there are rigid, well defined, distinctions between the approaches in terms of their function and what they regard as relevant information; also that there is unanimity of attitude to the importance of the internal criteria, and their interpretation, within any particular approach. Neither of these is true. The inevitable scope for personal interpretation of basic concepts and for alternative attitudes to what constitutes a reasonable way of representing information and processing it militates against this. In practice the boundaries between the approaches are blurred: it may be by no means obvious whether the function of a particular technique is inferential or decision-making, whether or not a particular interpretation of probability is implicit to any method or approach. Neither are there any truly objective grounds for assessing the merits of internal criteria, such as, for example, the *unbiasedness* of point estimators, the *minimax* principle for optimal choice of decision rules, or the use of *conjugate* prior distributions.

Later we shall develop the different approaches in sufficient detail for a cogent discussion of the arbitrariness of criteria or principles, and will be able to appreciate the subtleties of the controversy and criticism which occurs at all levels, and in respect of a vast range of ideas, in inference and decision-making. We shall also have need to broaden the range of approaches to consider other emphases, such as those embodied in the *likelihood approach* to inference. For the moment, however, we do need to recognize that many aspects of the study of statistical inference and decision-making must involve the personal judgement of those who instigate or develop the different approaches. This means that criteria may be correspondingly 'arbitrary' in depending on such personal attitudes for their justification, and also that criticism and controversy are inevitable.

We have already remarked on the oversimplicity of the classification given in the previous section and this is illustrated by the impression it may give of the *classical* approach. One may be tempted to view this approach as an embryo form on which more structure is built by the other approaches, which incorporate means of processing prior information and/or utilities, as well as sample data. The historical development outlined in the next section would appear to support this view. But this is too simple a picture. The failure of classical methods to accommodate prior information or consequences in any formal way did not arise from its failure to *recognize* these alternative forms of information. We shall see later that it was a reasoned and conscious choice by the initiators of the classical approach. They argued that such information could seldom be objective or detailed enough to form part of a formal theory of statistical inference. Instead they aimed at producing a theory which would be universal in application, free from subjective assessments, and based on information which would always exist in quantifiable form, namely *sample data*. Yet it is this very attitude that has stimulated most criticism of classical methods, along with its dependence on a frequency interpretation of probability.

We can illustrate the nature of such criticism, as well as demonstrate that even the function of a procedure is not immediately self-evident, by considering the example of an *hypothesis test*. At one level this may be regarded as an inferential procedure, in so far as it is a statement about the value of some parameter and a probabilistic assessment related to the appropriateness of this statement. But almost inevitably the test is prompted by the need to take alternative actions in some situation; its outcome stimulates a specific action. In this sense it is a decision-making procedure. But we might ask, 'Where have we incorporated an assessment of the consequences of different actions?' The reply that these cannot often be quantified may not be entirely satisfactory, especially if they *can* be quantified in the current situation.

Distinguishing the notion of *significance testing* as a purely inferential procedure stopping short of any need to accept or reject a hypothesis helps to clarify this issue in respect of philosophical (and, to an extent, practical) considerations. We shall return to this matter. Then again the claim of 'objectivity' of the classical approach, in being based on a frequency interpretation of probability, is not accepted by all. We shall see that many critics would claim that the interpretation of the results of an hypothesis test, or the information provided by a *confidence interval*, really rest on a *subjective* view of probability (and that from this viewpoint their interpretation is incorrect). At a more practical level, criticism is made of the 'arbitrariness' of the choice of significance levels for the test, and of the principle of drawing conclusions in terms of 'tail-area probabilities'.

The Bayesian approach is no more free of external criticism, centred (as implied above) on the intangibility of prior information, but again also focusing on specific techniques and criteria. Dissatisfaction is expressed with an approach which demands the formal use of prior information even when this is vague, imprecisely formulated, subjective or even non-existent. Then again, Fisher (1960, and elsewhere) was notable for his outright rejection of the whole concept of 'inverse probability', which underlies Bayesian inference, as an appropriate concept for a theory of statistics.

Critical debate does not operate only across the boundaries of the different approaches. There is no lack of internal dispute about the interpretation of concepts, and relevance of criteria, within a single approach. In Bayesian inference, for example, there has been much discussion on what constitutes an apt formal description of *prior ignorance*. Then again, on the interpretation of the probability concept we find a distinct contrast between the attitudes of von Mises (1964) and Lindley (1965). Both support a Bayesian approach to inference; von Mises from the frequency viewpoint alone, Lindley insisting on a subjective (degree-of-belief) interpretation. von Mises (1964), for instance, says

> Bayesian approach is often connected with a 'subjective' or 'personal' probability concept. *Our* probability concept is and remains 'objective'. (p. 332);

whilst Lindley (1965*b*) feels that a frequency interpretation cannot be achieved in general without a deal of 'mental juggling' (p. xi).

These different examples are not meant to be comprehensive, or representative. If it seems unfair to omit decision theory from criticism this is easily remedied. As a single example we have the debate on whether the concept of *utility* is appropriate for the formal expression of consequences; whether different outcomes can be measured on a *common* scale. As Pearson[9] asks,

> How, for example, as occurs in medical research, are we to balance in precise numerical terms the possible chance of saving life against the waste of a limited supply of experimental effort?

How are we to react to all this criticism and controversy? Firstly, to see that it is inevitable; secondly, to welcome it!

Controversy is inevitable in the field of statistical inference and decision-making. So many aspects of the subject depend upon personal opinions, attitudes and assessments, that this is bound to generate personalistic criticism or justification. Critical comment appears in all areas of the subject and operates at various levels. It is directed towards the 'arbitrariness' of criteria or concepts— but what is 'arbitrary' to one person may be 'self-evident' to another. It is basically this dilemma of 'what is one man's meat is another man's poison' that stimulates and perpetuates the criticism and controversy.

This controversy is healthy in so far as it generates a *constructively* critical attitude to the subject: a constant reappraisal. Constructive criticism in any subject is the prime stimulus for its development. This is certainly true of statistical inference and decision-making; witnessed by the dramatic developments that have occurred over the last 60 (and in particular, 30) years in this area, in a climate of continual critical debate.

The most important point to recognize, however, is that we can never expect any universal agreement on what constitutes a proper and correct approach to statistical inference. Subjective and personal elements must always remain. Preference for one approach or another must always incorporate subjective value judgements. It is not the function of this book to take sides; to support particular attitudes or condemn others. Its aim is quite the opposite; namely to develop some understanding of what the primary differences are in the different approaches, and to pinpoint the areas where the development of a particular approach becomes 'arbitrary', in that it depends on such personal, philosophical, attitudes and interpretations.

But whilst the personalistic element in the study of statistics acts as a stimulus for its development, it also has inherent danger. Human nature being what it is, there is always the risk that bigoted and extreme attitudes develop, and that argument descends to the emotional level of an interdenominational dogfight. Reasoned comparison of the different approaches then becomes impossible. Fortunately, little expression of this is apparent, but we might remark on the tendency there has been for factional groupings to develop—rather like gentlemen's clubs—whose members label themselves and others in distinctive styles. This is rather subtly illustrated by a remark by Bartlett[10],

> I am not being altogether facetious in suggesting that, while non-Bayesians should make it clear in their writings whether they are *non-Bayesian Orthodox* or *non-Bayesian Fisherian*, Bayesians should also take care to distinguish their various denominations of *Bayesian Epistemologists*, *Bayesian Orthodox* and *Bayesian Savages*.

This divisionalization is harmless enough in providing a shorthand system for expressing otherwise complex combinations of attitude and interpretation. But it does present a real confusion to the 'uninitiated', who are inevitably bemused and bewildered by all the 'name calling'. It is hoped that a secondary function of this book will prove to be its service as a dictionary in this respect!

Whilst some published work on basic ideas in inference does take a 'hard line' in attempting to demonstrate the inevitability of one approach or its superiority over another, it is vital to acknowledge the large effort which is being directed towards constructive evaluation. This takes many forms. Some seeks to *reconcile* the different approaches and show them as different (or dual) facets of a common objective. Other published work stresses the *relevance of different approaches* to a given problem depending on the particular circumstances and the availability of different types of information, and a further category *reassesses* or *re-interprets* the methods of one approach both internally and within the framework of another.

References to published material on the foundations, methods and inter-comparison of the different approaches to inference and decision-making will be given where appropriate in the later chapters. It is useful at this stage, however, to mention a few works which contain some fairly non-technical comment classified under the different emphases described in the previous paragraph.

A plea for the 'unifying role' of Bayesian methods at the foundations and applications level is made by de Finetti[11]. Hamaker[12] responds by asking if Bayesianism is a 'threat to the statistical profession' and this confrontation is continued by Hamaker[13] and de Finetti[14]. Moore[15] takes issue with Hamaker[12] on the question of the practicability of Bayesian decision theory. Cox[16] provides interesting commentary on the need to recognize an important role for the various approaches to statistical inference; see also Barnard[17]. Attempts at reconciliation, reassessment and re-interpretation between the different approaches appear in work by Bernardo[18], Bartholomew[19], De Groot[20], Pratt[21] and Thatcher[22]. See also Seidenfeld (1979).

The literature is not lacking in conciliatory comment, and it is right to conclude this section on such a note.

Pearson[23]

> I believe in the value of emphasizing the continuity as well as the difference in what have been the broad lines of development of our subject. I have the impression that by showing how the same situation is being tackled by alternative approaches the whole subject gains in richness in a way it would not if the exponents of one line set out to discredit another line by saying it was followed in error!

Plackett[24]

> The origins of many techniques of statistical inference which are now widely used can be traced back for 150 or 200 years. During most of this time, there have been differences of opinion about the validity of the methods proposed, and it seems to me unlikely that there will ever be a general agreement, or that arguments will be discovered which conclusively show the irrelevance of highly developed lines of thought. My view is that the differences of opinion should be patiently explored, and that the amount of agreement already in existence should be emphasized. We are dealing with ideas which go rather deeper than logic or mathematics and the important point is, not that we should start from the same assumptions, but that we should reach essentially the same conclusions on given evidence, if possible.

1.7 HISTORICAL COMMENT AND FURTHER REFERENCES

It is perhaps useful at this stage to give a brief indication of the historical development of ideas in probability theory and statistics. Further comment is made in relation to specific topics in the later chapters—in particular, Chapter 3 considers in historical sequence the awakening of alternative philosophical concepts of *probability*, as the language of statistics.

A detailed and comprehensive history of *probability theory* up to the beginning of the nineteenth century is given by Todhunter (1949, first published 1865). See also David (1962), Maistrov (1974). The development of the probability concept, and of its use, ranges from the tentative origins of the subject in the work of Cardano, Kepler, etc., in the sixteenth century, through its application to games of chance notably in the correspondence between Pascal and Fermat over a problem in the throwing of dice raised by Antoine Gombaud, the chevalier de

Méré, in the mid-seventeenth century, to its subsequent refinement and formalization in the following 150–200 years in the work of the Bernoulli brothers, de Moivre, Euler, Bayes and Laplace. The emphasis up to this time was on the 'classical' concept of probability based on the idea of 'equally likely outcomes'. The frequency concept, although appearing in vestigial form at this stage (e.g. in the work of de Moivre), did not find real expression before the middle of the nineteenth century (Venn). It awaited the arrival of von Mises early this century for its incorporation in a logical mathematical framework—the first example of an organized logico-mathematical basis for the probability concept.

The construction and systematic development of an axiomatic mathematical model for probability (using set and measure theory), free from the constraints of a particular philosophical viewpoint, was originated by Kolmogorov in the 1930s and continues to this day.

Alternative philosophical views of probability, of the logical or subjective type, have again found formal expression and study throughout this century (with increasing fervour over the last 30 years). Definitive ideas originate from the work of Keynes, Carnap, Jeffreys, etc., in one area (objective, logical concepts), and Ramsey, de Finetti, Good, Savage, etc., in another (personal, subjective). The implications of these alternative attitudes for statistical inference and decision-making have also been extensively studied over the same period, evidenced by the growing literature in the professional journals and the activities of the statistical organizations.

Whilst the pioneers and modern interpreters of probability theory were, and are, largely motivated by inference and decision-making interests, the bulk of the organized work on statistical theory has been produced over only the last 60 years or so. Bartlett (1962) attributes the first examples of 'statistical investigations in the modern sense' to Graunt in 1622, and to the exchanges between Pascal and Fermat at about the same time. But specific study of statistical techniques and concepts did not appear until much later: for example, in the introduction of 'inverse probability' by Bayes in 1764, 1765, the 'theory of errors' by Lagrange around 1760, and the criterion of 'least squares' by Gauss, Laplace and Legendre in the early nineteenth century. Further concepts began to appear during the nineteenth century. Bartlett (1962) describes as 'the great theoretical step' of the last century the development of the idea of correlation ('. . .1846. . . Bravais, but . . . [on] a wider basis by Galton in 1888'; p. 16).

Up to that time, however, there was nothing that could be described within the terms of this book as an organized 'approach' to a general theory of statistical inference or decision-making. This appears first during the period 1920–35 as what has been termed above classical inference. Around the turn of the century, and up to 1920, interest had focused on the application of probability and statistics to biological problems, and on the growing need for an organized study of experimentation in the agricultural and industrial spheres. Stimulated by the activity and enthusiasm of Edgeworth, Galton, Karl Pearson, Yule and others, there arose a period of detailed study of the principles of statistical analysis and of their logical bases. This bore fruit in the work of Fisher, Neyman and E. S. Pearson, who were largely responsible for an organized theory of estimation and hypothesis testing, centred on the ideas of likelihood and sufficiency. This was a period of rapid development of concepts and techniques which has continued to the present day, producing the current complex methodology of statistical inference based on the Fisher, Neyman and Pearson origins. Textbooks in this area are vast in number, but perhaps the most comprehensive survey of the current position is that of Kendall and Stuart (1977, 1979, 1976).

Within this single approach controversy was not lacking; the most notable example was the debate on the relative merits of *fiducial* and *confidence* methods of interval estimation.

The early work of Fisher, Neyman and Pearson is described in various publications. Perhaps the most accessible are Fisher's books (1970, 1966, 1959; first published 1925, 1926, 1956), his collected papers (1950), Bennett (1971–74) and Fisher's biography (Box, 1978); and the collected papers of Neyman and Pearson (1967), and of Pearson (1966). See also Edwards[25], Pearson[26] and Le Cam and Lehmann[27]. Other detailed historical commentary on this period (and indeed on a variety of other topics and periods) is given in the series of papers under the heading 'Studies in the History of Probability and Statistics' in the journal *Biometrika* over the last 25 years. Currently about 40 papers have appeared in this series; a selection is presented in Pearson and Kendall (1970) and Kendall and Plackett (1977). The latter volume also contains many contributions to the history of probability and statistics which have appeared elsewhere than in the *Biometrika* series.

The interest and activity surrounding Bayes' exposition of the use of the 'inverse probability' concept as an inferential aid, seen in the work of Laplace, waned during the nineteenth century under the criticism of Venn, Boole and others. Interest did not revive until the 1930s; even then only in the face of outright and persistent rejection of the concept by Fisher. It is probably fair to attribute the development of a formal system of inference based on the seed sown by Bayes to a period of merely the last 40 years or so, under the influence of Ramsey, Jeffreys, Lindley, Savage, Good, etc., with their variety of interpretative attitudes. In this respect *Bayesian inference* appears as a modern approach: Lindley (1965*b*) in the introduction to his text on this subject sees it in this light,

> The main difficulty in adopting . . . . a new approach to a subject (as the Bayesian is currently new to statistics) lies in adapting the new ideas to current practice. . . . A second difficulty is that there is no accepted Bayesian school. The approach is too recent for the mould to have set. (p. xi.)

A detailed review of this approach is given by Lindley (1971*b*).

It is difficult to separate the development of Bayesian methods of inference (*per se*) from the extensive application, over the same period, of the Bayesian attitude in the study of the wider objectives of *decision theory*. The incorporation of quantitative assessments of the consequences of actions under uncertainty found formal expression in the work of Wald in the 1940s. His book (1950) marked the start of continuing activity in decision theory, covering a whole variety of interpretative and conceptual problems as well as the development of specific techniques and applications. This is exemplified in the detailed mathematical presentation by Raiffa and Schlaifer (1961) and the illuminating introductory texts by Chernoff and Moses (1959), Aitchison (1970) and Lindley (1971a).

It is no historical accident that developments in statistical inference and decision-making reflect, and parallel, activity on the concept of probability and the refinements of probability theory as a deductive tool. The frequency basis of the classical approach to inference occurs alongside von Mises' formalization of the frequency concept of probability. The recent work on Bayesian inference and decision theory appears at a period of fervent activity on the logical and subjective views of probability. It is meaningless to enquire what comes first 'the chicken or the egg'. The statistical procedures need appropriate probability tools for their interpretation and employment; the development of alternative attitudes to probability must in turn colour statistical reasoning. The chicken could not exist without the egg, and vice versa.

In concluding this introduction it is appropriate to single out a few references to other works of differing, but relevant, emphasis. Some textbooks on statistics at an intermediate level do aim at a representative treatment of the subject and are not entirely restricted to a *single* approach. These include Schlaifer (1959) and Ferguson (1967). Examples in the literature of 'down-to-earth' *comparative* discussion of inference and decision-making, at varying levels of detail and emphasis, include Barnard[17], Bartlett (1962), Birnbaum[28], Cox[3], Lehmann (1962, Chapter 1), Plackett[24] and Smith[2]. In particular Plackett provides a concise and informative review of the different approaches. Amongst more detailed presentations of the 'foundations of statistics' we have Godambe and Sprott (1971), Good (1950), Hacking (1965), Hogben (1957), Jeffreys (1961) and Savage (1954).

All books referred to in this chapter are listed in the Bibliography at the end of the book.

References
The following publications were referred to by number throughout the chapter

1. Savage, L. J. (1961). *The Subjective Basis of Statistical Practice*. Report, University of Michigan, Ann Arbor.
† 2. Smith, C. A. B. (1965). 'Personal probability and statistical analysis' (with Discussion), *J. Roy. Statist. Soc. A*, **128**, 469–499.
† 3. Cox, D. R. (1958). 'Some problems connected with statistical inference', *Ann. Math. Statist.*, **29**, 357–372.

4. Lindley, D. V. (1977) 'The distinction between inference and decision', *Synthese*, **36**, 51–58.

5. Kerridge, D. F. (1968). Discussion on Marshall, A. W. and Olkin, I. (1968), *J. Roy. Statist. Soc. B*, **30**, 407–408.

6. Deming, E. W. (1965) 'Principles of professional statistical practice', *Ann. Math. Statist.*, **36**, 1883–1900.

7. Buehler, R. J. (1959). 'Some validity criteria for statistical inferences', *Ann. Math. Statist.*, **30**, 845–863.

8. Ramsey, F. P. (1931/1964) 'Truth and Probability'. In Kyburg and Smokler (1964) pp. 61–92. Reprinted from *The foundations of Mathematics and Other Essays*. Kegan, Paul, Trench, Trubner: London, 1931.

†9. Pearson, E. S. (1962). Prepared Contribution to Savage (1962).

10. Bartlett, M. S. (1965). Discussion on paper by Pratt[21].

†11. de Finetti, B. (1974). 'Bayesianism: its unifying role for both the foundations and applications of statistics', *Int. Statist. Rev.*, **42**, 117–130.

†12. Hamaker, H. C. (1977). 'Bayesianism; a threat to the statistical profession?', *Int. Statist. Rev.*, **45**, 111–115.

13. Hamaker, H. C. (1977). 'Subjective probabilities and exchangeability from an objective point of view', *Int. Statist. Rev.*, **45**, 223–232.

14. de Finetti, B. (1979). 'Probability and exchangeability from a subjective point of view', *Int. Statist. Rev.*, **47**, 129–136.

†15. Moore, P. G. (1978). 'The mythical threat of Bayesinism', *Int. Statist. Rev.*, **46**, 67–73.

†16. Cox, D. R. (1978). 'Foundations of statistical inference: the case for eclecticism' (with Discussion), *Austral. J. Statist.*, **20**, 43–59.

†17. Barnard, G. A. (1972). 'The unity of statistics: the Address of the President', *J. Roy. Statist. Soc. A*, **135**, 1–14.

18. Bernardo, J. M. (1979) 'A Bayesian analysis of classical hypothesis testing', in Bernardo, De Groot, Lindley and Smith (1980).

19. Bartholomew, D. J. (1965). 'A comparison of some Bayesian and frequentist inferences', *Biometrika*, **52**, 19–35.

20. De Groot, M. H. (1973). 'Doing what comes naturally: interpreting a tail area as a posterior probability or as a likelihood ratio', *J. Amer. Statist. Assn.*, **68**, 966–969.

21. Pratt, J. W. (1965). 'Bayesian interpretation of standard inference statements' (with Discussion), *J. Roy. Statist. Soc. B*, **27**, 169–203.

22. Thatcher, A. R. (1964). 'Relationships between Bayesian and confidence limits for predictions' (with Discussion), *J. Roy. Statist. Soc. B*, **26**, 176–210.

†23. Pearson, E. S. (1962). 'Some thoughts on statistical inference', *Ann. Math. Statist.*, **33**, 394–403.

†24. Plackett, R. L. (1966). 'Current trends in statistical inference', *J. Roy. Statist. Soc. A*, **129**, 249–267.

†25. Edwards, A. W. F. (1974). 'The history of likelihood', *Int. Statist. Rev.*, **42**, 9–15.

†26. Pearson, E. S. (1974). 'Memories of the impact of Fisher's work in the 1920s', *Int. Statist. Rev.*, **45**, 5–8.

†27. Le Cam, L. and Lehmann, E. L. (1974). 'J. Neyman—On the occasion of his 80th birthday', *Ann. Statist.*, **2**, vii–xiii.

†28. Birnbaum, A. (1962). 'On the foundations of statistical inference', *J. Amer. Statist. Assn.*, **57**, 269–326.

CHAPTER 2

An Illustration of the Different Approaches

In Chapter 1 we have discussed, in general terms, the purpose and nature of a statistical enquiry. We noted how the availability of different types of information may affect the form of statistical analysis which we conduct; in contrast, it was also seen how our philosophical attitudes and practical purpose may colour what we regard as relevant information, and the manner in which it should be processed. Such distinctions lead to different basic approaches to statistical inference and decision-making. It is convenient to illustrate the general discussion of the opening chapter by considering a single practical example in some detail. We shall approach this example entirely without preconceptions, imagining that we know nothing of the existence of particular techniques or prevailing attitudes to statistical inference or decision-making. Looking at things with this fresh eye we shall find it natural to ask certain types of question about the practical situation, and to develop on intuitive grounds specific methods of analysing the available information to throw light on these questions. We will thus be able to illustrate and further crystallize the basic distinctions of concept, attitude and methodology that arise in the different approaches to statistics.

2.1 A PRACTICAL EXAMPLE

One of the products made by the Trans-Western Electrical Engineering Co. (TWEECO, for short) is a special type of electronic component for assembly in a particular piece of hospital equipment. This is a high quality product demanding a complex automatic manufacturing process to meet the required standards. A basic fault can sometimes arise during this complex process; this renders an individual component quite useless so that it must be discarded. The fault is readily detected by an appropriate test.

A second company, The Hospital Equipment Co. (THECO), assembles the hospital equipment which utilizes the component and receives the component in large batches from TWEECO. The manufacturing process is given a regular overhaul at monthly intervals, and the batches consist of the output from

28

uninterrupted production runs between one overhaul and the next. Many such processes are running simultaneously, each producing this component, so that a large number of batches of the product are produced each year. During a comprehensive final test of the piece of hospital equipment after assembly by THECO, the basic component fault, if present, will be revealed and the component will need to be replaced before the equipment is released.

In this situation a variety of problems arise concerning the economical (or efficient) operation of the system and the quality of the final product. In so far as these factors are affected by the particular component under study, there are essentially two measures of the quality of the component which are important. For a given batch of components these are

(i) the **proportion** of components suffering from the basic manufacturing fault (the **proportion defective**),

(ii) the distribution of the lifetimes of the useful components, or as a summary measure, its mean (the **mean lifetime**).

Suppose that, as a result of the complex nature of the manufacturing process, the basic fault is by no means uncommon, although it is anticipated that a majority of the components will not suffer from the fault. It is obviously desirable that the proportion defective is as low as possible so that a minimum amount of expensive re-assembly work must be carried out by THECO, also that the mean lifetime of effective components is high so that the hospital equipment will give trouble-free service over a long time.

As far as the proportion defective is concerned, it may be worthwhile for TWEECO to operate some inspection procedure to try to ensure that batches with a high proportion defective are not sent on to (or used by) THECO. Whether or not such a procedure is desirable, and if so what its form should be, will depend on many factors including the net profit to TWEECO of acceptable components, and the costs to both companies of dealing with faulty ones. Also of relevance is the expense of conducting the inspection procedure, the 'reliability', 'accuracy' or 'usefulness' of such a procedure in statistical terms and the typical level of the proportion defective; also perhaps how this varies from batch to batch.

Concerning the mean lifetime of the components, THECO might like to be able to quote in the equipment specification some value for this quantity. This is useful for comparison purposes—perhaps to demonstrate the superiority of THECO's equipment over someone else's. Also a knowledge of the mean lifetime of the component is important to both companies to ensure that the design of the equipment is 'balanced': that is to say, that no single component (for example, the one under discussion) is either liable to fail much sooner than other components and thus dominate the performance of the equipment, or to last much longer than any other so that it is 'over-designed' and perhaps more expensive than it need be.

TWEECO as the manufacturer of the component, or THECO as its user, thus needs to be able to make informed comment on the two aspects of component behaviour: the proportion defective in a batch and the mean lifetime of effective components.

Suppose for the moment that we know nothing about existing techniques of statistical analysis, or different approaches to inference or decision-making. Let us try to follow the general path suggested in Chapter 1. We need first to construct a reasonable probability model for the situation, incorporating appropriate probability distributions and associated parameters; then to seek relevant information on which to base inferences about the unknown parameters or to indicate rules for action in the manufacturing and marketing process. We need to consider precisely *how* the information might be used to these ends: that is, how to contruct methods to analyse the information. Finally, we must give thought to what constitutes a reasonable *interpretation* of the results we obtain.

We can readily build a simple model here. We shall assume that the quality of a particular component is *independent* of the quality of any other components either in the same or in some other batch; however, the proportion of defective components, θ, *may be expected to vary from batch to batch*. Similarly, we shall regard the lifetimes of effective components as statistically independent. On the other hand, we assume that there are practical grounds for supposing that *the distribution of lifetimes of effective components does not vary from batch to batch*. Each batch is thus characterized by the proportion of defective components, θ, and the lifetime distribution of the effective components. Let us go further and assume that the form of this common lifetime distribution is known apart from the value of some single parameter, say its mean. We have now arrived at an elementary model for the system. In this model we take the random variable X to represent the lifetime of an effective component. It is a continuous, positive, random variable which has a probability density function $f_\lambda(x)$ which is known apart from its mean, denoted by λ.

We have thus represented the practical situation by a simple probability model involving just two parameters, θ and λ.

To know the value of λ is to have a tangible measure of the quality of usable components; this is valuable both for representing the behaviour of the piece of equipment incorporating the component and also to determine if the component is appropriately designed in relation to other components. To know the value of θ for a particular batch currently under study is to have a measure of the quality of that batch of components; knowing how θ varies from batch to batch provides a further assessment of the overall success of the manufacturing process. This knowledge about θ for the current batch, or from batch to batch, must be important in any decision to use or scrap a particular batch of components, or maintain or modify the style of manufacture of the components, respectively. But of course, neither λ nor θ (for some particular batch currently being examined) will be known.

The task of the statistician is to seek and to process relevant information about the practical situation to cast light on the values of λ and θ. The distinction between inference and decision-making is well marked in this problem—for example, on the one hand THECO may want to quote a value for λ on the equipment specification; on the other hand, TWEECO may have to decide whether the current batch of components is up to standard and can be passed on to the assembly process at THECO.

The value of a statistical analysis in this situation depends ultimately on the adequacy of the model. We have incorporated some quite specific assumptions: notably, the statistical independence of the properties of individual components, the constancy of the lifetime distribution of effective components from batch to batch and the specific form of this distribution. Ideally these assumptions need to be justified before any attempt is made to use the model. This would require a full-scale statistical investigation in its own right, and is beyond the framework of our present discussion. We shall adopt the model uncritically.

The whole question of model validation is a major one. All we will say here is that, in any real-life study of such a problem, it is often not feasible to carry out a thorough validation. It is unlikely that adequate information would be available, and the model might at best be justified on a combination of subjective and quasi-objective grounds. Independence might be justified by arguments about the physical properties of the manufacturing process, similarly the assumption of a constant lifetime distribution. The form of the lifetime distribution may be accepted because of 'empirical experience of similar situations', theoretical arguments about what distributions constitute reasonable models for component lifetimes in general, and the flexibility of form of the family of distributions $\{f_\lambda(x)\}$ for different λ, apart from any actual information obtained from the situation being investigated.

2.2   SAMPLE DATA AS THE SOLE SOURCE OF INFORMATION: THE CLASSICAL APPROACH

Suppose a current batch of the electronic components is available and we draw a random sample of them, from the large number in the batch, and examine their individual properties. A standard test is available to check for the basic manufacturing fault and this is applied to all the components in the sample. As a result we find that a number r, out of the n components in the sample, are defective. The effective components, which have been in no way damaged by the standard test, are now given a life-test to determine their lifetimes. A set of data of this type, with $n = 200$, is given in Table 2.1. Each component in the sample contributes either as a basic manufacturing fault (denoted*) or as a lifetime x_j ($j = 1, \ldots, n-r$). The number of faulty components in our data is $r = 60$.

What do these data tell us about the parameters, θ and λ, in the model? Since the data arise from a single batch they throw light on the value of θ specific to that

Table 2.1   Sample data for TWEECO components

*	58·8	79·2	61·9	73·1	*	*	78·5	82·2	83·0
*	93·6	*	75·0	96·0	97·9	76·3	84·1	102·4	84·5
95·3	100·6	*	76·3	*	110·7	96·2	73·3	63·5	*
83·1	78·6	*	*	75·5	80·9	*	94·5	86·0	77·1
97·1	77·8	*	*	*	87·2	94·3	90·1	105·4	*
*	*	74·3	109·9	87·0	82·2	66·3	78·3	69·6	*
83·4	66·9	*	74·9	*	93·0	84·9	*	67·0	*
*	*	*	72·5	80·4	60·5	86·3	69·2	*	84·2
66·0	85·6	99·6	68·6	67·1	77·8	84·8	*	*	85·6
61·7	66·1	91·5	*	*	67·5	*	*	*	*
*	*	64·8	64·9	63·4	79·9	*	*	90·5	75·7
67·9	*	78·2	84·8	59·3	77·3	*	78·7	*	86·5
68·3	84·7	84·7	82·0	*	*	81·6	68·5	*	82·2
69·1	94·3	103·0	*	77·3	95·5	64·3	83·1	71·1	70·9
80·2	*	68·2	*	81·0	109·0	*	103·2	87·0	85·8
87·1	81·6	62·2	94·5	73·5	69·8	74·2	*	76·7	98·2
*	73·5	97·3	*	97·1	55·2	80·3	93·1	74·2	60·5
78·9	*	90·1	78·0	91·1	*	80·3	71·0	*	*
83·8	96·3	66·5	*	86·0	*	76·0	80·4	*	63·0
*	87·7	86·0	72·7	*	109·5	60·9	86·4	82·9	86·3

batch. We shall call this value θ_0. Furthermore, the assumed homogeneity of the lifetime distribution means we can regard the lifetime data as representative of the manufacturing process as a whole, and not batch-specific, and we can use them to draw inferences about λ. Note that the sample information on batch *quality* (that is, whether individual components are defective or effective) is available as soon as the sample has been inspected. Consequently it can be used straightaway to infer the value of θ in the current batch; or indeed, as the basis for some decision such as whether or not to supply this batch to THECO. In contrast, the sample information on the *lifetimes of effective components* will not be available until life-tests have been conducted. The lifetimes given in Table 2.1 are in appropriate units (say maximinutes, MM) and will typically not have been obtained until long after the current batch has been dealt with (even allowing for some realistic accelerated ageing process in laboratory tests).

2.2.1   Batch Quality

Consider first of all what the data tell us about θ *for this batch*, that is about θ_0. Each component in the sample is noted to be either defective (D) or effective (E). In a typical situation we have a sequence of n independent qualitative observations, e.g.

DEEEDED . . . DE.

The model declares that a proportion θ_0 of the components in the batch are defective. Adopting the *frequency* concept of probability this amounts to saying that any individual component drawn at random from the batch has constant probability θ_0 of being defective. So

$$P(D) = \theta_0; \qquad P(E) = 1 - \theta_0.$$

Note that this does not conflict with the fact that any individual component has a *determined* quality, either defective or effective! The probability measure here relates to the values taken by the relative frequency of defective components from the batch, in samples of ever increasing size.

The assumptions of independence and constant probability suggest intuitively that the order of Ds and Es in our sample is irrelevant—that all that matters is the *number* of Ds and Es: that is, r and $n - r$. If this is so then n 'pieces of information' have been reduced to just two without loss of relevant information. This idea of reducing the extent, *but not the content*, of the data is an important one particularly in the *classical* approach, but also in *Bayesian* inference. Termed **sufficiency**, it is an idea we must discuss and develop in detail later (Chapters 5 and 6).

So what we know about θ_0 is that the data yielded r defectives out of n components examined. The model implies that r must be an observation of a random variable R having a **binomial distribution** with parameters n and θ_0; we shall write $R \sim \mathbf{B}(n, \theta_0)$.

We might try to use this to draw inferences about θ_0. The binomial distribution $\mathbf{B}(n, \theta_0)$ has mean $n\theta_0$ and variance $n\theta_0(1 - \theta_0)$. Thus *on average R is $n\theta_0$*; so if $\tilde{\theta} = R/n$,

$$E(\tilde{\theta}) = E(R/n) = \theta_0, \qquad (2.2.1)$$

where $E(\cdot)$ is the **expectation** operator. Our observation r is a typical value from the distribution of R. In view of (2.2.1) it seems reasonable to use r/n as a 'shrewd guess' at the actual value of θ_0, that is as an **estimate** of θ_0. We say that R/n is our **estimator** of θ_0; *it is the value this takes in the data that is the actual estimate*. From the data in Table 2.1 we estimate θ_0 by $\tilde{\theta} = r/n = 0.30$.† This illustrates the idea of **point estimation** in the **classical approach** to statistical inference.

Why is this a reasonable method of estimation? The argument above suggests only one criterion, namely that the *expected value* of the estimator is equal to the quantity it is estimating. Such an estimator is said to be **unbiased**.

---

† Strictly speaking we need to distinguish in our notation between the *random variable* $\tilde{\theta}(R)$ and the *value it takes*, $\tilde{\theta}(r)$, for a particular sample. Later, when discussing basic principles and theory in some detail, this distinction is made. For the moment, however (and wherever the level of treatment, or immediate context, renders this unnecessary), no formal distinction will be drawn between the *potential* and the *realized* values. In such cases the same symbol $\tilde{\theta}$ (or $\bar{x}$, or s^2, say) may be used unambiguously to denote an estimator, or an estimate, as required.

But the estimator θ has other properties in this situation. We know that its sampling variance is

$$\text{Var}(\tilde{\theta}) = \frac{\theta_0(1-\theta_0)}{n} \qquad (2.2.2)$$

so that as n becomes larger the variance becomes smaller. This means that the larger the sample size, n, the less we should *expect* the unbiased estimator $\tilde{\theta}$ to differ from θ_0 and we might conclude that *increase in the sample size leads*, in this sense, *to more accurate estimators of* θ_0. More particularly we can easily show that as $n \to \infty$, so $\tilde{\theta} \to \theta_0$, in probability. (This is a simple consequence of *Chebychev's inequality*.) Such an estimator is said to be **consistent**, a further intuitively appealing criterion for a point estimator. It is easy to see that consistency is not an inevitable property for an estimator. Suppose that, irrespective of the sample size, n, we choose to estimate θ_0 by the quality of the *first* component sampled; that is by 0 if this component is effective and by 1 if it is defective. Obviously such an estimator is unbiased, but its sampling variance is $\theta_0(1-\theta_0)$ *independent of n*. It is thus not a consistent estimator.

Finally, having remarked on the unbiasedness and consistency of $\tilde{\theta}$, we might ask if $\tilde{\theta}$ is the *best* unbiased, consistent, estimator of θ_0 based on the sample (r, n). This raises the question of what we mean by 'best'! One possibility is to seek an unbiased consistent estimator with *smallest variance*. In fact, it can be shown (§5.3.2) that there is no unbiased consistent estimator of θ_0 with variance less than (2.2.2). *In this respect $\tilde{\theta}$ is best*, and the best we can do to 'pinpoint θ_0' in the example is to use the point estimate $\tilde{\theta} = 0.30$ for θ_0.

Later, in Chapter 5, we shall consider in detail these various criteria for classical point estimators; their respective importance, incidence and interrelationships. We will face such questions as whether the sampling variance is the most useful measure of accuracy of estimation, whether unbiasedness and consistency are essential, whether or not sufficiency, consistency and unbiasedness are unrelated concepts, and whether we can recognize situations where best estimators exist, and identify them. Similar detailed study must be made of other aspects of the classical approach to statistics, which will be merely intuitively, and somewhat superficially, justified at this stage.

Before proceeding to consider other techniques and concepts of classical statistical inference applied to the present problem, we must pause to reflect on the interpretation of point estimators. In saying that $\tilde{\theta}$ is unbiased, expressed formally by (2.2.1), we have in mind a distribution of possible values of $\tilde{\theta}$; the **sampling distribution** of $\tilde{\theta}$. This must be viewed as having as 'collective' (in von Mises' terms) the set of values of $\tilde{\theta}$ which might arise from repeated random samples of n components drawn from (then replaced in) the current batch. Of course, no such repeated sampling is envisaged, but it must be conjured to provide the interpretative backing for the classical approach. Alternatively, a conditional argument can be invoked, and the distribution of values of $\tilde{\theta}$ viewed as arising as single values of r/n each from a different batch where the parameter θ_0 *happens to have the same value as it does for the current batch.* Neither view is entirely satisfying intellectually to some people; but some such attitude is inevitable in the classical approach, which defines its concepts, and assesses performance, in terms of a *postulated sequence of similar situations.* Consistency, and the use of the sampling variance to measure the accuracy of $\tilde{\theta}$, are clear illustrations of the use of such *aggregate* measures. We shall return to this point for a fuller discussion at a later stage (Chapter 5).

Returning to the problem of the TWEECO components we can go further in our inferences about θ_0. We might ask: 'How close is our estimate, $\tilde{\theta} = 0.30$, to the true value of θ_0 for the batch?' Some clue to this is given by (2.2.2). For $n(1 - \theta_0)$ and $n\theta_0$ both of reasonable size the binomial distribution $\mathbf{B}(n, \theta_0)$ may be approximated by a normal distribution with the same mean and variance as the binomial, i.e. by a **normal distribution** with mean $n\theta_0$ and variance $n\theta_0(1 - \theta_0)$, denoted by $\mathbf{N}[n\theta_0, n\theta_0(1 - \theta_0)]$. The conditions are appropriate here in that it may be assumed that both $n(1 - \theta_0)$ and $n\theta_0$ are greater than 40. So we have, approximately,

$$\tilde{\theta} \sim \mathbf{N}[\theta_0, \theta_0(1 - \theta_0)/n]. \tag{2.2.3}$$

The **standard error** (S.E.) of $\tilde{\theta}$, $\sqrt{[\theta_0(1 - \theta_0)/n]}$, is unknown since θ_0 is not known; but an *estimated* standard error (E.S.E.) may be obtained by substituting $\tilde{\theta}$ for θ_0. In this way we can use (2.2.3) to make probability statements involving θ_0. We have E.S.E. $(\tilde{\theta}) = 0.0324$, so that there is a probability of approximately 0.95 that $|\tilde{\theta} - \theta_0| < 1.96 \times 0.0324 = 0.0635$. This tells us that there is little chance (about 0.05) that $\tilde{\theta}$ is further from θ_0 than 0.0635.

But is this really as comforting as it might at first seem? The probability 0.95 is measured in relation to a sequence of repeated situations in each of which a sample of 200 components is drawn from the batch. In the long term about 95 per cent of the resulting values of $\tilde{\theta}$ will be within 0.0635 of the true value. This certainly provides some aggregate assurance, *but we do not in fact know if $\tilde{\theta} = 0.30$ is one of the 95 per cent within 0.0635 of θ_0, or one of the 5 per cent which are not!*

Aggregate assurance may be comforting to the statistician, viewed in the context of his wider professional activities. It may even satisfy TWEECO in the present problem to know that a large proportion of the conclusions it draws about consecutive batches will be correct. However, it provides little comfort to TWEECO as far as the present *isolated* batch is concerned, nor indeed to the interested party in any practical problem where a *single* inference is to be drawn with no obvious reference to a sequence of similar situations.

The same dilemma arises in the attempt *formally* to couple estimation with an assessment of the accuracy of the estimator, through the idea of an **interval estimator**.

From (2.2.3) we have

$$P\left\{\frac{|\tilde{\theta}-\theta_0|}{\sqrt{[\theta_0(1-\theta_0)/n]}} < z_\alpha\right\} = 1-\alpha,$$

where z_α is the double-tailed α-point of the standardized normal distribution. That is, if $Z \sim N(0, 1)$ we have

$$P\{|Z| < z_\alpha\} = 1-\alpha.$$

In particular if $\alpha = 0{\cdot}05$

$$\frac{|\tilde{\theta}-\theta_0|}{\sqrt{[\theta_0(1-\theta_0)/n]}} < 1{\cdot}96 \tag{2.2.4}$$

is true with probability 0·95. Inverting this as an interval for θ_0—either approximately by replacing θ_0 in the denominator by $\tilde{\theta}$, or precisely by determining the roots of an appropriate quadratic equation—yields what is called a **two-sided 95 percent confidence interval** for θ_0. Consider the more precise method. From (2.2.4) we can write

$$|\tilde{\theta}-\theta_0| < 1{\cdot}96 \sqrt{[\theta_0(1-\theta_0/n]}$$

or

$$Q(\theta_0) = (1+3{\cdot}84/n)\theta_0^2 - (2\tilde{\theta}+3{\cdot}84/n)\theta_0 + \tilde{\theta}^2 < 0. \tag{2.2.5}$$

Thus the confidence interval has the form

$$\alpha(\tilde{\theta}) < \theta_0 < \beta(\tilde{\theta}) \tag{2.2.6}$$

where α and β are, respectively, the smaller and larger roots of the quadratic equation

$$Q(\theta_0) = 0,$$

with $Q(\theta_0)$ as defined in (2.2.5). The region (2.2.6) provides an interval estimator of θ_0: an interval which contains θ_0 with probability 0·95. But again the measure of precision refers to repeated sampling, in the sense that 95 *per cent of the intervals, obtained by substituting in (2.2.6) values of $\tilde{\theta}$ obtained from different samples, will contain θ_0 in the long run.*

From the data of Table 2.1, (2.2.6) yields a two-sided 95 per cent confidence interval for θ_0 as $(0{\cdot}241 < \theta_0 < 0{\cdot}367)$. [In contrast the approximation obtained by replacing θ_0 by $\tilde{\theta}$ in the denominator of (2.2.4) is $(0{\cdot}236 < \theta_0 < 0{\cdot}364)$.] Likewise we can obtain, say, a 99 per cent confidence interval as $(0{\cdot}224 < \theta_0 < 0{\cdot}389)$ [or $(0{\cdot}217 < \theta_0 < 0{\cdot}383)$, respectively]. The **confidence level**, 95 per cent or 99 per cent, may be chosen at will, but we are forced to trade precision for confidence in the sense that the higher the confidence the wider the interval and hence the less precise is our statement about θ_0.

For the present example what does the confidence interval tell us about θ_0?

Consider the (more exact) 95 per cent confidence interval $0.241 < \theta_0 < 0.367$. We must beware of an appealing misconception! It is *not* true that $0.241 < \theta_0 < 0.367$ holds with probability 0.95. Either θ_0, a determined quantity not a random variable, is in this interval or it is not. All we know is that in the long run 95 per cent of such intervals obtained in similar circumstances to the present one will contain θ_0; 5 per cent will not. We have no way of knowing into which category the present interval falls! It is vital that we adopt the proper interpretation of the confidence interval. We shall pursue this later, and also consider criticisms directed towards the classical approach which hinge on such interpretative matters.

The two ways we have so far considered for processing the data on batch quality have been *inferential* in nature. They aimed to provide a fuller *description* of the underlying model for the current batch in terms of the value of θ_0. But it is likely that some *decision* on the current batch of components must be made, based on the data in the sample. Should the batch be sent on to THECO for use in the hospital equipment it is assembling? Should THECO accept it? With the type of catastrophic manufacturing fault possible in these components it is inevitable that a proportion of the components sent to THECO in any batch will be unusable. Presumably this is allowed for in the terms of the agreement between the two companies. But it is equally likely that some limit is placed on the numbers of defective components that THECO will tolerate. It may be, for example, that THECO agrees to accept batches provided that TWEECO reimburses the costs of locating and replacing defective components. Alternatively, a certain proportion of defectives may be accepted in a batch, but if the actual proportion exceeds this limit a heavy penalty is imposed. Let us consider this second case briefly.

Suppose it has been agreed between the two companies that batches of components which contain no more than, say, 24 per cent defectives must be accepted by THECO; whilst if there turn out to be more than 24 per cent defectives full costs of locating and replacing defective components must be met by TWEECO who must also pay a fixed penalty charge. TWEECO may well decide to carry out some *quality control* exercise on batches to safeguard against this double penalty. This can be achieved by taking a random sample of components from a batch, and accepting or rejecting the batch depending on the number of defectives in the sample. Whether to operate such a scheme, and what its form should be (e.g. whether to take a sample of predetermined size, if so of what size; or whether to sample components one at a time until a decision can be taken) depend really on cost factors such as the costs of sampling batches or of remedying rogue batches sent to THECO, as well as on the range of values of θ which might arise in different batches. These factors should ideally form part of the information which is examined in reaching a decision and we consider this in simple form later (§2.4). For the moment let us see what policy might be followed if we use the *sample data alone* from a sample of a fixed size n, with costs being only informally taken into account.

TWEECO obviously wants to avoid supplying the current batch if $\theta_0 > 0.24$, but at the same time cannot afford to reject too much of its production to this end. There are two extreme situations.

(a) The company might 'act safe' by assuming that $\theta_0 > 0.24$ for a given batch, and only supply the batch if the data give some reasonable indication that $\theta_0 \leqslant 0.24$.

(b) Alternatively, batches may be supplied to THECO on the assumption that $\theta_0 \leqslant 0.24$ unless, again, the data give evidence to the contrary.

Either of these policies may be effected by using what is known in the classical approach as an *hypothesis test*, or more specifically a *test of significance*. We illustrate this for policy (b) based on the data of Table 2.1.

Initially, we employ what is known as a **pure significance test** (Cox and Hinkley, 1974) to examine the inferential import of our data with respect to the prospect that $\theta_0 \leqslant 0.24$; that is, that the current batch is of suitable quality to supply to THECO. We term this prospect the *working hypothesis* and denote it by H. The aim is to assess the extent to which the data are consistent, or inconsistent, with H.

To examine H from this standpoint we need to consider some function of the data (a **test statistic**) which suitably reflects the value of θ_0. An obvious candidate here is our point estimator $\tilde{\theta} = r/n$. It has the property that the larger its value the stronger is the observed inconsistency with small values of θ_0, and in particular with H. We measure the extent of such inconsistency by the probability that the test statistic is as large as the observed value $\tilde{\theta}_{obs}$ were H true.

Specifically, we calculate

$$p_{obs} = P(\tilde{\theta} \geqslant \tilde{\theta}_{obs} | \theta_0 = 0.24),\dagger$$

which provides a measure (called the *observed significance level*, or *critical level*) of the inconsistency of our data with respect to the hypothesis $H_0 : \theta_0 = 0.24$. If p_{obs} is small (indicating inconsistency with H_0) it will in this situation be even smaller for other values of θ_0 within H, and thus indicates at least as much inconsistency with respect to the wider prospect H.

The observed value of θ from the data of Table 2.1 is 0.30. Using the normal approximation (2.2.3) we obtain $p_{obs} = 0.023$ (or 0.019 using a continuity correction), with the resulting inference that the data are highly inconsistent with H.

The pure significance test is not intended, in itself, to provide a basis for deciding whether or not to accept H, although clearly the smaller the observed significance level the less (in the sense of the greater the inconsistency of our data

---

$\dagger$ In probability theory the notation $p(\cdot|\cdot)$ refers to the probability of one event conditional upon another. Here it has the common wider interpretation, where the bar merely implies that any statement which follows it is assumed to be true. This distinction is often less meaningful in Bayesian inference, as we shall see later.

relative to H) is our credence in H. Indeed, if we were to take the view that any particular value of p_{obs} was small enough to justify *rejection* of H_0 (and hence of H), the observed significance level p_{obs} has the interpretation of the probability of *incorrectly rejecting* H_0 (or H). In contemplating accepting or rejecting H, we are now moving closer to decision-making (but without any overt consideration of costs or consequences).

Classical statistics incorporates a more formal mechanism (based on the interpretation above) for deciding whether to accept a working hypothesis H, or to reject it in favour of an *alternative hypothesis*, $\overline{H}$. Such a procedure is termed an **hypothesis test** or **significance test** (without any such epithet as 'pure' or 'simple'). For the current problem, H has the form: $\theta_0 \leqslant 0.24$; $\overline{H}: \theta_0 > 0.24$. We commence by specifying some small value α, the **significance level** of the test, as the maximum probability we will tolerate of *rejecting H when it is true*. This probability is maximized when $\theta_0 = 0.24$.

Again it will be large values of $\tilde{\theta}$ which cast doubt on H. Thus it suffices to determine some value c such that if we observe $\tilde{\theta} > c$ we reject H in favour of $\overline{H}$ (otherwise we accept H), with

$$P[\tilde{\theta} > c \,|\, \theta_0 = 0.24] = \alpha.$$

This implies, through the normal approximation (2.2.3), that we must take

$$c = 0.24 + 0.427 z_{2\alpha}/\sqrt{n}. \tag{2.2.7}$$

Such a procedure is called a **one-tailed level-α test of significance**. The value of α is open to choice. But so also is the rejection criterion. To reject H if $\tilde{\theta} > c$ is only one possibility; any other rejection criterion which has probability *no more than* α of being satisfied *when H is true* also yields a level-α test. We rest merely on the intuitive appeal of the current test for the moment, but will take up this point again later (§5.4). We must also return to the question of what role is played by the *alternative hypothesis* in the choice of a test.

If a maximum risk of 5 per cent of incorrectly rejecting H is regarded as tolerable in the light of the consequences of such wrong action, we would take $\alpha = 0.05$. For a one-tailed 5 per cent test in this case we must substitute $z_{0.10} = 1.645$ in (2.2.7) to obtain $c = 0.24 + 0.702/\sqrt{n}$. When $n = 200$, $c = 0.290$.

For the data of Table 2.1, $\tilde{\theta} = 0.30$, which would lead us to reject H on the basis of a 5 per cent test, and we should presumably not supply the batch to THECO.

If it was important to be even more sure of not incorrectly rejecting H we could take a smaller value of α, say $\alpha = 0.01$, and operate a 1 per cent test of significance. The data now provide insufficient evidence for rejecting H, since c has now become 0.31. The greater the assurance we require of being correct in rejecting H, the more extreme the data need to be in the sense that larger values of $\tilde{\theta}$ are needed for rejection.

Just what has been achieved by such a test? By making α arbitrarily small it is possible to ensure that there is a correspondingly small probability that a batch is unnecessarily held back (with the associated expense of scrapping, or fully inspecting, it). On the other hand, it is possible that there is a high risk that an *unacceptable* batch is supplied to THECO. In the extreme but trivial situation when $\alpha = 0$ no batches are held back. All, *including every unacceptable one*, are supplied and there is no control on batch quality.

However, unacceptable batches that are supplied involve TWEECO in the expense of remedial action, as well as the penalty charge. This is a second type of risk that must be studied; of *accepting H when it is false*. The test of significance above gives no direct consideration to this. It merely ensures that the *first* type of risk is kept within tolerable bounds. In this respect the test of significance is asymmetric in its regard of the two hypotheses; another point which we must consider in more detail later (§5.4). It is possible, however, to assess the risk of supplying an unacceptable batch. Suppose in some current batch the actual proportion defective is $\theta_0 = 0.26$, so that the batch is unacceptable. We would accept this batch on the basis of a 5 per cent test of significance on a sample of size 200 if $\tilde{\theta} < 0.290$. A simple calculation using the normal approximation (2.2.3) shows that the probability of accepting the batch is 0.833.

This is an astounding result! The cost of a reasonable guarantee that acceptable batches will not be suppressed is a very high probability of accepting unsatisfactory batches (with $\theta_0 = 0.26$). This could only be justified if the expense of suppressing a batch was enormous compared with the penalty of supplying an unacceptable batch, which is unlikely to be true. *In fact there has been no formal, explicit, consideration of consequences and costs of wrong actions in the construction of the test. This is characteristic of the test of significance*; such factors are considered only subjectively in the choice of what to take as the working hypothesis and in the choice of an appropriate significance level for action. Attempts to justify this lack of any formal regard to costs and consequences include the claim that it is seldom possible to quantify these factors, and hence that an objective approach to the problem should not involve them. Alternatively, it is sometimes argued that the test of significance is not a decision-making procedure but an inferential one, providing a description of the underlying probability process, and that costs are relevant only when we try to interpret this inference as a guide to action. This is nicely illustrated in our problem: in the distinction between an inferential conclusion to 'reject H' (i.e. to believe instead that $\theta_0 > 0.24$) and its behavioural implication that we 'reject the batch' (i.e. do not supply it to THECO). We shall see (§5.6) that much controversy centres on this matter of whether the test of significance plays an inferential, or decision-making, role.

The extreme attitude (b) leads to the test of significance described above. We have seen that it implies that production losses must far outweigh the cost of remedial action on sub-standard batches, if the procedure is to make real sense. The opposite extreme, as represented by (a), only makes sense for the opposite cost structure (where remedial action is far more costly than production losses).

This policy, (a), may also be implemented by a test of significance. All that is necessary is to interchange the working and alternative hypotheses. But this extreme is also unlikely to represent a realistic informal assessment of the cost structure of the problem. One wonders, then, if the test of significance is an appropriate vehicle for *making decisions on batch quality in this problem*. It is true that the second type of risk can be reduced by increasing the sample size, but the sample size is already 200 which may be as large as can be justified on grounds of cost (if not *too* large; again costs should ideally be incorporated in the analysis).

There is a compromise procedure which is possible. The two companies might co-operate in the construction of the inspection scheme for each batch in the following way. TWEECO agrees to accept (say) a 5 per cent risk that a batch with $\theta_0 = 0 \cdot 20$ is rejected by the scheme; in return THECO agrees to bear (say) a 5 per cent risk of receiving a batch with $\theta_0 = 0 \cdot 28$. We can then seek a critical value c_0, and sample size n, to ensure that a rejection procedure of the form

$$\text{'reject batch if } \tilde{\theta} > c_0\text{'}$$

has the agreed characteristics. The function

$$\phi(\theta_0) = P(\tilde{\theta} \leqslant c_0 | \theta_0) = P \text{ (accept batch of quality } \theta_0)$$

is called the **operating characteristic (OC) function** of the inspection scheme. We need to choose c_0 and n so that

$$\phi(0 \cdot 20) = 0 \cdot 95; \quad \phi(0 \cdot 28) = 0 \cdot 05.$$

See Figure 2.1, which shows the typical form of the OC function on the normal approximation. Such a scheme requires a sample size of 305 with $c_0 = 0 \cdot 238$. That is, we should reject a batch if a sample of 305 components contains 73 or more defectives.

It is a common practice for problems of this type to be examined by means of such a compromise procedure, without specific regard to costs of sampling, batch rejection or remedial action. See Wetherill (1969) for a detailed treatment of the

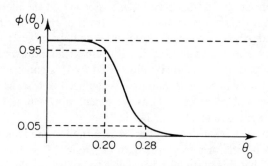

Figure 2.1 OC function for batch inspection

subject from the classical approach. Barnett and Ross[1] discuss the specific problem of computer acceptance testing from this standpoint. If reliable information is available on costs, or on the way in which θ varies from batch to batch, alternative methods incorporating the ideas of decision theory, or Bayesian inference, may be applied. See §§2.3, 2.4.

As in the case of estimation, any assessment of the properties of the classical test of significance must again be framed in terms of a 'sequence of similar situations'. Thus the significance level α provides an upper bound to the long-term proportion of batches, *satisfying the claims of the working hypothesis*, which will be rejected. This conditional interpretation is unavoidable. The quantity α neither measures the overall proportion of batches which are rejected, nor does it express the chances of incorrect rejection *for the current batch*, however appealing this latter interpretation might seem.

2.2.2   Component Lifetimes

We have briefly considered the use of classical methods in the study of the proportion of defectives in the current batch. The data in Table 2.1 also provide information on the lifetime distribution for effective components over the whole manufacturing process. We will briefly consider how methods of point estimation, interval estimation and hypothesis tests may be applied to this problem as well.

The model assumes a particular form, $f_\lambda(x)$, for the lifetime distribution: known apart from the value of λ, the mean of the distribution. Various distributions have been successfully employed to represent the lifetime distributions of industrial products and components. [See Barlow and Proschan (1965) and associated references.] These include the Weibull distribution and the gamma distribution. As special cases of the latter, both the exponential distribution and the normal distribution have been seen to provide useful empirical models. Davis[2] applies these distributions in a variety of situations, including human mortality, payroll errors, and failures of hand tools, of calculating machines, and of a variety of electronic components. In this spirit let us suppose that the lifetime distribution for the present problem is essentially of the form of the χ^2 distribution (that is, gamma with index 1/2) with mean λ which is taken to be large. It would in practice be more natural to assume a gamma distribution with unknown index. However, the basic principles are more easily illustrated by this constrained distribution which is not entirely unrealistic, and avoids the need to discuss the problems of estimating a second parameter. Thus we essentially assume that the lifetime distribution is approximately $N(\lambda, 2\lambda)$.

We might now ask if the data support such a model; and seek an estimate of λ. The histogram of lifetimes from the data of Table 2.1 is shown as Figure 2.2. Suffice it to say that this provides no evidence to contradict the $N(\lambda, 2\lambda)$ model. A simple probability plot would also be a useful informal aid to model validation.

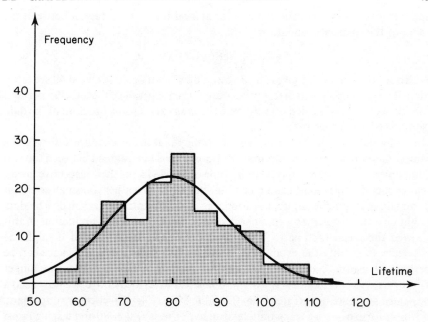

Figure 2.2   Histogram of lifetimes

How can we estimate λ? Again since λ is the population mean, we might take the sample mean $\bar{x}$. From the data we obtain the sample mean and sample variance, s^2, as

$$\bar{x} = 80 \cdot 80, \quad s^2 = 148 \cdot 19.$$

(It is comforting that s^2 is about twice $\bar{x}$ in value!) So it seems intuitively reasonable to estimate λ by

$$\tilde{\lambda} = \bar{x} = 80 \cdot 80.$$

Unfortunately our intuition is not entirely sound here. We can show, in fact, that $\bar{x}$, with sampling variance $2\lambda/n'$, is *not* the best unbiased consistent estimator of λ. (The sample size, n', is 140.) This exemplifies the fact that the sample mean is not always the best estimator of the population mean. We might have suspected some trouble here in view of the simple relationship between the mean and variance. Some function of the data reflecting variation about the mean possibly carries additional information about λ.

We might also consider a confidence interval for λ based on $\bar{x}$. We know $\bar{x} \sim N(\lambda, 2\lambda/n')$, so that

$$P\left\{ \left| \frac{\bar{x} - \lambda}{\sqrt{(2\lambda/n')}} \right| < 1 \cdot 96 \right\} = 0 \cdot 95$$

and we obtain a 95 per cent confidence interval for λ as the region between the roots of the quadratic equation

$$\lambda^2 - \{2\bar{x} + 7\cdot68/n'\}\lambda + \bar{x}^2 = 0.$$

For the current data this gives $78\cdot72 < \lambda < 82\cdot93$; but again (as we shall see later) there is a real sense in which this procedure does not produce the 'best' 95 per cent confidence interval; we do better to base the interval on some function of the data (some **statistic**) other than $\bar{x}$.

Finally, let us consider some simple policy-making procedure utilizing the sample data on component lifetimes. Suppose it has been agreed between the two companies that a mean component lifetime of 80 MM suits their respective needs; that is, such a component has a mean lifetime which is neither too small so that it dominates equipment failure, nor too large so that it is 'over-designed'. The data might be used to perform a spot check on the control of the mean lifetime; if this showed the mean lifetime to be markedly different from 80 MM it might be necessary to modify the production process. Again a test of significance can be performed as the classical expression of such a check. But it is no longer appropriate to conduct the type of *one-sided* test described above; we now need a *two-sided test* since departures from the value 80 in *either* direction are important. A simple form of such a test would take an hypothesis $H:\lambda = 80$ and test it against the two-sided alternative hypothesis $\overline{H}:\lambda \neq 80$.

Such a test has the following form. Suppose, *for the moment*, that our lifetime data constitute a *random* sample of 140 observations from $N(\lambda, 2\lambda)$. We know that, *if H is true*,

$$Z = \frac{\bar{x} - 80}{\sqrt{(160/140)}}$$

has a standardized normal distribution, $N(0, 1)$. We have only then to calculate the value of Z, the *test statistic*, and assess whether it is a reasonable value to have arisen from $N(0, 1)$, by determining the probability of obtaining a random observation from $N(0, 1)$ as large (or larger) in absolute value as $|Z|$. The corresponding, more formal, *level-α test of significance* has a rejection criterion: 'reject H if $|Z| > z_\alpha$'. The interpretation, and practical choice, of the significance level are subject to the considerations already discussed.

In the current example $|Z| = 1\cdot06$, which is far from being significant even at the 5 per cent level. Presumably no action is called for!

Before leaving this topic we must also note some further distinctions between the two-sided, and one-sided, tests. The hypothesis $H:\lambda = 80$ provides a *complete* specification of the probability model; the earlier hypothesis $H:\theta_0 \leqslant 0\cdot24$ provided only a *partial* specification. To stress this distinction the former type of hypothesis is termed a **simple hypothesis**, the latter a **composite hypothesis**.

Much of the literature on hypothesis testing refers to the basic working hypothesis as the **null hypothesis**, frequently with the restriction that this must be *simple* so that our working assumption provides a completely specified model. To adopt this attitude for studying the batch quality in our example means we must test a null hypothesis $H: \theta_0 = 0.24$ against an alternative hypothesis $\overline{H}: \theta_0 > 0.24$. This implies that the parameter θ_0 is restricted to the range (0.24, 1) which seems both unnatural and unrealistic. In view of this, and of the general inconsistency in the literature, we shall follow the practice of rejecting the term 'null' and allowing the working hypothesis to be simple or composite according to practical need.

Indeed, the use of a point hypothesis is at best a convenient abstraction! In terms of application, or import, small departures from $\lambda = 80$ cause us little practical concern. Indeed, we should really design the test in terms of how far λ needs to depart from the value 80 before this becomes materially important. Such a facility is available through consideration of the second type of risk mentioned earlier (we take this up again in §5.4). Finally, the lack of any *formal* consideration of costs is again unfortunate in its implications. It is possible that values of λ greater than 80 might have quite different consequences to values lower than 80. If we could quantify this difference it would provide additional important information, which cannot be accommodated through a test of significance. On the other hand, the *pure* test of significance maintains its function of representing the import of the sample data *alone*; some claim that this is as far as we should go.

This brief review of some of the *classical* means of data analysis applied to a particular problem is far from comprehensive either in terms of what methods might be used for such a problem in practice, or in terms of the interpretation of concepts or conclusions. Even within this simple treatment we see clues to further developments, to inadequacies or arbitrariness in the basic ideas and to interconnections between the different techniques. For example, we must later trace the relationship between confidence intervals and tests of significance, and give thought to questions of optimality in the choice of statistical procedures. Also there is the question of choosing the sample size to meet particular needs, or indeed of conducting a *sequential* study where we take observations one by one until we obtain adequate information for the purpose in hand.

Strictly speaking, we must note that our whole analysis of the lifetime data, whilst illustrative, is in one basic respect suspect! It implicitly assumes that the size of the random sample of lifetimes, 140, was predetermined. This is not true; the sample size is itself a random quantity because of the way the sample was chosen and this may well affect the analysis. In general, the nature of the sampling mechanism *is* relevant to the construction of methods of statistical analysis in the *classical* approach. In other approaches it is not so! We will consider this factor in more detail later.

2.3   RELEVANT PRIOR INFORMATION: THE BAYESIAN APPROACH

So far attention has been restricted to sample data as the sole form of explicit information on which to base inferences and actions in our problem. But other information may exist, as has been discussed in Chapter 1. The TWEECO example provides a clear illustration of this and acts as a useful vehicle for demonstrating some of the techniques of **Bayesian inference**; an approach specifically concerned with the combination of sample data and prior information in the inferential process. Again we shall consider separately the questions of batch quality and component lifetime. This distinction of practical objective leads to an interesting distinction of emphasis (if not basic attitude) within the Bayesian approach!

2.3.1   Prior Information on Batch Quality

We start with a study of batch quality, being concerned with saying something about the quality, θ_0, of a batch currently being produced and sampled. The sample data consists of the observation that there were r defective components out of a random sample of size n; in the actual data r is 60 and n is 200. But there is a very real opportunity for obtaining additional information in such a sampling inspection problem. The current batch, with its value of θ expressing batch quality, is only the latest in a long sequence of such batches. Earlier batches, each with their own specific value of θ, have been produced and processed. We can imagine a long sequence of batches terminating with the current one, each indexed by a pair of parameter values (θ, λ) in our model. The parameter θ is assumed to vary in value from batch to batch, the parameter λ to be fixed. See Figure 2.3.

Not only do the current data (r, n) reflect on θ_0, but so indeed would the previous values $\theta_1, \theta_2, \ldots, \theta_k, \ldots$ of the quality parameter, if they were known. We can think of the current value θ_0 as typical of the *frequency distribution* of values of the parameter which arise from batch to batch. We might go further and talk of a probability distribution of θ values, the **prior distribution**, with an unexceptionable frequency interpretation. The current value θ_0 may be regarded as arising at random from this prior distribution. In terms of this distribution we can talk about the **prior probability** that θ_0 is in a certain range of values.

Suppose the prior distribution has probability density function $\pi(\theta)$. Then $\pi(\theta)$, if known, provides essential information about the current value of θ in defining

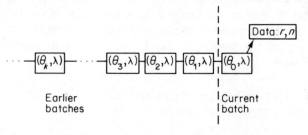

Figure 2.3

the probability framework from which it has arisen. The sample data add to this information in giving evidence on which specific value, θ_0, has arisen from the distribution $\pi(\theta)$. But there are two questions. Firstly, how can we know, precisely or inferentially, what $\pi(\theta)$ is? Secondly, how do we combine $\pi(\theta)$ and (r, n) to reflect on θ_0?

In this type of problem we are likely at least to have some previous *sample data* available to give an indication of the form of $\pi(\theta)$. How this might be utilized will be considered when **empirical Bayes' methods** are discussed later (§6.7). In our particular problem, however, it is to be expected that much more specific information on $\pi(\theta)$ is available. Any batches that are accepted by THECO will be subject to full inspection, by virtue of the components being used in assembling the hospital equipment of which they are part and which will be fully tested before being sold. For such batches the precise values of θ will thus be known. But it is possible that the precise values of θ will also be known for rejected batches—since these may have been fully inspected to salvage the effective components. So a large number of representative values of θ may well have accumulated and have been recorded. If sufficient in number these can give a fairly precise picture of $\pi(\theta)$. For illustration, suppose such records exist for the present problem, and in fact 498 previous θ values are available. The histogram of this information is shown as Figure 2.4.

Before considering how we might combine the information given by the prior distribution $\pi(\theta)$ with that yielded by the sample data (r, n), we note that $\pi(\theta)$ *in itself* opens up new avenues for enquiry about θ_0. It may be used to make direct probability assessments about what value, θ_0, has occurred in the current batch, quite apart from anything the sample data might tell us. Thus we can say that the value of θ for which $\pi(\theta)$ is a maximum is the *most likely value* to have arisen. Alternatively, we can measure the prior probability that θ_0 exceeds, or is less than, some particular value ϑ. We can even derive an interval in which θ_0 lies with some prescribed prior probability. This leads to an alternative type of confidence interval—a **Bayesian prior confidence interval**. (Terminology is not standardized on this topic; the terms **Bayesian interval estimate**, or **credible interval**, are sometimes used to distinguish the Bayesian, from the classical, concepts.)

To fix ideas suppose we know the exact form of $\pi(\theta)$. The curve superimposed on the histogram in Figure 2.4 is the probability density function of a **beta distribution** (of the first kind) with parameters 20 and 80. Empirically this seems a reasonable representation of the variation in θ values from batch to batch suggested by the histogram. Consequently we shall assume that

$$\pi(\theta) = \frac{99!}{19!79!}\theta^{19}(1-\theta)^{79}, \qquad (2.3.1)$$

which we will denote by $\mathscr{B}_1$ (20, 80). It is easily confirmed that $\pi(\theta)$ has a mode at $\theta_m = 19/98 = 0.194$. *In the absence of any other information* it seems sensible to take 0.194 as the best indication of the *current* value, θ_0, in that it is the most likely

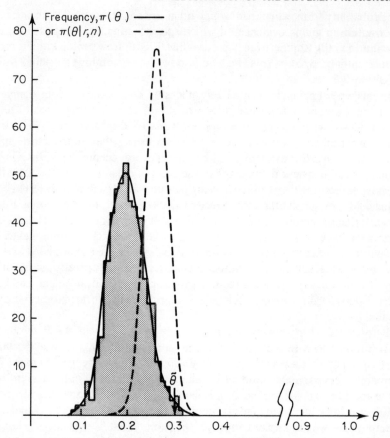

Figure 2.4   Histogram of batch qualities, with prior and posterior distributions

value *a priori*. Furthermore, we can obtain a direct assessment of the prior probability that the batch meets the standard required by THECO, that is that $\theta_0 \leqslant 0.24$. By means of a simple transformation this probability is obtained by inverse interpolation in tables of the **F-distribution** to be 0.835. This prior probability can equivalently be regarded as the long-run proportion of acceptable batches. A further extension is to derive a prior Bayesian confidence interval for θ_0. Transforming the lower and upper $2\frac{1}{2}$ per cent points for $F_{40,160}$ yields, via (2.3.1),

$$P\{0.128 < \theta < 0.283\} = 0.95,$$

so that (0.128, 0.283) can be regarded as a 95 per cent prior Bayesian confidence interval for θ_0. Note how on the one hand $\pi(\theta)$ describes variation of θ from batch to batch and how at the same time it represents our prior information about the value of θ for the current batch. In this latter respect $\pi(\theta)$ might be denoted by $\pi(\theta_0)$!

It seems reasonable to assume that the sample data will increase our knowledge of θ_0 over that provided by the prior distribution alone. But just how are we to make use of this extra information?

Our estimate of θ_0 from the sample data alone was $\tilde{\theta} = 0.30$. Viewed in relation to the prior distribution $\pi(\theta)$ shown on Figure 2.4, this seems untypically high. Intuitively we might expect that the combined information from the sample and the prior distribution would yield a *weighted* estimate of θ_0 somewhere between the most likely prior value of 0.194 and the sample estimate of 0.30. This is just what happens—but rather than obtaining merely a weighted *point estimate* of θ_0 we can weight the complete prior distribution by the information in the sample, to obtain an augmented *probability distribution* for θ_0 as an expression of the combined inference. Roughly speaking the sample data, $\tilde{\theta} = 0.30$, stretches $\pi(\theta)$ towards the higher values.

How is this achieved formally? We have only to apply a standard result in probability theory, namely **Bayes' theorem**, to obtain the *conditional distribution* of θ given the sample data (r, n)—hence the description of this approach as **Bayesian inference**. This conditional distribution, with density function denoted by $\pi(\theta|r, n)$, is called the **posterior distribution** of θ (or θ_0) and it provides the required measure of the combined information on the value of θ in the current batch given by both the prior distribution and the sample data. We have

$$\pi(\theta|r, n) \propto p_\theta(r, n)\pi(\theta), \tag{2.3.2}$$

where

$$p_\theta(r, n) = \theta^r(1 - \theta)^{n-r} \tag{2.3.3}$$

is the probability of obtaining the sample data from a batch of quality θ [or, in terms of variation in values of θ, the **likelihood function** of θ for the given sample (r, n)]. The constant of proportionality in (2.3.2) may be taken as

$$\left\{ \int_0^1 p_\theta(r, n)\pi(\theta)d\theta \right\}^{-1}$$

to make $\pi(\theta|r, n)$ a proper probability density function. *Note*: Since $\pi(\theta|r, n)$ describes our modified state of knowledge about the value of θ in the current batch, it might alternatively be denoted by $\pi(\theta_0|r, n)$.

In our example θ has the prior beta distribution (2.3.1) and it is easily confirmed that the posterior distribution is also of the same form. It is $\mathscr{B}_1(80, 220)$; in general a prior distribution $\mathscr{B}_1(l, m)$ is transformed by a binomial sample (r, n) to a posterior distribution $\mathscr{B}_1(l - r, m + n - r)$; see §6.6. So we have the distribution $\mathscr{B}_1(80, 220)$ as the augmented probabilistic assessment of what we know about θ_0 now that the sample data have been incorporated. As before we might choose to summarize the properties of the distribution. The most likely value, *a posteriori*, for θ_0 is now $79/298 = 0.265$; our estimate has been revised upwards by the unfavourable sample data in which $\tilde{\theta} = 0.30$. We find also that the posterior probability that $\theta_0 \leqslant 0.24$ has become approximately 0.147 compared with 0.835

a priori; and that a **posterior** 95 per cent **Bayesian confidence interval** for θ_0 is approximately (0·220, 0·309), compared with (0·128, 0·283). The decrease in $P(\theta_0 \leqslant 0·24)$, and in the width of the confidence interval, both similarly reflect the sample data in an intuitively reasonable manner. These features are evident in Figure 2.4 where the value of $\hat{\theta}$, and the posterior distribution, are both shown (the latter as a dotted curve).

At this stage it is worth noting certain of the features of the Bayesian approach as reflected through this example, also some distinctions of attitude or emphasis compared with the classical approach. A fuller discussion of these points will be given in Chapters 5 and 6.

(i) An immediate appeal in the Bayesian approach is the direct probability assessment it provides about the parameter θ. $\pi(\theta|r, n)$ measures in straight-forward probability terms our inference about θ. The Bayesian interval may be regarded as *a region within which θ lies with prescribed probability*, an interpretation we were unable to attribute to the *classical* confidence interval.

(ii) This direct probability assessment naturally suggests the *mode* of the distribution of θ as a summary measure for point estimation. In the classical approach it is more natural, on the concept of unbiasedness, to consider the *mean* of the sampling distribution to assess the value of a point estimator. In both cases the choice is somewhat arbitrary (later we shall see how extra consideration might also commend the *mean* for Bayesian estimation) but the different measures do seem to have the most immediate appeal. In either situation a case can be made for the use of the *median*!

The same spirit that suggests the mode as a point estimator also prompts a further condition on the choice of Bayesian confidence intervals. These should surely not *include* values of θ with lower probability (density) than any values of θ which are *excluded*. If the distribution of θ is not unimodal and symmetric, such a criterion may rule out the concept of exclusion in terms of 'tail-area probabilities' as used in the classical approach, and may lead to confidence *regions* rather than *intervals* (see §6.3).

(iii) The invariance of the distributional form for θ before and after sampling is interesting! Only the values of the parameters l and m are affected, being increased by the number of defective and effective components in the sample, respectively. It is as if our prior information is equivalent to a prior sample of size $l + m$ with l defectives. This concept of 'equivalent prior sample size' can be formalized, and is most useful. However, and this is most important, it does not arise universally but only for particular combinations of prior distributional form and sample data. If such a happy combination arises naturally, in the sense that such a prior distribution is a reasonable practical expression of the actual prior information available, then we will obtain a very simple, and valid, interpretation of the Bayesian analysis. Otherwise not! See §6.6.

(iv) The keystone to the Bayesian method is (2.3.2). Suppose we had sampled components *until r defectives were obtained*, needing n components to achieve this. In spite of this change in sampling method the posterior distribution is completely unaffected (after normalization) since the likelihood is unchanged. What this means is that the Bayesian approach takes no regard of the method of sampling, in direct contrast to the classical approach. Depending on attitude, this might be seen as an indication of the strength or weakness of the Bayesian approach!

(v) Finally we return to our first point, (i). This simplicity of interpretation is certainly attractive. But we must not be too hasty in our acclaim. The prior distribution $\pi(\theta)$, in view of (2.3.2), is a vital ingredient in the Bayesian method. In this example it was possible to give a detailed specification of $\pi(\theta)$ based on real knowledge. But this situation is far from representative, and in many cases we will be required to prescribe a specific form for $\pi(\theta)$ on perhaps only the flimsiest of information. Subjective elements may influence this choice and we shall need to consider the implications of not knowing $\pi(\theta)$ precisely. Then again, even if $\pi(\theta)$ is known precisely, there is the question of what is meant by the posterior distribution. In what sense is it a *probability* distribution? After all, θ has a determined, if unknown, value, θ_0, in the current batch!

Consider what happens if we were to obtain some more sample data in addition to the earlier observed (r, n). How would we employ Bayesian ideas to draw inferences in the light of further data (s, m)? Would we use as prior distribution the original $\pi(\theta)$, or the current posterior distribution $\pi(\theta|r, n)$? *The answer must depend on which batch (s, m) was drawn from.* If from the current batch with $\theta = \theta_0$, then $\pi(\theta|r, n)$ describes our state of knowledge about θ_0 before observing (s, m). (s, m) relate to that same batch and thus $\pi(\theta|r, n)$ would seem to be the appropriate *prior* distribution to use. If, however, (s, m) comes from a *new* batch, and $\pi(\theta)$ is truly a representation of how θ varies from batch to batch, it seems that $\pi(\theta)$ must again be the appropriate prior distribution when using (s, m) to draw inferences about θ in the new batch. This is a rather crucial distinction between repeated sampling from a single batch, and use of distinct samples from successive batches. It also has interesting implications with regard to the interpretation of the probability concept in the various prior, and posterior, distributions! See §6.8.1 below for further consideration of this matter.

2.3.2 *Prior Attitudes about Component Lifetimes!*

Can we apply a similar method of inference for λ? In principle, yes! However, rather than developing this in detail we shall consider just one aspect of the study of λ on the Bayesian approach. In considering θ we worked by reference to an accumulation of previous values which constituted a frequency distribution from which the current value had arisen as a 'typical member'. No such model is available for λ, which relates to the manufacturing process as a whole. It is unique (see Figure 2.3) and does n‹ t naturally figure in any actual (or even conceptual) manner as a representative of a more general population. It seems that we must abandon any direct frequency interpretation for the prior distribution of λ. This raises the further problem that we may know nothing about λ *a priori*; or at best have just some general indications from the physical structure of the process. But the Bayesian approach demands that a specific distribution is introduced. Such questions of interpretation and the formal expression of **prior ignorance** will be discussed later (Chapter 6). We consider here just one possible way in which subjective prior information might be utilized.

It is possible that information about λ, apart from that given by the sample data, is available in the form of technical knowledge of the nature of the manufacturing process; also in the form of feedback of information from hospitals using the equipment. Neither source is likely to yield very specific information. Such information as does arise will be unlikely to be quantitative but probably exists largely in the form of personal impressions of the technical, or service, staff. Nonetheless, it may be possible to get such staff to agree to a common quantitative expression of their prior *beliefs* about λ. This will provide at best a rough picture of the situation, amounting perhaps to a statement of the 'most likely value of λ' and of values of λ 'most likely to bracket the actual value'. For example, it may be agreed that 'the most likely value of λ is 75 MM' and that 'it is unlikely that λ is more than 4 MM away from 75 MM'. But this *must* be expressed in the form of a specific prior distribution for λ, and a somewhat arbitrary choice of distributional form must necessarily be made, based in part, perhaps, on mathematical expediency.

Suppose for illustration we assume λ to have a prior normal distribution, $N(\mu, \sigma^2)$. A possible interpretation of the expressed prior views about λ is to choose $\mu = 75$ and to take 71 and 79 as lower and upper $2\frac{1}{2}$ per cent points. This produces an approximate prior normal distribution, $N(75, 4)$, which can be augmented by the sample data, using (2.3.2), to yield the posterior distribution of λ. Our model claims that the data comprise random observations from $N(\lambda, 2\lambda)$, but for ease of illustration we shall suppose instead that this is specified as $N(\lambda, 160)$, which does not seem unreasonable for the data of Table 2.1. It is easy to confirm that this leads to a normal *posterior* distribution with mean 79·514 MM and variance 0·889 $(MM)^2$ (see §6.5). Again we might summarize this *augmented inference* by saying that the most likely value of λ, a posteriori, is 79·51 MM or that (77·66, 81·36) constitutes a 95 per cent posterior Bayesian confidence interval.

The speculative nature of this process hardly needs to be stressed. Qualitative subjective prior beliefs about λ have been combined with an arbitrary assignment of the prior distributional form of λ to yield a specific prior distribution. The form and propriety of the resulting posterior distribution in this example rests heavily on the validity of the prior distribution and obviously every effort must be made to ensure that we do have a reasonable representation of our prior knowledge. (On other occasions this may be less worrying if the prior information is essentially 'swamped' by the extent of the sample data.) There have been some studies made of the construction of prior distributions from *subjective* prior information. See, for example, Winkler[3-5], Savage[6], Murphy and Winkler[7], O'Carroll[8], Tversky[9], Suppes[10] and Hendrickson and Buehler[11]; also Chapter 3 of Raiffa and Schlaifer (1961).

Finally, just what is the nature of the probability concept on which a prior distribution such as that for λ is based? A frequency interpretation is no longer tenable here, and a 'degree-of-belief' interpretation seems to be required. This feature of the Bayesian approach leads to some critical comment from those unprepared to entertain anything but a frequency view of probability for the purposes of statistical inference!

2.4 COSTS AND CONSEQUENCES: SIMPLE DECISION THEORY IDEAS

There are other items of information that may exist in this problem, apart from those that have found tangible expression in the classical or Bayesian methods of inference described above.

In considering tests of significance on the quality of the current batch we noted that the relative expenses involved in scrapping a batch rather than supplying it to THECO, or in remedying the effects of a sub-standard batch supplied to THECO, may be relevant to the choice of what type of test to operate. These cost considerations arose only in an informal way, and were allowed to affect the issue only if they were extreme in one direction or the other. But it may be possible to be fairly precise in quantifying such costs as those of component manufacture, examination and testing of individual components or of any action arising out of supplying a batch of inadequate quality. This will be highly relevant information in a problem such as the one under study where we may reasonably assume the ultimate concern of the two companies to be a financial one; expressible in terms of maximization of profits.

We might hope, then, to use statistical methods which incorporate such information. The basic idea of **decision theory** is to derive rules for action, in a practical problem, which *objectively* incorporate the gains and losses arising from different possible actions: 'by your actions shall ye be judged'! We can illustrate some simple aspects of decision theory for the present problem, but we must bear in mind that not all situations are as straightforward as this one. A commercial problem resting, as it does, almost exclusively on *financial* cost considerations is likely to prove more amenable to the quantification of gains and losses than problems in other areas. Social, medical and psychological situations involve quite different concepts of gain and loss. These may be difficult even to define, let alone quantify. The intangibility of the loss structure is a basis for some critical comment, as we have already seen (§1.6), and is a matter we must return to. Indeed, even the evaluation of monetary gains and losses is not without its anomalies; see §4.7.

However, for the moment we will ignore any conceptual or interpretative difficulties and examine how financial cost considerations may be explicitly included in our analysis of the TWEECO–THECO operation, following fairly closely work by Wetherill and Campling[12] on the use of decision theory in sampling inspection.

To illustrate ideas we shall make some very simple assumptions about the consequential costs of defective components. In some convenient monetary units we shall assume that it costs an amount a to produce one component; that THECO accepts all components supplied to them and pays an amount, $a + b$, for each component, but charges an amount c for the inconvenience of dealing with each component subsequently found to be defective. This means that TWEECO makes a net *gain* of b units for each *effective* component that it supplies, sustains a net *loss* of a units for any component not supplied (irrespective of its quality), whilst *defective* components which are supplied provide a net return (positive or negative) of $b - c$ units. (This cost structure is not the only possible one. An alternative structure has already been suggested in §2.2. The actual choice must reflect the real 'facts of life' of the problem under investigation.)

Using this cost structure we can express its implications for individual components either in terms of net losses or net gains to TWEECO. The choice of whether to use *losses* or *gains* is arbitrary since we must expect either of these to take both positive and negative values depending on circumstances. For the moment we shall adopt the more usual convention of measuring losses rather than gains, although consideration of gains leads more naturally into the general discussion of the concept of **utility** later, in Chapter 4.†

For individual components we have the following **table of net losses** to TWEECO (sometimes called a **loss table** or, if dealing with gains, a **payoff table**); depending on the quality of the component and the action taken in respect of it by TWEECO.

Table 2.2

Quality	Action	
	Supply to THECO	Scrap
Effective	$-b$	a
Defective	$c - b$	a

† In Chapters 4 and 7 fuller reference is made to the lack of consistency in the literature in the *loss* or *gain* formulation of decision theory, and to some of the resulting ambiguity of terminology.

Obviously it is only the *relative* net losses which are relevant to the decision as to whether or not to supply a component, so that we can operate with a revised loss table which assumes zero losses for scrapped components.†

Table 2.3

	Action	
Quality	Supply to THECO	Scrap
Effective	$-f$	0
Defective	g	0

Where $f = a + b$ and $g = c - a - b$.

It is apparent that g must be positive. The alternative would imply that THECO is still prepared to make a positive payment even for a defective component. This is clearly unreasonable; also in this situation there would be no problem of choice of action for TWEECO since it would then be in their interest to supply components irrespective of their quality.

Suppose now that decisions are to be made for whole batches rather than individual components. The possible actions for each batch are to supply, or scrap, that batch; we will want to choose one or the other of these actions in the light of the cost information given in Table 2.3, and any information available on the overall batch quality as measured by the proportion defective, θ.

Consider first what happens if θ is known. We can measure the loss to TWEECO of supplying, or scrapping, a particular batch of quality θ_0 by what we might term the **batch-average loss**, per component, of these two actions. These are merely the *expected* losses, per component, with respect to variations in the quality of the components in the batch. Thus if we supply the batch the batch average loss, per component, is

$$L(\theta_0) = -f(1 - \theta_0) + g\theta_0. \tag{2.4.1}$$

On the other hand, if we scrap the batch each component contributes a zero loss irrespective of its quality, and the batch-average loss, per component, is

$$M(\theta_0) = 0. \tag{2.4.2}$$

But $L(\theta_0)$ can be rewritten as

$$L(\theta_0) = k(\theta_0 - \theta_B), \tag{2.4.3}$$

† Note that we are implicitly assuming that the batch *has been manufactured* and are then seeking the best action to adopt for that batch. The subsequent discussion (and in particular the concept of 'breaking-even') is pursued from this standpoint. The more basic question of deciding whether or not to manufacture a batch requires a different (more complicated) model. Both have their relevance for different situations.

where $k = f + g$ and $\theta_B = f/(f + g)$. So we see that the batch-average loss, per component, of supplying the batch varies linearly from $-f$ units when $\theta_0 = 0$ (i.e. all components effective) to g units when $\theta_0 = 1$ (i.e. all components defective). The decision whether to supply, or scrap, the batch will obviously rest on whether $L(\theta_0)$ is negative or positive, in view of (2.4.2). If θ_0 were known any prior information on variations in batch quality, or sample data, would of course be superfluous. Thus in such a *no-data* decision theory problem we should *decide to supply the batch if* $\theta_0 \leqslant \theta_B$ *and to scrap it if* $\theta_0 > \theta_B$. For obvious reasons $\theta_B = f/(f + g)$ may be regarded as, and is commonly termed, the **break-even quality.**

For illustration, suppose the manufacturing costs per component are 5p, and THECO pays 6p for each component with a subsequent penalty charge of 24p for any defective component. Then $f = 6$, $g = 18$, so that the break-even quality is given by $\theta_B = 0.25$, and any batch with less than 25 per cent defective components should be supplied to THECO.

Of course, it is unrealistic to assume that the actual quality of a batch will be known precisely. In practice we will *not* know θ_0 and we must utilize any prior information or sample data to throw light on this unknown value. Suppose for the moment that we have no prior information about θ but do possess sample data (r, n) from the current batch. We might assume that collecting these data involved some expense, say at a rate, per observation, of one unit on our cost scale. Other cost schemes for experimentation are also feasible. For example there might be a fixed overhead cost irrespective of sample size with additional constant costs per observation, or again costs per observation may reduce as the sample size increases. For illustrative purposes, however, we adopt the simple scheme of a constant unit cost per observation with no 'overheads'.

Consider what happens if we base our decision to supply or scrap the batch on the sample data alone: *by choosing some value γ and scrapping the batch if $r > \gamma$, otherwise supplying the batch to THECO.* (Compare this approach with the significance test criterion in §2.2.1.) Such a **decision rule** is intuitively in accord with our desire to scrap batches of sufficiently low quality, but we need to find some sensible measure of its usefulness. In fact, this is again easily expressed in terms of the table of losses per component (Table 2.3), the batch quality, θ_0, and the sample size, n.

We need to specify the size of the batch if we are to compare trading profits (or losses) per batch with costs of sampling. Suppose each batch contains the same large number, N, of components. The loss that arises from operating the decision rule for a particular batch depends on the batch quality, θ_0, and the value of r in relation to γ. In view of the batch-average losses, per component, given in (2.4.1) and (2.4.2), the overall loss for the whole batch due to this decision rule splits into two parts, as

$$L_n(\theta_0, r) = \begin{cases} NL(\theta_0) + n & (r \leqslant \gamma), \\ NM(\theta_0) + n & (r > \gamma). \end{cases} \qquad (2.4.4)$$

From (2.4.2) and (2.4.3) this gives

$$L_n(\theta_0, r) = \begin{cases} Nk(\theta_0 - \theta_B) + n & (r \leqslant \gamma), \\ n & (r > \gamma). \end{cases} \qquad (2.4.5)$$

Note how an actual *gain* will arise only if θ_0 is sufficiently less than θ_B to counteract the sampling costs, n.

It is envisaged that we operate this decision rule repeatedly for deciding what action to take with different batches. Consider a sequence of batches of the same quality θ_0. In some cases the sample data will yield a number of defectives, r, in excess of γ; in other cases not. Taking an average over the distribution of r we obtain an *average loss for the batch of quality* θ_0 when using this decision rule. This is known as the **risk function** for the decision rule and may be denoted $R_{n,\gamma}(\theta_0)$. We have

$$\begin{aligned} R_{n,\gamma}(\theta_0) &= E_r\{L_n(\theta_0, r)\}^\dagger \\ &= \{N L(\theta_0) + n\} P(\text{supply batch of quality } \theta_0) \\ &\quad + \{N M(\theta_0) + n\} P(\text{scrap batch of quality } \theta_0) \\ &= \{Nk(\theta_0 - \theta_B) + n\} P(r \leqslant \gamma | \theta_0) + nP(r > \gamma | \theta_0) \\ &= Nk(\theta_0 - \theta_B) P(r \leqslant \gamma | \theta_0) + n. \end{aligned} \qquad (2.4.6)$$

This is again a *conditional* concept, involving only those batches of some fixed quality θ_0; as we remarked for the use of classical methods generally in §2.2.

The risk function, as a function of θ_0, measures the long-run expected loss from operating this decision rule on batches of quality θ_0. Since we will want to consider how $R_{n,\gamma}(\theta_0)$ varies from batch to batch of different qualities, it is again expedient (as in §2.3) to drop the subscript on θ and denote the risk function by $R_{n,\gamma}(\theta)$. What we would like to know is, for a given sample size n, what value to choose for γ in order to produce in some sense the best decision rule of the type under consideration. But it is no simple matter to define what is best in this context.

We see this by considering qualitatively the way in which $R_{n,\gamma}(\theta)$ varies with θ for different values of γ; shown in typical form in Figure 2.5 for γ near to zero, at some intermediate value between 0 and n, and near to n (with an appropriate choice of the values of the basic parameters a, b, c, n and N). The way in which the shape of $R_{n,\gamma}(\theta)$ varies with γ is easily confirmed from (2.4.6). In particular, $R_{n,\gamma}(\theta_B) = n$, whatever the value of γ. Ideally we should like to choose a decision rule (i.e. value of γ) with a risk which is lower for all values of θ than that of any other rule of the type being considered. But Figure 2.5 indicates that this is not possible; we cannot obtain a 'best' decision rule irrespective of the value of θ. If θ happens to be small we seem best advised to choose γ near to n; that is, only to scrap batches if almost all the observed components are defective. But whilst this minimizes the risk when θ is small we are also maximizing the risk if θ turns out to

---

† The subscript r on the expectation operator implies that we are considering the expectation over the distribution of values of r.

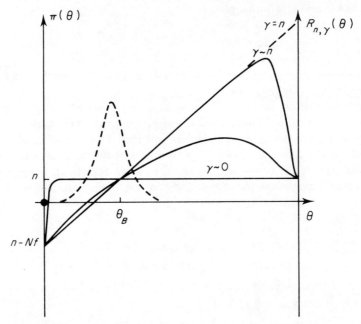

Figure 2.5   Risk function, $R_{n,\gamma}(\theta)$, and prior distribution, $\pi(\theta)$

be large. In contrast, to choose γ near to 0, that is to accept batches only if the sample contains no defective components, minimizes the risk for large values of θ but provides a very poor return if θ is in fact small.

Not knowing θ, how are we to resolve this dilemma? A policy sometimes adopted is to choose that decision rule for which *the maximum possible risk is minimized*. This yields what is called the **minimax procedure**. From Figure 2.5 we see that in the present example this corresponds with choosing $\gamma = 0$; that is, only supplying batches where the sample contains no defective components. In guarding against the worst that can happen the minimax policy implies an extremely pessimistic attitude. This may occasionally be justified but must inevitably be paid for in terms of poor returns in favourable circumstances. We see this highlighted in the example where, although we minimize our maximum risk, *we more or less ensure that no gain can be achieved* unless θ is very close to zero.

Unfortunately, decision theory *per se* offers no alternative means of distinguishing between the risk functions to choose a best decision rule, other than in the rare eventuality of one rule having a risk function uniformly (in θ) lower than any other. Such a **dominant** rule, if one exists, is obviously the one to choose. At the opposite extreme, any decision rule for which the risk function is uniformly higher than that of *some* other rule may be immediately rejected; we would hope to be able to 'weed out' such **inadmissible** decision rules.

All values of γ in our example lead to admissible decision rules, and TWEECO might surely expect to do better than adopt the minimax procedure.

One way of doing this is to take account of what is known about the values of θ which might arise from batch to batch. If it so happened that we knew that nearly all values of θ were greater than θ_B then the choice of γ would be obvious. We see from Figure 2.5 that in such circumstances we would be best advised to choose $\gamma = 0$; that is only to supply batches in which no defective components were observed. In contrast, if nearly all values of θ were expected to be less than θ_B we should scrap a batch only if the sample contained all defective components. This amounts to saying that it makes no sense to base our actions on the implications of values of θ which are unlikely to be encountered.

Formalizing this attitude; we must aim to use any information we have about what values of θ might arise in practice to refine our choice of decision rule. In our problem quite specific information about θ is available in the form of the prior distribution (2.3.1), and it seems sensible to impose *Bayesian* ideas on the decision theory framework. We could weight the risk function by the prior distribution, and measure the usefulness of a decision rule by the **expected risk**

$$r_n(\gamma, \pi) = E_\theta \left\{ R_{n, \gamma}(\theta) \right\}$$
$$= \int_0^1 R_{n, \gamma}(\theta) \pi(\theta) d\theta, \tag{2.4.7}$$

that is, by considering the expected value of $R_{n, \gamma}(\theta)$ with respect to the prior distribution $\pi(\theta)$. The criterion of choice now amounts to adopting that decision rule (value of γ) for which $r_n(\gamma, \pi)$ is a minimum.

The choice of the posterior *mean* value of $R_{n, \gamma}(\theta)$ might seem somewhat arbitrary; other summary measures of location of $R_{n, \gamma}(\theta)$ with respect to $\pi(\theta)$ might appear equally appealing. We shall return later (§7.3) to the question of why the mean value is used.

The prior distribution of θ in the TWEECO problem is shown dotted on Figure 2.5 and its qualitative effect may be assessed from the diagram. Obviously, if $\pi(\theta)$ is highly concentrated around some value much less than the break-even value, θ_B, we will be led to a best procedure (in the class under consideration) based on a choice of γ close to n; if $\pi(\theta)$ is highly concentrated at a point in excess of θ_B the best procedure will utilize a value of γ close to 0. Apart from these extreme situations, however, we must expect the best procedure to involve a choice of γ intermediate between 0 and n. The evaluation of γ to minimize $r_n(\gamma, \pi)$ for the present problem, where $\pi(\theta)$ is taken to be $\mathscr{B}_1(l, m)$ and $R_{n, \gamma}(\theta)$ to have the form (2.4.6), is not difficult for a particular assumed sample size, n. Using the so-called (more useful) **extended form** of decision theory analysis (see Chapter 7) the appropriate value turns out to be

$$\gamma = \theta_B(n + m + l) - l, \tag{2.4.8}$$

reflecting in a natural way the influence of the information given by the sample data and prior distribution. Through the influence of θ_B in (2.4.8) the loss structure is seen to be crucial. In the problem being discussed we have $\theta_B = 0.25$, $n = 200$, $m = 80$, $l = 20$, so that $\gamma = 55$. We should thus supply any batch for which our sample contains at most 55 defectives; otherwise scrap it. The extent to which this best procedure improves on any other choice of γ, say γ', is measured by the respective expected risks, min $r_n(\gamma, \pi)$ and $r_n(\gamma', \pi)$, given by (2.4.7). The evaluation of $r_n(\gamma, \pi)$, for any γ, involves some tedious calculations but is by no means prohibitive on a computer.

We have considered only a very special type of decison rule, based on scrapping the batch if $r > \gamma$ in samples of a fixed size n. Even within this limited framework we would need, in practice, to go beyond the optimal choice of γ for a given value of n to consider the simultaneous choice of both the sample size n and the critical number of defectives γ. That is, we need to choose both n and γ to minimize $r_n(\gamma, \pi)$, with resulting expected risk, $\min_{n,\gamma} r_n(\gamma, \pi)$. In principle this double optimization is straightforward, although the calculations are tedious (if not prohibitive). See Barnett[13].

But there is no obvious reason why this particular type of decision rule should be best. Ideally we should attempt an unconstrained choice of optimum decision rule from amongst *all* possible ones. Many will be immediately rejected ('supply the batch if $r > c$, otherwise scrap it'; 'supply the batch only if r is even'). Whilst there is no immediately obvious reason why the best choice should fall into the class previously considered, *this does in fact turn out to be so*. Again see Barnett[13]. That rule which has minimum expected risk amongst all possible decision rules is called the **Bayes' decision rule**; its expected risk, the **Bayes' risk**.

Rather than adopting a fixed sample size procedure of the type described, we might consider an alternative procedure where components are sampled *sequentially* from the batch and their cumulative import assessed until, on some appropriate criterion, a decision can be made about the batch. As in well known, classical sequential procedures may improve on fixed sample size ones. The same is true of sequential *decision* procedures although their construction may be complicated, sometimes prohibitively so, from the point of view of application. See De Groot (1970), Chapter 12, for further discussion and references to fundamental work on sequential decision theory, also Wetherill and Campling[12] consider specifically the sequential approach in sampling inspection problems.

Let us consider briefly what has been achieved by adopting a decision theory approach in the study of batch quality. It has been possible to utilize all three sources of information; prior knowledge of variations in quality from batch to batch, sample data and consequential cost considerations. In giving direct consideration to this latter factor we are able to construct specific rules for action rather than restricting ourselves to *inferential* statements about batch quality such as were obtained through the Bayesian approach in §2.3. Undoubtedly any derived decision rule will rest heavily on the loss structure which is adopted, and its usefulness will depend critically on this loss structure being appropriate to the problem in hand. This is at one and the same time the strength and weakness of decision theory; that it allows the consequences of different actions to be a central factor in the choice of what specific action to take, but requires these consequences to be reliably and accurately quantified. Some problems may not admit such accurate quantification as the one we have considered.

Finally, an important distinction between decision theory and other approaches appears in the way in which the different possible actions are compared. In expressing this comparison in terms of expected losses or risks there is no probability measure attached to the results. One decision rule is better than another in so far as it has a lower expected loss. Such a direct comparison is appealing but not always immediately interpretable, particularly when the losses are not monetary or economic. Furthermore, the averaging process involved in producing the Bayes' procedure is a complicated two-stage one. It involves a conditional aggregate concept (as in the *classical* approach) for dealing with the sample data, and in addition a weighting by a prior probability distribution which may have had to be *subjectively* specified. What this really means in practical terms needs to be properly appreciated!

2.5 COMMENT AND COMPARISONS

Our study of the TWEECO–THECO problem has in its various facets thrown up certain clear distinctions of approach, and suggested some specific methods of handling different types of information. We can use this experience to make some simple comparative comments.

(a) *The Classical Approach.* Designed specifically for the *processing of sample data alone*, this approach makes no provision for the *direct* use of prior information or of consequential costs or benefits. On the other hand, it is apparent that such extra information will affect the choice of what hypothesis to test, or of significance levels, in an informal manner. This informality may be felt to lead to the imprecise use of prior or consequential information; but, in contrast, if the information itself is insubstantial or subjective it might be thought best not to be forced into an arbitrary quantification. The approach is neither entirely 'inferential' nor entirely 'decision-making'. Point estimation is undoubtedly inferential, but hypothesis testing must often play an 'action guidance' role. Classical procedures are constructed and assessed through the idea of a 'sampling distribution' based on a *frequency* probability concept. Any measures of their success thus involve aggregate, or long-term, considerations. As such no direct assurance is possible of the truth of a *particular* inference or decision.

(b) *The Bayesian Approach.* Sample data is now augmented by prior information, through the use of Bayes' theorem, in the drawing of inferences about a probability model. These inferences now have a direct probability interpretation, with the advantage that the accuracy of 'particular' inferences may be immediately assessed. When the prior information is detailed and may be expressed in terms of a *frequency*-based probability distribution few obvious interpretative difficulties arise (but see §6.8). But prior information is seldom as comprehensive as this. Nevertheless, the Bayesian approach *demands* a full quantitative statement of prior information even if the actual prior information is insubstantial, subjective or for that matter non-existent. The virtues of this approach in general will need to be judged in terms of how well we can construct detailed statements of the prior information, on how critically the resulting inferences depend on these statements and on the probability interpretation of the results. In particular, an entirely *frequency*-based view of probability cannot be maintained and alternative attitudes will need to be adopted in some contexts.

(c) *Decision Theory*. This third approach is entirely 'decision-making' or 'action guiding' in spirit. To this end it requires the statement of a set of possible actions and an assessment of the losses or gains which would arise from such actions in different circumstances. Sample data (and possibly prior information) are processed to describe the prevailing circumstances, and the approach aims at producing a 'best' action in terms of the pattern of losses and gains. The value of a prescribed policy for action is not measured in probability terms but in terms of the loss structure. As in the classical approach, however, aggregate or long-term considerations affect the choice of what is a 'best' policy. So no immediate assurance is possible, only the knowledge that, *viewed in relation to a series of decisions in the same situation*, the policy which prompts a current action is a sensible one. The choice of a 'best' policy for action depends strongly on the loss structure. The strength of this approach rests on its ability to take overt consideration of this loss structure and in extending the inference function of the other two approaches to one of actual decision-making. Its success depends on the validity of the quantitative losses and gains which are used. The mere specification of the set of possible actions is not always a simple matter, neither is the definition of what constitutes a 'best' policy for action. This latter feature seems most straightforward when prior information is available; but then the specification problems and interpretation aspects of the Bayesian approach must also be considered.

References

1. Barnett, V. D. and Ross, H. F. (1965). 'Statistical properties of computer acceptance tests', *J. Roy Statist. Soc. A*, **128**, 361–393.
2. Davis, D. J. (1952). 'An analysis of some failure data', *J. Amer. Statist. Assn*, **47**, 113–150.
3. Winkler, R. L. (1967). 'The assessment of prior distributions in Bayesian analysis', *J. Amer. Statist. Assn.*, **62**, 776–800.

† 4. Winkler, R. L. (1967). 'The quantification of judgment: some methodological suggestions', *J. Amer. Statist. Assn*, **62**, 1105–1120.

† 5. Winkler, R. L. (1971). 'Probabilistic prediction: some experimental results', *J. Amer. Statist. Assn*, **66**, 675–685.

6. Savage, L. J. (1971). 'Elicitation of personal probabilities and expectations', *J. Amer. Statist. Assn*, **66**, 783–801.

† 7. Murphy, A. H. and Winkler, R. L. (1977). 'Reliability of subjective probability forecasts of precipitation and temperature', *Appl. Statist.*, **26**, 41–47.

† 8. O'Carroll, F. M. (1977). 'Subjective probabilities and short-term economic forecasts: an empirical investigation', *Appl. Statist.*, **26**, 269–278.

† 9. Tversky, A. (1974). 'Assessing uncertainty' (with Discussion) *J. Roy. Statist. Soc. B.*, **36**, 148–159.

10. Suppes, P. (1974). 'The measurement of belief' (with Discussion), *J. Roy. Statist. Soc. B*, **36**, 160–191.

11. Hendrickson, A. D. and Buehler, R. J. (1972). 'Elicitation of subjective probabilities by sequential choices', *J. Amer. Statist. Assn*, **67**, 880–883.

12. Wetherill, G. B. and Campling, G. E. G. (1966). 'The decision theory approach to sampling inspection' (with Discussion), *J. Roy. Statist. Soc. B*, **28**, 381–416.

13. Barnett, V. (1974). 'Economic choice of sample size for sampling inspection plans', *Appl. Statist.*, **23**, 149–157.

CHAPTER 3

Probability

At this stage we must dig somewhat deeper into the subsoil of inference and decision-making by examining two truly basic concepts, those of **probability** and **utility**. The concept of **probability** is crucial to any study of our subject. Whilst many of the ideas of inference and decision-making may be developed and discussed with only an informal (essentially undefined) attitude towards probability, we have already observed how certain *criteria* and *desiderata* in statistical theory imply a particular probability viewpoint. In reverse, specific views on the basic nature of probability have been seen to colour the approaches to inference or decision-making which incorporate such views. **Utility** (as a means of placing a quantitative measure on some action in relation to its consequences) also plays a central role in the construction and interpretation of statistical ideas; whether it enters purely informally as in the choice of significance levels or working hypotheses in the classical approach, or formally as an integral part of the quantitative information to be assessed by decision theory methods.

This chapter and the next deal respectively with probability and utility, although we shall see that it is not possible to dissociate these concepts completely.

3.1 TYPES OF PROBABILITY

As for our main theme of statistical inference and decision-making, so the variety of approaches to *probability* is vast and complex. Each writer on the subject has his own individual slant, often developed in great detail. Whole volumes are devoted to a single viewpoint and its implications. See, for example, von Mises (1957), Reichenbach (1949), Jeffreys (1961), Savage (1954), de Finetti (1974, 1975). Many (e.g. von Mises, or de Finetti) can countenance only one specific view of probability, whilst others (Bartlett, Koopman, Carnap) would claim that probability has different roles to play in different circumstances and that no single attitude is adequate. Then again, the *circumstances* under which the probability concept has meaning may be deliberately restricted (see §1.3); or restrictions may be placed on the extent to which probabilities are *numerical* quantities rather than merely qualitative relationships. See Borel[1], or Koopman[2], in this respect.

We shall consider the broad spectrum of views of probability, bearing in mind that our prime interest is in the way in which different views of probability interrelate with different modes of statistical reasoning. Such a review, of necessity, grossly oversimplifies the true situation. The simple classification which is adopted is by no means the only possible one, but serves as a useful framework for understanding the different ways in which the probability concept is viewed and applied. The reader who wishes to explore ideas in greater depth will find the references given in the text useful. These references have been selected from the enormous literature on this topic on grounds of readability and accessibility; as a result they are not always the earliest relevant references but they provide a useful springboard for detailed study.

There are some *summaries* of the variety of attitudes to probability. These include the introductory remarks by Kyburg and Smokler (1964), Savage[3], de Finetti[4], Atkinson[5] and the opening chapters of Good (1950, 1965) and Carnap and Jeffrey (1971). Most are quite brief, being intent on introducing a more detailed study of some *particular* attitude. The exceptions to this are Atkinson[5], the more accessible (if somewhat slanted) survey by de Finetti[4], and, in particular, the detailed comparative study of different theories and concepts of probability by Fine (1973). Many different methods of classifying probability viewpoints have been advanced (see, for example, Good, 1950, de Finetti[4] or Fine, 1973) but we shall adopt a simple *subjective/logical/frequency* system and take as a starting point some of the distinctions drawn by Good (1965, Chapter 2). Within this simple system we shall distinguish just four basic views of probability, as shown in the following table.

Probability viewpoint	Basis
'Classical'	Symmetry considerations; 'equally likely outcomes'
Frequency (or **frequentist**)	Empirical; relative frequencies in 'repeatable' situations
Logical	Objective; intrinsic 'degree-of-belief' as a logical measure of implication
Subjective	Personalistic; individual assessment of 'rational' or 'coherent' behaviour

It is easy to see why different views of probability are entertained and how different circumstances influence the relative claims for the separate views.

In our everyday lives the words 'probability', 'likelihood', 'chance' constantly arise: 'In all probability I shall stay at home tonight', 'There is little chance that the U.N. can influence the Ankardia crisis', 'It is unlikely that I shall remain in this job for long'. Such statements are made with little (if any) regard to a *conscious* interpretation, let alone any measurement, of the uncertainty concept. The words are used in a largely conventional way. At most they express a *personal* conviction that the proposition under consideration (staying at home tonight, etc.) is neither inevitable nor impossible. The emphasis in the remark may suggest a leaning towards the one or other extreme of some 'inevitability–impossibility scale'. In this sense we see an informal attempt to introduce into our human affairs a system of 'partial belief' specific to the individual 'I'. But equally apparently 'I' will often need to *take action* on the basis of such feelings. This is so when 'I' decide to cross a crowded road ('It is probably safe to cross now!'), or take a shot at goal in a football match rather than trying to manoeuvre into a more favourable position. The probability concept implicit in such human assessments or actions is described by Good (1965) as **psychological probability** and has been discussed by Cohen (1960).

In trying to set up a formal definition of probability from this standpoint we must allow for the fact that individuals differ (and even individually are not always consistent) in interpreting their personal assessments of uncertainty; in acting on their personal 'degree-of-belief' in some eventuality. An individual's assessment may vary with his mood even in apparently similar tangible circumstances. A footballer who has just scored a goal to put his team in the lead may confidently take another pot-shot from 40 yards in the spirit that there is nothing to lose, but on another similar occasion may prefer to try to dribble the ball into the goal *in spite of the fact he has nothing to lose by taking a pot-shot from 40 yards.*

Then again, we detect an 'irrationality' in human behaviour which may render the individual liable to be exploited, or perhaps unable to effect a preference in certain circumstances. This is easily demonstrated by actual behaviour, or simulated in laboratory experimentation. Exploitation can be demonstrated in some betting situations, where a person may be quite prepared to accept a series of bets where the stakes, and odds, imply that he is bound to lose *whatever happens*. Inability to effect a preference appears, trivially, in the common hesitancy with which so many of us face rather mundane alternatives with no immediately discernible relative advantages—(chips or mash!). It is tempting to equate no preference with equality of preference, but is this necessarily what is felt in such circumstances? No preference often stems from essentially no relevant information, or ignorance of the prevailing circumstances, and we shall see later the difficulties that arise in expressing this concept in formal terms (see §3.2, but particularly §6.4).

In spite of, indeed spurred on by, such observable 'irrationality' in human behaviour an important *behaviourist* school of thought on probability has developed. This promotes a concept of **subjective probability** and provides a framework for processing (usually) numerical representations of the *degrees-of-belief* that a person holds for different eventualities on the basis of his environmental experience. We shall consider the details of this approach later (§3.5), but its motivation and direction are easily summarized.

The attitude is essentially this! Probability acts as an intuitive stimulus to the individual in his assessment of situations involving uncertainty (i.e. less than certainty). A formal theory of probability (whether numerical or quasi-logical) must respect this personal basis, and serve to define the probability concept in terms which admit *conscious* (rational, formal) processing of this intuitive stimulus—perhaps as an aid to its better application. In this latter respect it becomes difficult to distinguish the subjective probability concept *per se* from some attempts to construct a theory of decision-making—the concept of *utility* becomes the implicit link as we shall see in the next chapter. A person's intuitive probability assessments are completely bound up with his environmental and psychological experiences. Thus only *conditional* probabilities (conditioned by this experience) have any meaning in probability theory (see Lindley, 1971a, or Good, 1950). Individuals are irrational in their processing of uncertainty basically because they are ignorant of how this should be achieved. If only enlightened through formal probability theory they would (according to the *subjective* viewpoint)

(i) appreciate that all alternatives can be ordered in terms of their relative positions on the inevitability—impossibility scale,

(ii) be *consistent* in their judgements; for example, if A is felt to be more likely than B, and B than C, then C cannot be felt more likely than A,

(iii) be *coherent* in not being prepared to accept a series of bets under which they *must* lose. (Some would say 'may lose'.)

Subjective probability theory represents the behaviour of the individual who is 'enlightened' in these respects. It deals with a conceptual 'rational man' and shows that this enlightenment leads logically to a system of axioms for probability theory essentially the same as those adopted in most approaches; although their interpretation is quite different. Only quite late in the development of the *subjective* attitude have methods been advanced for assigning actual numerical values to an individual's personal probabilities for some eventuality, usually involving him in a process of introspection through the consideration of different hypothetical bets (at different odds) concerning the eventuality in question, or allied ones. It is fully anticipated that these numerical values may vary from one person to another. After all, their background experiences are likely to be different and it is *the individual* who must assess the situation and produce *his* evaluation (perhaps as a basis for action). On the other hand, situations arise where it will be reasonable to assume a deal of overlap in the experiences of different individuals; in such cases it will hardly be surprising if their probability assessments virtually coincide.

de Finetti stresses the *personal* subjective nature of probability.

> My thesis . . . is simply this:
>
> ## PROBABILITY DOES NOT EXIST.
>
> The abandonment of superstitious beliefs about . . . Fairies and Witches was an essential step along the road to scientific thinking. Probability, too, if regarded as something endowed with some kind of objective existence, is no less a misleading misconception, an illusory attempt to exteriorize or materialize our true probabilistic beliefs.
>
> In investigating the reasonableness of our own modes of thought and behaviour under uncertainty, all we require, and all that we are reasonably entitled to, is consistency among these beliefs, and their reasonable relation to any kind of relevant objective data ('relevant' in as much as subjectively deemed to be so). This is Probability Theory.

(de Finetti, 1974, p. x)

This *rationale* is the broad basis of subjective probability, in its various formulations.

As an alternative to the *subjective*, or *personal*, view of probability we have what is termed the **logical** concept of probability (called the **necessary** view by Savage; see Savage[3]). Preceding subjective probability in terms of the history of its *formal* development, it aims at representing different degrees in the relationship between one proposition and another, or between a proposition and a body of evidence. Motivated by the principles of mathematical logic in which a proposition A either implies, or refutes, another proposition B, it modifies the relationship between A and B to admit a *degree of implication* in real life situations. Thus if A is some propositional information (or set of statements) describing a particular situation and B is some other statement relating to that situation, we measure by probability the *extent* to which A supports B—which in general is intermediate between the extremes of logical implication, or complete denial. In distinction to the idea of subjective probability, this measure [conventionally represented on the scale $(0, 1)$] is assumed *unique*. There is one and only one degree of implication of B afforded by A, *logically* determined by A and B. This degree of implication, $p = P(B|A)$, is a *formal* property of A and B. Probability becomes a formal concept relating A and B, in a sense somewhat similar to that of *mass* or *velocity* as a formal property of a physical body.

This view has certain implications. Again probability is an entirely conditional concept. Furthermore, it is not expected that we will know $P(B|A)$ in any particular context. Individuals may, by various means, attempt to throw light on the value of $P(B|A)$. In their support they may invoke empirical evidence (information gleaned from appropriate experimentation) or subjective impressions or experience. The extent to which they are unsuccessful in determining the value of $P(B|A)$ is the reflection of the inadequacy of their empirical, or subjective evidence. Good (1965) summarizes this in the following way:

> A logical probability, or . . . 'credibility' [in the terminology of early writers] is
> a rational intensity of conviction, implicit in the given information, and such that
> if a person does not agree with it he is wrong. (p. 7.)

(Note how different this latter attitude is to the *subjective* view!) With this attitude statistical analyses are seen as the means of measuring, or assessing the effects, of this 'inadequacy'.

It is apparent that the logical view demands the introduction of a concept of 'ignorance' or 'no information', with the difficulties which (we shall see) this generates.

Such a formal view of probability leads to rules for its logical consistency akin to the usual probability axioms which are encountered in almost all practical approaches to probability.

Even earlier in terms of detailed development, and more limited in its range of application, we have the **frequency** view of probability. Nonetheless, this approach has been more widely discussed and used than any other. It provides the basis, and interpretative framework, for classical statistical methodology. As such it is encountered as the prime attitude to probability inherent in the practical application of statistics in a wide range of different disciplines. The major element in this view of probability is that the concept should be an 'objective' one, in being divorced from any conideration of *personal* factors, and amenable to practical demonstration through experimentation. This causes the frequency view to differ fundamentally from the behaviouristic (personalistic) or logical attitudes outlined above, since there is now only a restricted class of practical situations in which it is reasonable to define a probability concept. By 'reasonable' is meant; where such a concept can be unambiguously applied, free from the 'intangible, ill-defined or immeasurable factors inherent in a subjective view of probability' and where unique probabilities exist which can be demonstrated empirically. The frequency view claims that to meet these aims probability can only be defined and used in situations which (potentially at least) are *able to be repeated over and over again under essentially identical conditions.*

Thus the possibility that the poet Homer was blind is not one which is amenable to such probability arguments, nor (perhaps) is the question of whether the footballer decides to 'take a pot-shot at goal'. On the other hand the quality of an industrial component coming off the production line, or the life expectation of a school teacher, might be discussed in probability terms by reference to successive components of the same type produced under the same circumstances, or to the set of 'similar school teachers'. Furthermore, factual data may be obtained in each case to throw light on the values of any relevant probabilities.

In fact the only information regarded as relevant to probability assessments in any situation comes from observing *outcomes* in repeated realizations—what we have previously called *sample data*. Probabilities have unique (if unknown) values determined by the nature of the situation under study. As such probability is an unconditional concept (cf. the subjective or logical views), having no concern for the different environmental experience of different individuals, nor for any circumstantial evidence relating to the situation. This uniqueness, however, leans heavily on the assumption that basic conditions do not change, and that we really can recognize situations where this is so.

The frequency approach rests on two observable features of the behaviour of outcomes from repeated realizations.

Firstly, it is a fact that outcomes vary from one repetition to another in an unpredictable manner. This is what is meant by the term 'random variation'. Secondly, it is noted as an empirical fact that out of such short-term chaos a sort of long-term regularity emerges. This regularity shows itself in the following way. Suppose there is some particular eventuality, A, that interests us. We take repeated observations noting those occasions on which A occurs. Then the ratio of the number of times A occurs, n_A, to the total number of repetitions, n (the **frequency ratio**, or **relative frequency**, of occurrence of A, n_A/n) seems to tend to some limiting value as $n \to \infty$.

Without attempting to place any mathematical construction on this purely empirical observation of the long-term stability of relative frequencies, its adoption as an essential feature of practical situations involving uncertainty leads to an *empirical* (**frequency**) view of probability. This takes the following form. As an empirical fact we accept that relative frequencies stabilize. Their 'limiting values' if known provide a full description of the situation under study. We assign to A a number $P(A)$ called the probability of A, designed to *represent* the 'limit' of the relative frequency, n_A/n.

In view of the interpretative significance of $P(A)$ in terms of relative frequency, there are certain simple and obvious requirements we must place on the values which can arise for $P(A)$. For example, if A and B cannot occur simultaneously (they are **mutually exclusive**) it seems reasonable to demand that the probability that either of them occurs is merely the sum of their separate probabilities, *since this is true of relative frequencies*. Such arguments lead to a system of formal axioms, as the basis for developing a detailed body of probability theory. The proof of the frequency pudding is attributed to its eating! The simple axioms lead logically to the conclusion that n_A/n must tend, as $n \to \infty$, to a limiting value in a well defined mathematical sense. Furthermore, this limiting value is $P(A)$ itself, thus recapturing in a *formal* manner the *empirical* feature that prompted the theory.

But, as with the subjective view of probability, there are many implicit assumptions underlying the frequency argument which need to be brought into the open. We shall examine these later and consider (for both the frequency, and subjective, approaches) the criticisms of impracticality that they engender.

One further approach to probability that needs mention largely for its historical interest is what is (somewhat confusingly in the present context) known as the 'classical' view. Fundamental to this approach is the idea that we can recognize situations in which there is a finite number of equally likely outcomes which exhaust all possibilities (they are **exhaustive**). In such situations the *probability of a particular eventuality*, A, *is defined to be the ratio of the number of outcomes whch support, or imply,* A, *to the total number of possible outcomes.* Obviously certain difficulties arise on this approach. To restrict attention only to situations where equally likely outcomes arise would seem greatly to limit the application of probability ideas. 'Equally likely' is itself a concept which is most readily (if not exclusively) represented in probability terms, so that the definition appears somewhat circular. If 'equally likely' is not represented in this way we need to know what else is meant by it, and how we *recognize* it. We shall return to these points.

However, before embarking on this, there is one more topic which needs to be aired. This century has seen a great deal of activity in the rigorous *mathematical* formulation of the probability concept and in its associated theoretical development. Pre-eminent amongst the innovators was Kolmogorov who published in 1933 (Kolmogorov, 1956) one of the earliest rigorous axiomatic treatments of probability theory. Why does it not figure then in our above classification? The answer is simple. Whilst undoubtedly a milestone in terms of the *mathematical* development of probability theory, it makes no contribution to the *philosophy* of the subject. Indeed, Kolmogorov (1956) rejects any questions of purpose or interpretation.

> The theory of probability, as a mathematical discipline, can and should be developed from axioms in exactly the same way as Geometry and Algebra. This means that after we have defined the elements to be studied and their basic relations, and have stated the axioms by which these relations are to be governed, all further exposition must be based exclusively on these axioms, independent of the usual concrete meaning of these elements and their relations. (p. 1.)

Employing the theories of measure and integration, the elaborate machinery of probability theory is developed in detail from the generally recognized basic probability axioms. Certainly, understanding of many of the complexities of probability theory requires some such formal, and highly mathematical, approach. Modern teaching practice reflects this also in its frequent use of set theory as basis. But as far as understanding the *conceptual* nature of probability is concerned such an approach gives no help—it does not claim to.

Fine (1973) remarks:

> The Kolmogorov axioms . . . provide neither a guide to the domain of applicability of probability nor a procedure for estimating probabilities nor appreciable insight into the nature of random phenomena.
> The apparent utility of the Kolmogorov theory is in fact due to its supplementation in practice by interpretative assumptions, many of which often go unstated. (p. 83.)

The tongue-in-cheek comment

> . . . there is no problem about probability: it is simply a non-negative, additive set function, whose maximum value is unity.
>
> (Kyburg and Smokler, 1964, p. 3.)

places in context the mathematical, in contrast to the conceptual, nature of probability. de Finetti[4] expresses this well in the following passage (some italics have been added, for emphasis):

> There are myriad different views on probability, and disputes about them have been going on and increasing for a long time . . . we note a seeming contradiction; it may be said with equal truth the different interpretations alter in no substantial way the contents and applications of the theory of probability and, yet, that they utterly alter everything. . . .
> Nothing changes for the mathematical theory . . . thus a mathematician not *conceptually* interested in probability can do unanimously acceptable work on its theory, starting from a merely axiomatic basis. And often nothing changes even in practical applications, where the same arguments are likely to be accepted by everyone *if expressed in a sufficiently acritical way.*
> . . . [Nonetheless, it] would be a most harmful misappraisal to conclude that the differences in interpretation are meaningless except for pedantic hairsplitters or even worse, that they do not matter at all. . . . The various views not only endow the same formal statement with completely different meanings, but a particular view also usually rejects some statements as meaningless, thereby restricting the validity of the theory to a narrowed domain, where the holders of that view feel more secure. (pp. 496—497.)

The broad spectrum of conceptual or motivational attitudes is represented by the earlier remarks in this section. What is important is that such attitudes should support formal expression in terms of axioms similar to those used in the mathematical developments. Otherwise (depending on our viewpoint) the conceptual attitudes are groundless, or the mathematical model sterile.

Fortunately, the various conceptual attitudes do lead to axiomatic schemes akin to that of Kolmogorov (apart perhaps for the relevance of his axiom of complete additivity to the subjective approach), and indeed go further sometimes in suggesting the need to introduce new concepts into the formal mathematical model (such as de Finetti's idea of exchangeability).

We have now given a convenient summary of three major directions (**subjective, logical, frequency**) in the definition of probability, together with an embryo concept (the **'classical'** view) which interrelates with them. Of course, the classification we have used is by no means complete, nor is it the only possible one. See Fine (1973), Good (1950, §1.4) or de Finetti[4]. The following sections present various modifications of attitude, and further details, within the three major areas and provide references for more detailed study. Having set the scene in what seemed a natural order of development, we now return to the (reverse) historical order in further considering the different views of probability.

3.2 'CLASSICAL' PROBABILITY

In spite of attempts to handle the idea of a random event even in very early times, David (1962) attributes the first corect (reported) probability calculations to the sixteenth century Italian mathematician, Cardano. Such calculations, based on an idea of equally likely outcomes, attracted the interest of many other eminent mathematicians in the succeeding two centuries. This interest was directed predominantly, if not exclusively, to games of chance involving cards, dice and so on. Huygens in 1657 provided rules for calculations involving dice games, for the types of problems being considered by Pascal, Fermat and others. At this stage no attempt was made to *define* probability, interest centred merely on its *evaluation*. Inherent in all these primitive applications of probability was an assumption of a framework of 'equally likely' outcomes. The earliest definition of probability, given tentatively by de Moivre in 1718 but more explicitly by Laplace at the beginning of the nineteenth century, adopts this framework as basic to the probability concept. Laplace's 'classical' definition of probability: *as the ratio of the number of outcomes favourable to the event to the total number of possible outcomes, each assumed to be equally likely*, figured as the accepted attitude to probability until early this century. But within the context of current attitudes it retains merely an historical interest, as the stimulus for the early development of some quite complex rules in the calculus of probability, which now find justification in alternative definitions of the probability concept itself.

The 'classical' view of probability cannot stand up to serious scrutiny for a variety of reasons. A detailed and compelling rejection was provided by von Mises (1957) in 1928. We must consider briefly why this view of probability is unsatisfactory. Dissatisfaction hinges of the preoccupation with 'equally likely' outcomes.

(a) *What is meant by 'equally likely'*? The phrase is synonymous with 'equally probable', a condition which apparently needs to be assessed *ab initio* in terms of a prevailing probability concept. In this respect the definition appears to consitute a vicious circle. But a more liberal attitude might be to regard 'equally likely' outcomes as recognizable in terms other than those of probability. Borel (1950), for example, claims that everyone has a 'primitive notion' of what is meant by 'equally likely' thus removing the circularity of the definition. If this is so the 'classical' definition is merely restrictive rather than circular, relating only to these situations where a basic framework of 'equally likely' outcomes exists. On this viewpoint two further questions must be resolved.

(b) *How do we recognize 'equally likely' outcomes?* We must assume that some prior knowledge supports such a framework. But in what way? Two distinct principles have been advanced, neither of them particularly acceptable nor widely accepted. The first is an appeal to the symmetry or homogeneity of the experimental situation. A coin is physically symmetric so why should a head or a tail be favoured? Likewise a six-faced die! Surely we must accept that the outcomes are 'equally likely'! This attitude is sometimes called the **'principle of cogent reason'** or the 'principle of indifference' (Keynes, 1957; Fine, 1973) and again we must beware of the circularity of such an argument. On what precise grounds does physical symmetry imply equi-probability? But than again we do not really believe in the perfect symmetry of the coin or die—the faces are differently figured, the manufacturing process by no means perfect! At best, then, we are considering an idealized coin or die to which our real one approximates; that is, a so-called *fair* or *unbiased* one. Fine (1973) makes the interesting observation that the ancient form of die (the *astragalus*, or heel bone, usually of a sheep; see David, 1962) was far from symmetric. Its development into the present day, symmetric, cubical form may well reflect empirical experience that such a form is needed for 'fairness'. Thus equiprobable outcomes in throwing a die may be as much justified by empiricism (observed behaviour) as by any such principle as 'cogent reason' or 'indifference'.

The alternative attitude is the so-called **'principle of insufficient reason'** first formally advanced by Bayes† (see Chapter 6). This argues as follows. If we have no reason to believe that one or another face is more likely to arise, then we should act as if (that is, assume) that all are 'equally likely'. The concept now becomes a subjective one, as a quantitative description of the state of being ignorant. Substantial objections can be raised to this 'principle' also; the matter is taken up in some detail in §6.4.

(c) *How restrictive are equally likely outcomes?* Suppose on one or other of the grounds advanced in (b) we accept that a particular die is equally likely to show any of its six faces. As von Mises (1957) remarks:

> It is obvious that a slight filing away of one corner of an unbiased die will destroy the equal distribution of chances. Are we to say that now there is no longer a probability of throwing a 3 with such a die, or that the probability of

† Indications of a similar idea appear much earlier in the work of Jacob Bernoulli.

throwing an even number is no longer the sum of the probabilities of throwing a 2, 4 or 6? . . . (since there is no probability without equally likely cases.) (p. 69.)

There are, of course, much more serious restrictions which arise in this spirit. On the 'classical' approach probability would appear to find application in at most a small range of artificial problems related to unbiased coins or dice, or fair deals of cards, and the like. The gamut of human enquiry in sociology, agriculture, genetics, medicine and so on, would seem to be outside the realm of probability theory. Attempts to extend the range of application of the 'classical' definition (from that of Laplace onwards) seem singularly unconvincing—often an implicit frequency view of probability is imported into the argument.

de Finetti[4] stresses and illustrates the difficulties encountered under our headings (b) and (c). Fine (1973) examines appeals to invariance principles, or use of entropy arguments from information theory, as bases of support for the 'classical' definition of probability, but finds them unsatisfactory.

Finally, we should refer briefly to attempts that have been made to reconcile the 'classical' view with, or relate it to, alternative views of probability, particularly the frequency or subjective ones. von Mises spends much time discounting the value of the 'law of large numbers' as the natural link between 'classical' probability and the results of empirically based probability arguments, claiming that it is the latter which is the valid starting point for a theory of probability. Then again, (as with Fine's remark about the *astragalus*), many have argued that the 'principle of cogent reason' with its appeal to symmetry is really a frequency concept. We incline towards an equal assignment of probability to a head or a tail with a typical coin because of our accumulated, inherited, experience of what happens with such coins in the long run, rather than because the coin appears essentially symmetric (if we are applying the 'principle of cogent reason'), or does not appear asymmetric (if we are applying the 'principle of insufficient reason').

3.3 THE FREQUENCY VIEW

. . . we shall see that a complete logical development of the theory on the basis of the classical definition has never been attempted. Authors start with the 'equally likely cases', only to abandon this point of view at a suitable moment and turn to the notion of probability based on the frequency definition. (p. 61.)

. . . When the authors have arrived at the stage where something must be said about the probability of death, they have forgotten that all their laws and theorems are based on a definition of probability founded only on equally likely cases. . . .

. . . we may say at once that, up to the present time [1928], no one has succeeded in developing a complete theory of probability without, sooner or later, introducing probability by means of the relative frequencies in long sequences. There is, then, little reason to adhere to a definition which is too narrow for the inclusion of a number of important applications and which must be given a forced interpretation in order to be capable of dealing with many questions of which the theory of probability has to take cognizance. (p. 70.)

(von Mises, 1957)

Whilst paying lip service to the 'classical' definition of probability, many of the early writers were implicitly adopting frequency (or subjective) principles when it came to describing the numerical evaluation of probability in everyday life. This is evidenced in Laplace's attempts in 1812 (Laplace, 1951) to apply probability to such diverse topics as human mortality, possession of human sensation by animals, the credibility of witnesses or the downfall of military powers. One of the earliest attempts at an *overt* **frequency** definition of probability is found in the work of Venn in 1866 (see Venn, 1962). Concerned with counterbalancing a growing preoccupation with subjective views of probability stemming from ideas of James Bernoulli and more particularly Augustus de Morgan (1847) (see the later critical essay, Venn[6]), Venn formalized the idea of expressing probability in terms of the limiting values of relative frequencies in indefinitely long sequences of repeatable (and identical) situations. He claimed that probability must be a measurable, 'objective', concept and that the natural basis for such a definition was statistical frequency. He provided a systematic, if non-mathematical, treatment of probability from the frequency viewpoint. However, Venn did not consider the formal conditions under which such a definition makes sense, or the mathematical structure of the basic concept and its rules of behaviour: the so-called 'calculus of probability'.

An attempt to construct a sound mathematical basis for the *frequency* view of probability did not appear until the work of von Mises in the 1920s. His approach, in rejection of the classical view, strictly limited the class of situations in which the probability concept has relevance. In von Mises (1957) we find:

> The rational concept of probability, which is the only basis of probability calculus, applies only to problems in which either the same event repeats itself again and again, or a great number of uniform elements are involved at the same time. . . . [In] order to apply the theory of probability we must have a practically unlimited sequence of uniform observations. (p. 11.)

In his lengthy development of the frequency viewpoint (1957), von Mises explains with great care the conditions under which probability may be defined. This development is discursive. A formal axiomatic mathematical treatment is presented in other publications, for example von Mises[7] provides a brief summary in English of earlier more detailed work in the original German editions. Central to von Mises' ideas is that of a **collective**;

> . . . a mass phenomenon or an unlimited sequence of observations fulfilling the following two conditions: (i) the relative frequencies of particular attributes within the collective tend to fixed limits; (ii) these fixed limits are not affected by

any place selection. That is to say, if we calculate the relative frequency of some attribute not in the original sequence, but in a partial set, selected according to some fixed rule, then we require that the relative frequency so calculated should tend to the same limit as it does in the original set.

(von Mises, 1957, p. 29)

The condition (ii) he describes as the **principle of randomness**; the limiting value of the relative frequency *when this condition holds* is defined as the *probability* of the attribute 'within the given collective'.

On this framework a theory of probability is developed in which new collectives are formed from initial ones, and have probability assignments which derive from those in the initial collective. The new collectives are constructed on the application of one or more of von Mises' 'fundamental operations'. These amount to no more than the familiar manipulative rules of modern probability theory.

The present day expression of basic probability theory from the frequency viewpoint uses different words to those used by von Mises, with its concern with sample spaces, mutual exclusion, independence, conditional probability and so on. In essence, however, it adds little to the earlier construction, other than an economy of statement coupled with an implicit acceptance of certain features that von Mises (in historical context) found it necessary to elaborate.

It is worth picking out one feature of current terminology, the idea of an **event**, for brief mention; not in contrast with earlier work in the frequency area but because we shall find a real distinction here between the *frequency* and *subjective* approaches. On the *frequency* view, an event is just one of the *potential* 'observations' in the 'collective'. Thus, if we throw a six-faced die the occurrence of an even score (2 or 4 or 6) is an 'event'. Any observation may or may not support such a description; we say that this event does or does not occur at any particular stage in the collective. We talk about the probability of this event as the probability that it occurs as a *typical* observation; in the sense of the limiting value of the relative frequency of even scores in the unlimited sequence of identically operated throws of the die.

Subjective probability, we shall observe, cannot admit such a liberal view of an 'event'. An 'event' becomes a *particular* 'event'. We need to talk about the probability that, say, the seventeenth outcome in some specific set of throws of the die is even. It is the 'seventeenth outcome . . . set of throws' which is now the 'event' whose probability can be discussed. The subjective approach has no equivalent concept of an unlimited sequence of similar situations. In response to the enquiry 'What is the probability of an even score?' it replies 'When? At what throw? By whom? What has gone before?' Probability is personalistic and conditional on immediate precedent as well as global environment.

But let us return to the *frequency* view of probability. As in the 'classical' approach, there is an implied restriction of the class of situations to which the probability concept may be applied, and we are bound to ask again how serious is this restriction. Certainly less so than for 'classical' probability! With its empirical basis, the frequency view of probability inevitably relates to a vast range of situations of practical interest; the wide acceptance and application of statistical methods based on this approach is beyond dispute. Fields of application range over most spheres of human activity (from Agriculture to Zoology) as is witnessed by the continuing use of *classical* statistical methods, which are firmly based on the frequency concept of probability. Certain areas are not fair game, as von Mises himself was quick to exclaim (apart from any critics of the approach). These are mainly individual, personal, or behaviouristic—the authorship of the Gospels is one example.

The response to this restriction depends on one's attitude. Some statisticians would claim that situations which do not admit 'repetition under essentially identical conditions' are not within the realm of objective statistical enquiry, others that no single view of probability is enough and different circumstances may legitimately demand different attitudes. Some go so far as to see this restriction as an inherent fallacy of the frequency approach, declaring that the concept of a collective is untenable: that we can never really assert that an unlimited sequence of essentially identical situations exists. Life is too short to demonstrate it, our prior knowledge never adequate to assert it. It is seen as unrealistic to base an *empirical* concept of probability on an *infinite* sequence of realizations. Furthermore, a common counter-blast to any criticism of the subjective approach is to declare that nothing is more *subjective* than the construction of a collective. A vicious circle if ever there was one!

Fine (1973, Chapter IV) describes in detail the various sources of dissatisfaction with the frequency viewpoint. The implicit assumption of the *independence* of separate trials in the 'unlimited sequence of similar situations' is questioned on the grounds of its pragmatism (its lack of empirical support) and of its extravagance—the much weaker condition of *exchangeability* (see §3.5 below) is a better representation of the 'collective' and adequate for the determination of central results in the probability calculus. As Kolmogorov (1956) remarks:

> . . . one of the most important problems in the philosophy of the natural sciences is . . . to make precise the premises which would make it possible to regard any given real events as independent. (p. 9.)

The distinction between *essentially finite experience* and probability defined in relation to *indefinitely large numbers of realizations* (limiting values of relative frequencies) is another crucial basis for dissatisfaction. After detailed examination Fine (1973) reaches extreme conclusions:

> Observation of any finite number of experiments would tell us nothing about limits of relative frequencies. (p. 94.)

whilst in reverse,

> ... a relative frequency interpretation for [probability] ... does not enable us to predict properties of actual experimental outcomes. (p. 103.)

Fine sees the basic difficulty residing in what is essentially von Mises' *principle of randomness* with its implicit assumptions of independence, identity of distribution and ill-defined notion of the irrelevance of 'place selection'. He adduces a circularity of argument akin to that in the 'classical' definition of probability:

> ... even with the relative-frequency interpretation, we cannot arrive at probability conclusions without probability antecedents. (p. 102.)

One direction in which we are lead by the conflict between 'finite experience' and 'infinite sequences' of outcomes is to seek to define *randomness* in the context of *finite* sequences of outcomes. Some interesting proposals have been made by Kolmogorov[8], Martin-Löf[9] and others (see Fine 1973, Chapter V) based on the notion of **computational complexity**: essentially the difficulty in describing the sequence in some programming language on a computer (where both the 'language' and the 'computer' are appropriately defined in formal mathematical terms).† The greater the 'complexity' the more 'random' the sequence. Such a *notion* has been used for the partial development of alternative forms of the frequency-based probability concept. See, for example, Solomonoff[10] and Willis[11].

We must remark on the use of the word 'objective' in relation to views of probability. In the early literature it was used to distinguish the *frequency* approach from alternative views based on *degree-of-belief* or *credibility*, which were at the time neither well distinguished one from the other nor clearly defined. Nowadays the term 'objective' has a more specific meaning. It describes any view of probability which does not depend on the personal feelings or actions of an individual. In this respect the 'classical' view is also an objective one, as is the so-called logical view which we consider next. (See de Finetti[4] for a more detailed classification of *objective* and *non-objective* approaches to probability.) But this does not seem a completely adequate basis for classification, since the logical view also has affinities with the subjective approach. They both aim to represent degree-of-belief; the latter an individual's *personal* assessment of the situation, the former a rational (inevitable, unique, correct) degree-of-belief induced by, and logically stemming from, a certain body of evidence—a degree-of-belief any

---

† Barnard, G. A. (1972) *J. Roy. Statist. Soc. A*, **135**, 1–14, gives a simple explanation of the principles involved in this notion of randomness.

'rational' person should have, justified by the evidence and unrelated to his personal make-up. So in a sense the logical view straddles the subjective and frequency views, in being 'objective' but expressing 'degree-of-belief'.

An alternative to von Mises' (and others') belief that the inherent restriction in the frequency approach is not only inevitable, but desirable, is to be found in the work of another frequentist, Hans Reichenbach. In his book, *The Theory of Probability* (1949), first published in 1934, he attempts to demonstrate that even isolated, behaviouristic, eventualities can be given a frequency interpretation in terms of a wider 'reference class'. His ideas have affinities with the logical view advanced by Jeffreys, but in respect of extending the domain of frequency-based probability appear contrived (he himself admits to 'ambiguities') and have not found substantial following. See also Salmon (1966).

Thus we see in the frequency view of probability an attempt at an empirical, practically oriented, concept: one that is very widely accepted as the basis for a substantial battery of (classical) statistical techniques and methods but which from the philosophical and practical standpoints is (in common with *any* view of probability) by no means free from critical comment.

> The form of this definition restricts the field of probability very seriously. It makes it impossible to speak of probability unless there is a possibility of an infinite series of trials. When we speak of the probability that the Solar System was formed by the close approach of two stars . . . the idea of even one repetition is out of the question; . . . But this is not all, for the definition has no practical application whatever. When an applicant for insurance wants to choose between a policy that offers a large return if he retires at age 65, as against one that offers return of premiums if he retires before 65 and a smaller pension if he retires at 65, his probability of living to 65 is an important consideration. The limiting frequency in an infinite series of people is of no direct interest to him. The insurance company itself is concerned with a large number of cases, but up to a given time even this number is finite, and therefore the probabilities are meaningless according to the definition.
>
> Again, what reason is there to suppose that the limit exists? . . . The existence of the limit is in fact an *a priori* assertion about the result of an experiment that nobody has tried, or ever will try.
>
> (Jeffreys, 1973, pp. 194–195)

> Thus the criterion based on the notion of frequency is reduced, like that based on equiprobable events, to a practical method for linking certain subjective evaluations of probability to other evaluations, themselves subjective. . . .
>
> (de Finetti[12], pp. 116–117)

3.4   LOGICAL PROBABILITY

> . . . in practice no statistician ever uses a frequency definition, but . . . all use the notion of degree of reasonable belief, usually without even noticing that they

are using it and that by using it they are contradicting the principles they have laid down at the outset.

(Jeffreys, 1961, p. 341)

> The idea of a reasonable degree of belief intermediate between proof and disproof is fundamental. It is an extension of ordinary logic, which deals only with the extreme case. . . . The problem of probability theory is to find out whether a formal theory is possible. Such a theory would be impersonal in the same sense as in ordinary logic: that different people starting from the same data would get the same answers if they followed the rules. It is often said that no such theory is possible. I believe that it is possible, and that inattention to the progress that has actually been made is due to an active preference for muddle.

(Jeffreys[13])

The **logical** view of probability stands in distinct contrast to the '*classical*' or *frequency* (empirical) views. These latter were developed in the desire to express aspects of the real world—to serve as models for practical problems, from card games to medical research. This practical motivation is reflected in their basic nature, and the associated probability theory developed from these viewpoints is designed *to be applied*. Probability is seen as a means of quantifying the uncertainty which is observed in practical phenomena (albeit within a restricted range of problems).

The emphasis in the **logical** view is quite different; probability is regarded as a concept which modifies, and extends the range of application of, formal logic. Instead of declaring that two propositions A and B stand in relation to each other (if at all) in one or other of two ways, namely A implies B or A refutes B, the concept of probability is introduced to express a *degree* of implication of B afforded by A. When applied to a body of knowledge, E, about a situation, and a potential outcome, S, probability expresses the extent to which E implies S. Probability is always *conditional* in form. It is the *rational* degree-of-belief in S afforded by E; or what has been called the '*credibility*' of S in the face of E or *degree of confirmation* (Carnap). This is not expressing an *empirical* relationship but a unique (impersonal) logical one. Kyburg and Smokler (1964) summarize the attitude in the following way:

> The essential characteristic of this view, in most of its formulations, is this: given a statement, and given a set of statements constituting evidence or a body of knowledge, there is one and only one degree of probability which the statement may have, relative to the given evidence. A probability statement is *logically* true if it is true at all, otherwise it is *logically* false. Probability statements are purely formal in tne same way that arithmetical statements are purely formal. (p. 5.)

Early ideas leading to the logical view of probability are to be found in Edgeworth[14], but the names usually associated with this view are those of Keynes, Jeffreys and Carnap (see Jeffreys, 1961, first published 1939; and Carnap and Jeffrey, 1971; and Carnap, 1962, first published 1950). Arising from its basis in logic rather than empirical experience it is natural to find in the literature a considerable preoccupation with the formal structure of probability: what is its nature as a system of logic, what axiomatic framework supports it. Consideration of the application of probability to practical problems becomes a secondary matter, in attention, rather than importance [see Jeffreys (1961), who devotes great effort to constructing a Bayesian statistical theory on the logical attitude to probability]. One result of this is that the probability concept need not necessarily give a *numerical* measure to the 'rational degree of belief' in some eventuality. Indeed, in Keynes' original work (Keynes, 1921) probabilities are only partially ordered. Degrees-of-belief in some outcome, on the basis of *different* bodies of evidence, may or may not, according to Keynes, be amenable to direct comparison. Within the class of *comparable* probabilities, numerical assignments may be made, but this represents only a special class of situations. It is only in this class that probability becomes the subject of detailed mathematical analysis, or quantitative application. When numerical values are assignable, Keynes adopts a *frequency* basis for their calculation. [It might be mentioned that Good (1965) claims that Keynes subsequently 'nobly recanted' from the logical view, in favour of a personalistic, *subjective*, attitude to probability; also Borel[1] appears to view Keynes' proposals from the outset as subjective, rather than *objectively* logical.]

Jeffreys (1961, first published 1948; and elsewhere) differs in two fundamental respects from Keynes. His probabilities, again expressing rational degrees-of-belief, are always completely ordered and have numerical values. Furthermore, frequency is denied any relevance even in the calculation of probabilities, this being achieved by updating current probability assessments in the light of new information, by means of Bayes's theorem. The 'principle of insufficient reason' (see §3.2) is given central prominence, and much importance placed on quantifying the state of ignorance.

> If we assert a probability for a law, presumably a high one, on some experimental evidence, somebody can ask 'what was the probability before you had that evidence?' He can then drive us back to a position where we have to assess a probability on no evidence at all. . . . A numerical statement of a probability is simply an expression of a given degree of confidence, and there is no reason why we should not be able to say within our language that we have no observational information. The idea that there is any inconsistency in assigning probabilities on no evidence is based on a confusion between a statement of logical relations and one on the constitution of the world.

(Jeffreys[13])

We take this matter up in more detail in §6.4.

Jeffreys views the quantification of probabilities as important, since probability theory is to him the vital tool in the exploration of scientific theories. The development of a method of statistical influence is vital. His is not the purely philosophical interest of Keynes, it is more pragmatic.

> Keynes defined probability as a degree of belief because it then provided an extension of the abstract ideas of logic; Jeffreys adopted this definition because he felt it was the one best suited to the needs of scientific induction. . . .
>
> (Atkinson[5])

But whilst placing great emphasis on the numerical nature of probabilities, Jeffreys obviously felt that quantification was merely conventional (utilitarian) and not fundamental to the formal development of probability theory from the logical viewpoint. He devotes much attention to this formal development and presents (Jeffreys, 1961) a detailed axiomatic system within which any considerations of actual *numbers* as probabilities are intioduced merely as 'conventions'.

Carnap (1962) is not alone in finding the need for more than one view of probability to meet the variety of interpretative and applicative roles that the concept is called on to fulfil. He regards both the frequency and logical views as valid, but applicable in distinct circumstances. In both, probability is assumed to be numerical valued, the logical probability expressing degree-of-belief in a propositional statement, the frequency probability applied to practical problems which accommodate the idea of an infinite sequence of similar experiments. The two coincide in value when both apply to the same situation: for example, in relation to the number of similar experiments in which a particular statement is true.

Predominant amongst criticisms of the logical view of probability, from whatever platform, is a dissatisfaction with the basis of obtaining a numerical valued probability. If probability represents the unique appropriate rational degree-of-belief in some outcome, just what can be meant by assigning a number to it; how do we measure it? Subjectivists further criticize the uniqueness assumption and cannot accept the thought of some idealized rational (non-personal) being to whom the probability assessment applies. Probability must be a *personal* assessment, from their standpoint, and relate to the individual. This means that different individuals legitimately attach different probabilities to the same outcome. Savage[3], who refers to the logical view as the 'necessary' concept of probability, since the probability of S, on evidence E, is a 'logical necessity' (intrinsic feature) of the relationship of S and E, sees logical probability as merely the modern expression of the classical view and to have been 'thoroughly and effectively criticized' and 'not now active in shaping statistical opinion'. This essay[3], though highly slanted [the frequency view has 'drastic consequences for the whole outlook of statistical theory'; the work of Jeffreys is 'invaluable in developing the theory of (*personalistic*) Bayesian statistics'], is nonetheless a stimulating digest of the different attitudes to probability. See also Fine (1973, Chapter VII) for a detailed critique of the *logical* view.

To the frequentist the universality of application of logical probability is anaethema; and the adoption of the frequency approach as a computational device where appropriate appears to him to place the cart a very long way in front of the horse!

> Another probability concept has been of historical importance and still fascinates some scholars. This I call the necessary (or logical, or symmetry) concept of probability.
>
> According to this concept, probability is much like personal probability, except that here it is argued or postulated that there is one and only one opinion justified by any body of evidence, so that probability is an objective logical relationship between an event A and the evidence B. The necessary concept has been unpopular with modern statisticians, justifiably, I think.
>
> (Savage[15])

> ... logical probability may serve to explicate an objective version of classical probability, ... we are less inclined to agree that logical probability is the proper basis for estimating empirical (say relative-frequency-based) probability.
>
> There may be a role for logical probability in rational decision-making, although the form it will take is as yet unclear.
>
> (Fine, 1973, pp. 201–202)

3.5 SUBJECTIVE PROBABILITY

> I consider that in the last resort one must define one's concepts in terms of one's subjective experiences.
>
> (Good, 1950, p. 11)

> One can only conclude that the present theory is necessarily incomplete and needs some theory of subjective probability to make it properly applicable to experimental data.
>
> (Smith[16])

> I will confess, however, that I ... hold this to be the only valid concept of probability and, therefore, the only one needed in statistics, physics, or other applications of the idea. ... Personal probability at present provides an excellent base of operations from which to criticize and advance statistical theory.
>
> (Savage[3])

It is revealing (and in no way contrived) that these comments on **subjective** probability are expressed in a very personal style. This is of the essence of subjective probability, that it is concerned with individual behaviour; with the expression of preferences among different possible actions by the individual, and with the way in which he forms judgements. A probability concept is developed which is *specific to the individual* in the sense that it relates to the accumulated *personal* experience *he* brings to bear in assessing any situation. No restriction is placed on the range of application of probability theory. It applies as much to the throwing of a die or the quality of an industrial component as to the authorship of a classical work or to the choice of a marriage partner. In every case probability measures a particular individual's degree-of-belief in some eventuality *conditional* of his relevant experience. It may vary from one person to another in that their experiences may be different; this is accepted and expected. In contrast to the logical (or necessary) view, it postulates no *impersonal, rational* degree-of-belief which is seen as a unique inevitable measure of the support given by prevailing circumstances to some outcome of interest, independent of the observer. A limited *rational subjective* theory which assumes such a true underlying measure as the goal at which the imperfect individual is aiming seems to find little current support. de Finetti[4] (also de Finetti, 1974, 1975) rejects such a view in his fine survey of different approaches to probability, and develops a detailed case for the strictly *personal* concept; likewise Savage[17].

Shades of opinion within the subjective approach, and detailed developments of its implications, are seldom concerned with justifying the personal behaviouristic basis. They are directed towards producing a formal probability theory on such a basis, and with describing how the *quantitative* expression of personal probabilities may (or may not) be elicited. We shall briefly review the variety of attitudes to these topics.

The idea of subjective probability goes back a long way. It was informally advanced by Bernoulli in *Ars Conjectandi* (1713) when he talked about probability as the 'degree of confidence' that we have in the occurrence of an event, dependent upon our knowledge of the prevailing circumstances. Another early formulation was provided (somewhat obscurely) by de Morgan in 1847.

> By degree of probability we really mean, or ought to mean, degree of belief. . . . Probability then, refers to and implies belief, more or less, and belief is but another name for imperfect knowledge, or it may be, expresses the mind in a state of imperfect knowledge.
>
> (See Kyburg and Smokler, 1964, p. 9)

Although de Morgan defined the subjective concept, declared that it was amenable to numerical expression, and could be detected in the individual through his personal feelings, no formal theory was advanced until quite recent times. Both Ramsey and de Finetti, independently and almost simultaneously (in the period 1925–35), addressed themselves to the construction of a formal definition of subjective probability and an associated manipulative theory. A most detailed exposition of the subjective viewpoint, with extensive discussion of the motivational, conceptual and computational aspects, is given by de Finetti (1974, 1975).

Being concerned with personal behaviour it was inevitable that ideas were expressed in behaviouristic terms, and that rules should be laid down to delimit what constitutes reasonable ways of expressing preferences or forming judgements. Whether these really apply to the individual in his complex (and often irrational) behaviour is a point we shall return to. For the moment we consider some aspects of the theory that has been developed, and the methods advanced for assigning numbers to an individual's beliefs.

The key to de Finetti's formal definition of probability is found in his remark:

> It is a question simply of making mathematically precise the trivial and obvious idea that the degree of probability attributed by an individual to a given event is revealed by the conditions under which he would be disposed to bet on that event.
>
> (de Finetti[12], p. 101)

With this emphasis on behaviour as expressed through 'betting' situations, the probability of an event E (for an individual; that is, conditional on his personal experiences) is obtained as follows. It is 'the price he is just willing to pay for a unit amount of money conditional on E's being true. . . . an amount s conditional on E is evaluated at ps'. This is to say that if I believe a coin to be unbiased I would exchange $\frac{1}{2}$ a unit of money for *1 unit of money when a head occurs*. My inclination to do this expresses my subjective probability of a head as $\frac{1}{2}$.

The general study of subjective probability places much emphasis of this quantification of personal belief in terms of betting behaviour, and the subsequent development of a complete theory of probability often leans heavily on this. Thus, for example, if one attributes in this sense a probability p to an event E, then a probability $1 - p$ should be attributed to the denial of E. Otherwise it could happen with an appropriate combination of bets that one will lose money *whether or not* E *happens*. Subjective probability theory assumes that such 'irrational' behaviour does not arise; it requires the individual to be **coherent** in his bets. In his discussion of operational definitions of probability, de Finetti (1974) defines coherence in the context of choosing a value $\bar{x}$ as a estimate of a random quantity X under the following circumstances:

(i) If you choose $\bar{x}$, you *must* accept *any* bet with *gain* $c(X - \bar{x})$ for arbitrary positive or negative c, chosen by an opponent.

(ii) If you choose $\bar{x}$, you suffer a *penalty* proportional to $(X - \bar{x})^2$.

Under these two possible conditions *coherence* is expressed as follows:

(i) 'It is assumed that You do not wish to lay down bets which will *certainly* result in a loss for You' (p. 87).

(ii) 'It is assumed that You do not have a preference for a given penalty if You have the option of another one which is *certainly* smaller' (p. 88).

In addition it is assumed that the individual is logically **consistent**† in his reaction to whether two events are more, less or equally probable, one to another. As one example of this we demand that if E_1 is more probable than E_2 which is more probable than E_3, then E_1 is more probable than E_3. With such simple requirements of *coherence* and *consistency* an axiomatic probability theory is developed; the formal expression of these concepts is found to be equivalent to the usual axioms of probability theory.

Thus we again arrive at the common rules for the calculus of probabilities, but with quite obvious interpretative differences. All probabilities are *personal* and *conditional*. The *events* to which they refer are isolated outcomes, a particular spin of a particular coin by a particular individual, rather than global in the frequency sense. An event does not have a unique probability, it may differ for different people. All that is essential is that any set of probabilities attributed to events by an individual should be *coherent*, and not violate the demands of logical *consistency*.

But what happens if an individual is incoherent? Does this mean that the theory is inadequate? On the contrary, the attitude expressed is that a person, on having his incoherence pointed out to him, will modify his opinions to act coherently in a desire to behave logically in accord with subjective probability theory. Thus the approach is a *normative* one. It aims to show how individuals *should* behave, not how they *do* behave. This attitude is crucial to the subjective viewpoint which maintains that it does not aim to model *psychological* probability behaviour. de Finetti[12] made this clear:

> It is . . . natural that mistakes are common in the . . . complex realm of probability. Nevertheless . . ., fundamentally, people behave according to the rules of coherence even though they frequently violate them (just as . . . [in] arithmetic and logic). But . . . it is essential to point out that probability theory is not an attempt to describe actual behaviour; its subject is coherent behaviour, and the fact that people are only more or less coherent is inessential. (p. 111, footnote.)

The 'betting' basis for subjective probability figures widely in writings on the subject. It is given special recognition by Smith[16] who in his individual treatment of the subject, talks about 'personal pignic probability' from the latin *pignus* (a bet). One feature of the use of bets to measure probability needs to be remarked on. We have talked of exchanging p units of money for one unit of money if the event E happens. But what is our *unit*, a farthing, or $75 000? This would seem relevant. Someone who would wager \$1 against \$1 on the spin of a coin, may well not play the same game with \$1000 versus \$1000. If I posses only \$1000 I may not play *any* betting game with \$1000 as stake. How then are we to define probability in betting

---

† Some confusion of terminology exists in this area. Certain writers use the term *consistency* to describe what is here labelled *coherence*. Consistency has a traditional meaning in logic as 'non-contradictory' and the distinction between the two words will be maintained, as recommended by Kyburg and Smokler (1964, p. 12).

terms? The *relative values* we attach to losing or gaining some sum of money vary with the sum of money involved. The *absolute value* to an individual of a sum of money varies in no unique way with its literal value. What it is worth to him, its *utility*, is crucial to his betting behaviour. Thus subjective probability, with an emphasis on the *monetary* basis appears inevitably entwined with *utility theory* (which we shall consider in the next chapter). de Finetti recognized this and thought originally to avoid anomalies by considering bets with 'sufficiently small stakes'. This was somewhat unconvincing. But he subsequently remarks that when he became aware (in 1937) of Ramsey's work, which takes as the primitive notion that of *expected utility* and derives probability from it, he had greater sympathy with this as an operational approach to the enumeration of probability. Thus although betting situations still figure widely in subjective probability discussions the use of utility rather than money is predominant.

As remarked in Chapter 2, there has been much attention in the literature to the matter of how individuals do, or should, face the problem of setting numerical values on their subjective (personal) probabilities. Various empirical studies have been reported in different areas of application. For example, Murphy and Winkler[18] and Murphy[19] are concerned with subjective probability forecasts of temperature and rainfall; Hoerl and Fallin[20] consider horse-racing; Winkler[21], the outcomes of (American) football games; O'Carroll[22], short-term economic forecasts. The general principles for quantifying subjective probabilities, and specific practical methods, are considered by Savage[23]; Hampton, Moore and Thomas[24]; Smith[16] and Tversky[25]. The assessment of probabilities by groups, rather than by individuals, has also been examined; see, for example, Hampton, Moore and Thomas[24], and De Groot[26]. An interesting recent paper by Lindley, Tversky and Brown[27] faces up to what we should do when quantified subjective probabilities exhibit *incoherence*. Two approaches are considered. In the 'internal' approach a subject's 'true' coherent probabilities are estimated from his incoherent assessments. The 'external' approach has an external observer modifying *his* coherent probabilities in the light of the subject's incoherent views.†

Let us now briefly consider the parallel (somewhat earlier) development by Ramsey[28], published in 1931*. He, also, stressed the centrality of concepts of coherence and consistency, and again believed that the appropriate way to measure degree-of-belief was through the observation of overt behaviour in betting situations. The emphasis, however (as also with de Finetti, 1974, 1975), is on the use of 'expected returns' rather than betting odds as an attempt to reduce the emotive content of the probability concept and to avoid the monetary bias in assigning numerical values. Ramsey's approach was as follows. Faced with various 'propositions' (events) about some situation of interest the individual expresses the personal probabilities he attributes to these propositions by his

† French, S. (1980) *J. Roy. Statist. Soc. A*, **143**, 43–48, finds some implications of this work 'counter-intuitive' and suggests a remedy.
* Dated 1926 in Kyburg and Smokler (1964).

reaction to different 'options' (bets) offered to him. Thus an option might be to risk a loss if a die shows a 'six' or the *same* loss if his favourite football team loses its next match. His acceptance or refusal of this option reflects the relative probabilities he assigns to a 'six', or to his team losing. If, furthermore, he believes the die to be unbiased this provides a reference level for the latter probability. Note how the *amount* of the loss has become irrelevant, the option refers merely to identical utilities. The theory of probability is developed axiomatically on this type of consideration. The main axioms are that options are totally ordered, and that there always exists an **'ethically neutral proposition'** as a yardstick. This latter requirement means that there is a proposition such that the individual is indifferent to two options which promise to return α when the proposition is true and β when false, or β when true and α when false, with $\alpha \neq \beta$. Such a proposition is defined to have degree-of-belief, $\frac{1}{2}$.

With the addition merely of some consistency axioms, Ramsey shows that a utility function must exist, and that this in turn yields a formal concept of probability providing numerical values for degrees-of-belief. A detailed and careful modern development of subjective probability on largely similar lines is given by Savage (1954) who is at pains to point out the difficulty with *monetary* bets (overcome in Ramsey's approach) in that the stakes must either be infinitesimally small, or else we must assume the utility function for money to be linear (see next chapter). In a more mature statement of his views of subjective probability Savage[15] expresses some dissatisfaction with his earlier (1954) work and presents a detailed reappraisal. This places emphasis on the dual concepts of *subjective probability* and *utility* as the principle components of 'a general theory of coherent behaviour in the face of uncertainty'. This work contains an extensive bibliography on the 'foundations' of statistics and probability. An entertaining elementary presentation of the linked themes of probability and utility, for coherent decision-making, is given by Lindley (1971a).

The historical and attitudinal perspective on subjective probability is ably provided by Kyburg and Smokler (1964) through their introductory remarks and their reproduction of critical works by Venn, Borel, Ramsey, de Finetti, Koopman and others. This book also presents an extensive bibliography of relevant material.

We cannot leave this brief review of subjective probability without allusion to one or two further matters. The first of these concerns the important concept of **exchangeable events**. Framed originally by de Finetti in 1931 it is crucial to the development of statistical ideas from the subjective viewpoint.

Exchangeability is essentially an expression of symmetry in the probabilistic behaviour of events. Consider a sequence of experiments where E_j denotes the potential (random) outcome of the jth experiment. In *frequency* terms physical independence of the experiments induces statistical independence of the E_j in the sense that

$$P\{E_j | E_{j-1}, E_{j-2}, \ldots\} = P\{E_j\}.$$

In subjective terms physical independence has no such probabilistic implication: indeed, knowing the outcomes of (independent) experiments up to the $(j-1)$th will be expected to *modify* the subjective probability assigned to E_j. Also, it is argued that assumptions of independence are extreme and can seldom be justified from tangible knowledge of the situation under study. Instead it is proposed that (when appropriate) we adopt the weaker, more supportable, assumption that the E_j are **exchangeable** in such a situation, in the sense that

$$P\,(E_{j_1}\,E_{j_2}\,\ldots\,E_{j_n}) = P\,(E_{k_1}\,E_{k_2}\,\ldots\,E_{k_n})$$

for any distinct subscripts $(j_1,\ldots,j_n)$ and $(k_1,\ldots,k_n)$, and all n. In particular $P\,(E_j)$ is the same for all j; the probability of the joint occurrence of a set of outcomes does not vary with the specification of the subset of experiments on which the outcomes occurred.

Thus in a sense *exchangeability* plays a role in the subjective approach akin to von Mises' *principle of randomness* in the frequency approach; it expresses a notion of good behaviour for sequences of events. Apart from serving as a basic tool in developing statistical ideas, it also acts as an interpretative link between the frequency and subjective views. For example, de Finetti[12] is able to derive some justification in subjective probability terms for relative frequency arguments in estimation, via this concept of exchangeability. (A theme also considered at length by Ramsey[28] from a different standpoint.) It is this analysis that leads de Finetti to the conclusion expressed at the end of §3.3. But perhaps the most important effect of exchangeability is its role in reconciling different subjective prior probability assessments in Bayesian inference; where if events are exchangeable resulting inferences are found to be essentially robust against different prior probabilities for different individuals. See de Finetti[12] for further details on this. Exchangeability is by no means restricted in its use to *subjective* probability. A similar notion appears in Carnap's *logical* probability proposals and Fine (1973, p. 86) employs exchangeability in seeking 'an invariant description of the outcomes of repeated experiments.'

All probability concepts need an axiomatic system to support the corresponding probability calculus. All embody some axiom of **additivity**. There are different views (between individuals, or approaches) on the appropriate form of the additivity axiom: on whether it should apply to a *countable* set of events, or just to a *finite* set. de Finetti (1974, pp. 116–119) argues forcefully, on conceptual and interpretative grounds, for *finite additivity* as the appropriate form in the development of probability theory, from the subjective standpoint.

As with all the views of probability described in this chapter, a variety of ramifications of the subjective proability viewpoint exist. We cannot consider these in detail other than to add to the names of Ramsey, de Finetti and Savage those of Koopman (some probabilities are measurable in betting odds terms, others are not, and are not necessarily completely ordered), Borel (pioneer of subjective probability; see Borel[1]), Good (who considers partially ordered subjective probabilities applied to Bayesian statistical analysis: the stimulus of much modern study of subjective probability), and C. A. B. Smith (favouring partially ordered probabilities leading to intervals rather than point values).

As with other views of the probability concept, subjective probability does not lack critics: who complain of its lack of 'objectivity', of the intangibility of a *numerical* basis for such probabilities, and of what some view as the almost theological leap of faith needed for its unilateral adoption.

Venn[6], in criticizing 'the prevailing disposition to follow de Morgan in taking too subjective a view', was voicing as early as 1888 what even now remain the major doubts on its basis and implementation. More recently van Dantzig[29] presented their modern expression in his impassioned rejection of the subjective view of probability put forward by Savage (1954) in *The Foundations of Statistics*. Central to such criticism is the declaration that statistical investigations advance knowledge of the world only to the extent that their conclusions *avoid* the personal preferences or prejudices of the investigator. To base techniques on a personal or subjective view of probability is a denial of this honest resolve, and renders such a view impracticable. Only an empirically based probability concept, it is declared, can achieve the 'objectivity' necessary to reduce personal factors to an insignificant level.

> To those who take the point of view that the external world has a reality independent of ourselves and is in principle cognizable, and who take into account the fact that probabilistic judgements can be used successfully to obtain knowledge about the world, it should be perfectly clear that a purely subjective definition of mathematical probability is quite untenable.
>
> (Gnedenko, 1968, p. 26)

> Most modern statisticians will hardly consider the conclusions they arrive at (like the statements made by art critics . . .) as being of a 'subjective' nature, i.e. depending on their mood, their taste, . . . [that where one] does not find a significant result, whereas his colleague does, . . . [this] is due to differences in their metabolism . . . . . . [On an extreme subjective view of probability] the statistician . . . gets a status like that of an ancient priest, whose statements the layman, unable to reproduce the statistician's order of preferences, has to accept without a possibility of criticism. Statistics, in as much as it would remain a science at all, would become a 'sacred and secret' science.
>
> (van Dantzig[29])

3.6 OTHER VIEWPOINTS

The range of attitudes expressed about the probability concept is vast. Each individual presents his own particular viewpoint. In the main, however, one can detect a *basic* attitude in the ideas advanced by any writer and attribute this to one of the broad categories described above. However, there are exceptions to every rule and a truly individual approach is that of **fiducial** probability proposed by

R. A. Fisher. Its stimulus, however, is the construction of a principle of statistical inference, rather than the desire to produce a fundamental view of probability itself. For this reason it is best regarded as an approach to inference and it will be discussed in that light at a later stage (in Chapter 8) rather than in this present chapter.

3.7 SOME HISTORICAL BACKGROUND

The probability concept has a chequered and poorly documented history. The appearance of different modes of thought have been often submerged in informality of expression, coloured by prejudice, inhibited by persecution or merely inaccessible in published form. Added confusion arises when we find the attribution of attitudes to historic personalities resting on the personal convictions of their modern commentators. This is well illustrated in relation to Bayes's[30] original enigmatic definition of probability:

> The *probability of any event* is the ratio between the value at which an expectation depending on the happening of the event ought to be computed, and the value of the thing expected upon its happening.

Pitman[31] views this as support for a *degree-of-belief* attitude to probability, whilst Fisher (1959) sees it as a clear statement of the *frequency* viewpoint

> . . . equivalent to the limiting value of the relative frequency of success. (p. 14.)

Obviously the element of chance, and attempts to represent it and measure it, have figured in some sense in the activities of man from the earliest times. See David (1962; *Games, God and Gambling*). But so crucial a concept has proved most elusive until quite recently. We have come far in terms of attitude from the days in which anyone who dabbled in probability theory might be labelled a religious heretic or sorcerer. But we have only in this century escaped from the (contrary) belief that the world is entirely deterministic, and that the probability concept had merely pragmatic relevance as a cover for our inadequate knowledge of the laws of determinism, or to cope with an unfortunate paucity of confirmatory experimental evidence. von Mises (1957) was certainly not hammering the final nail in this particular coffin when, in 1928, he wrote:

> The point of view that statistical theories are merely temporary explanations, in contrast to the final deterministic ones which alone satisfy our desire for causality, is nothing but a prejudice. Such an opinion can be explained historically, but it is bound to disappear with increased understanding. (p. 223.)

However, whilst nowadays accepting the relevance of probability in its own right as a facet of the world we live in, we are still far (perhaps further than ever) from reaching common agreement on the 'nature of the beast'. Deeper interest, study and discussion, only serve to elaborate the vastly different attitudes to probability which prevail. Little wonder then that its *origins* are difficult to trace!

Studies are available of some early developments; dealing with relevant historical personalities and the formal details of their work, though often with little regard to the philosophical interpretation of the views of probability they were promoting. Historical commentaries which aim at objectivity in this sense include Todhunter (1949, first published 1865), David (1962), Kendall and Pearson (1970), Kendall and Plackett (1977) and (perhaps) Maistrov (1974); but these attempt little comment on the thorny areas of controversy that have sprung up (mainly this century) between the objective and subjective views of probability.

Leaving aside questions about the basic nature of probability (and ambiguities of attributative interpretation, as illustrated above) we also find differences of opinion on the motivating forces for the study of probability in early times. Traditional belief that the origins of attempts to formalize the probability concept arose in the context of gambling and games of chance must be tempered by, for example, the view of Maistrov (1974) who sees the driving force to lie in broader contemporary social, economic and political affairs. The romantic image of probability theory arising out of the diminishing returns of the Chevalier de Méré in the gaming room (reflected in the famous correspondence between Pascal and Fermat) is altogether too simple a view of the origin of probability theory. See Kendall[32], David (1962) and the interesting fictitious Pascal side of the correspondence with Fermat, by Renyi (1972). Nonetheless gambling, and the betting instinct generally, has been a major stimulus from the beginning and continues to figure widely in discussions of the basic nature of probability, of how to measure it, in classroom illustration of simple properties of probability theory and as the subject of sophisticated research activity. In this latter respect see Dubins and Savage (1965), *How to Gamble if You Must. Inequalities for Stochastic Processes.*

We shall not attempt in this short chapter to emulate the existing objective studies of the 'pre-history' of probability, published tracts speak for themselves and are readily available. Gaps certainly remain, particularly in *interpreting* eighteenth and nineteenth century developments. More important, definitive historical treatment of the variety of attitudes which have appeared in modern times is much needed.

3.8 AND SO ...

What are we to accept as the true basis of probability? Does it matter what attitude we adopt? We have already seen that different views of probability are confounded with different approaches to statistical inference and decision-making. To this extent it is important to our main theme that we appreciate the variety of views of the probability concept.

... suppose someone attaches the probability one sixth to the ace at his next throw of a die. If asked what he means he may well agree with statements expressed roughly thus: he considers $1 a fair insurance premium against a risk of $6 to which he might be exposed by occurrence of the ace; the six faces are equally likely and only one is favourable; it may be expected that every face will appear in about 1/6 of the trials in the long run; he has observed a frequency of 1/6 in the past and adopts this value as the probability for the next trial; and so on. . . . each of these rough statements admits several interpretations. . . . only one . . . can express the very idea, or definition of probability according to this person's language, while the others would be accepted by him, if at all, as consequences of the definition.

(de Finetti[4])

3.9   AND YET . . .

The frequentist seeks for objectivity in defining his probabilities by reference to frequencies; but he has to use a primitive idea of randomness or equi-probability in order to calculate the probability in any given practical case. The non-frequentist begins by taking probability as a primitive idea but he has to assume that the values which his calculations give to a probability reflect, in some way, the behaviour of events. . . . Neither party can avoid using the ideas of the other in order to set up and justify a comprehensive theory.

(Kendall[33])

References

† 1. Borel, E. (1924). 'Apropos of a treatise on probability', *Revue Philosophique* Reprinted in Kyburg and Smokler (1964).
† 2. Koopman, B. O. (1940). 'The bases of probability', *Bull. Amer. Math. Soc.* Reprinted in Kyburg and Smokler (1964).
† 3. Savage, L. J. (1961). 'The foundations of statistics reconsidered', in *Proceedings of the IVth Berkeley Symposium on Mathematics and Probability*. Berkeley: University of California Press. Reprinted in Kyburg and Smokler (1964).
† 4. de Finetti, B. (1968). 'Probability: II Interpretations', in *International Encyclopedia of the Social Sciences*, Vol. 12. New York: Macmillan Co. and The Free Press.
  5. Atkinson, L. F. (1969). 'The concept of probability', internal publication of the Department of Mathematics, University of Western Australia.
† 6. Venn, J. (1888). 'The subjective side of probability', reprinted from Chapter VI of the 1888 edition of Venn's book *The Logic of Chance*, in Kyburg and Smokler (1964).
  7. von Mises, R. (1941). 'On the foundations of probability and statistics', *Ann. Math. Statist.*, **12,** 191–205.
  8. Kolmogorov, A. (1963). 'On tables of random numbers', *Sankhyā A*, **25,** 369–376.
  9. Martin-Löf, P. (1966). 'The definition of random sequences', *Information and Control*, **9,** 602–619.
 10. Solomonoff, R. (1964). 'A formal theory of inductive inference', Parts I and II, *Information and Control*, **7,** 1–22 and 224–254.
 11. Willis, D. (1970). 'Computational complexity and probability constructions', *J. Assn. Comput. Mach.*, **17,** 241–259.

† 12. de Finetti, B. (1937). 'Foresight: its logical laws, its subjective sources', *Annales de l'Institute Henri Poincaré*. Reprinted (in translation) in Kyburg and Smokler (1964).

13. Jeffreys, H. (1955). 'The present position in probability theory', *Br. J. Philos. Sci.*, **V**, 275–289.

14. Edgeworth, F. Y. (1910). 'Probability', in *Encyclopaedia Britannica*, 11th edn, Vol. 22. London: Encyclopaedia Britannica, pp. 376–403.

15. Savage, L. J. (1961). 'Bayesian statistics', lecture to Third Symposium on Decision and Information Processes, Purdue University.

† 16. Smith, C. A. B. (1965). 'Personal probability and statistical analysis', *J. Roy. Statist. Soc. A*, **128**, 469–499.

17. Savage, L. J. (1961). 'The subjective basis of statistical practice', internal publication of the University of Michigan, Ann Arbor.

† 18. Murphy, A. H. and Winkler, R. L. (1977). 'Reliability of subjective probability forecasts of precipitation and temperature', *Appl. Statist.*, **26**, 41–47.

† 19. Murphy, A. H. (1975). 'Expressing the uncertainty of weather forecasts', *The Statistician*, **24**, 69–71.

20. Hoerl, A. E. and Fallin, H. K. (1974). 'Reliability of subjective evaluations in a high incentive situation', *J. Roy. Statist. Soc. A*, **137**, 227–230.

† 21. Winkler, R. L. (1971). 'Probabilistic prediction: some experimental results', *J. Amer. Statist. Assn*, **66**, 675–685.

22. O'Carroll, F. M. (1977). 'Subjective probabilities and short-term economic forecasts: an empirical investigation', *Appl. Statist.*, **26**, 269–278.

† 23. Savage, L. J. (1971). 'Elicitation of personal probabilities and expectations', *J. Amer. Statist. Assn*, **66**, 783–801.

† 24. Hampton, J. M., Moore, P. G. and Thomas, H. (1973). 'Subjective probability and its measurement,' *J. Roy. Statist. Soc. A*, **136**, 21–42.

† 25. Tversky, A. (1974). 'Assessing uncertainty' (with Discussion), *J. Roy. Statist. Soc. B*, **36**, 148–159.

† 26. De Groot, M. H. (1974). 'Reaching a consensus', *J. Amer. Statist. Assn*, **69**, 118–121.

† 27. Lindley, D. V., Tversky, A. and Brown, R. V. (1979). 'On the reconciliation of probability assessments' (with Discussion), *J. Roy. Statist. Soc. A*, **142**, 146–180.

† 28. Ramsey, F. P. (1931/1964) 'Truth and probability', in Kyburg and Smokler (1964) pp. 61–92. *Reprinted from The Foundations of Mathematics and Other Essays.* Kagan, Paul, Trench, Trubner: London, 1931.

† 29. van Dantzig, D. (1957). 'Statistical priesthood (Savage on personal probabilities [1]]', *Statistica Neerlandica*, **11**, 1–16.

30. Bayes, T. (1963). *Facsimiles of Two Papers. Reprinted from the Philosophical Transactions of the Royal Society of London, Volume 53*, 1763. New York: Hafner.

† 31. Pitman, E. J. G. (1965). 'Some remarks on statistical inference', in Neyman and Le Cam (1965).

32. Kendall, M. G. (1956). 'Studies in the history of probability and statistics. II. The beginnings of a probability calculus', *Biometrika*, **43**, 1–14. Reprinted in Pearson and Kendall (1970).

† 33. Kendall, M. G. (1949). 'On the reconciliation of theories of probability', *Biometrika*, **36**, 101–116.

CHAPTER 4

Utility and Decision-making

... we tried to show that it was sensible to associate numbers with uncertain events. Now we show that the consequences of decision-making can also have numbers attached to them, and that these two sets of numbers combine to solve the problem and determine the best decision.

(Lindley, 1971a, p. 50)

In Chapter 1 the information to be processed in an inference or decision-making problem was distinguished into three elements: *sample data, prior information* and the *consequential benefits* (or disadvantages) of any potential action. It was quite clear in this earlier discussion that an assessment of consequences might be an important aid to the choice of action in the face of uncertainty. This applies whether we are to take formal regard of such an assessment in constructing policies for decision-making (as is the case in *Decision Theory*, to be discussed at length in Chapter 7) or whether it enters in a less formal way, as for example in the choice of the working hypothesis, or significance level, in the classical test of significance.

One approach to the assignment of *numerical* values to consequences is through the **theory of utility**. As subjective probability models how an individual reacts to uncertainty, so utility theory offers a model for the way in which individuals react (or should react) to different possible *courses of action* in a situation involving uncertainty. Starting from the empirical observation that an individual inevitably chooses some specific course of action, it claims that he does so (or should do so) in relation to how he *values* the different possibilities. This evaluation amounts to setting up some order of preferences for the different alternatives, and choosing the most preferred action. If such preferential action is to be *consistent* and *coherent* (on defined meanings of these terms within *utility theory*, similar to those explained above in respect of the subjective probability concept) it is shown to be implicit that the individual has assigned actual *numerical* values, **utilities**, to the different possible actions. In choosing a best action, he is merely following that course which has greatest utility.

Some aspects of emphasis and status of this theory must be made clear at the outset. Within utility theory it is not suggested that an individual necessarily

96

consciously constructs his utilities, scans these and takes that action with the largest numerical value. It merely claims that in setting up a 'sensible' preference scheme (one that is consistent and coherent) there is an *implied* set of utilities on which the individual is acting; it even goes further to propose procedures for determining the implicit utilities relevant to the individual in a particular situation. In practice, however, we observe that individuals are not always 'sensible' in their actions. They act on impulse, and may express combinations of attitude which are circular, or logically inconsistent. Within utility theory the standpoint which is adopted is to claim that such 'irrational' behaviour is a result of ignorance of the proper bases for decision and action, provided by *utility theory*. If only enlightened they will learn from their mistakes, and strive to act more rationally in future.

Not everyone accepts the model proposed through utility theory. Many would claim that human *personal* action cannot, and should not, be formalized in this way, but that the intangibility of a logical basis for human action in certain circumstances is part of its strength. This is of the essence of human personality! To answer that utility theory is only a basic framework, and that *any* model inevitably simplifies the true situation, is at most a partial resolution of the conflict of attitude here. Any fuller analysis, however, quickly takes us beyond statistics into the realms of philosophy or, even, religion. We shall not venture into these areas.

Instead we shall consider in this chapter some of the details of formal *utility theory*, as just one possible model to describe action under uncertainty. As such it forms part of the organized theory of inference and decision-making and is thus relevant to our theme. The fact that doubts exist on its validity or general applicability places it in no different category to many of the other topics we have discussed. To best assess such doubts we need to know what utility theory is about, and in this spirit we shall consider the behaviouristic basis of the theory, its terminology and postulates, the properties of utility functions and their proposed use in establishing criteria to guide the taking of decisions. We shall consider also the particular problems of establishing a utility function for *money*.

It is obvious from the above remarks that utility theory is entirely *subjective* in origin. This is not to say, however, that it is only applied in situations where personal factors are predominant. In certain situations (in industry, agriculture and so on) the major components in the utilities of different actions are clear cut and objective. If so there may be little dispute about the relevance, or general form, of utilities and their use in making decisions. Problems may still remain, however, concerning their specific quantification.

The idea of *utility* has already been encountered in the *subjective* approach to probability (Chapter 3). In the subjective formulation it was proposed that probabilities might be calculated by reference to betting situations. One difficulty was that the probability calculations seemed likely to change with the stakes of the envisaged bets. Ramsey dealt with this problem by proposing a combined theory of *probability* and *utility*, in which the twin concepts were intrinsically interrelated. Here we encountered a fundamental form of the **utility** concept.

Published material on utility theory ranges from elementary motivational treatments to detailed formal (mathematical) developments. The most accessible, attractively styled, expositions are to be found in textbooks on decision-making or *decision theory*. A leisurely and well illustrated introduction to the idea of utility is given by Lindley[1]. Perhaps the first axiomatic treatment is that of von Neumann and Morgenstern (1953; first published 1944); others, in rough order of their level of treatment, appear in Raiffa (1968), Lindley (1971a), Winkler (1972), Chernoff and Moses (1959), Fine (1973), Savage (1954), De Groot (1970). Many of these references give special attention also to the important topic of the utility of *money*; Cox and Hinkley (1974) highlight a range of practical considerations in the assessment of utility. The utility of money, and some of the special practical problems, will be considered later in this chapter. Two slightly confusing features of the existing treatments of the subject are the lack of uniformity in naming ideas and concepts (**reward**, **prospect**, etc.) and variations in the order in which topics are introduced (behavioural bases, axiomatics and so on). The terms used below (drawn from the multiplicity in use) seem natural, as does the order of development, but care will be needed in augmenting the following brief review by reference to other publications.

4.1   SETTING A VALUE ON REWARDS AND CONSEQUENCES

We start by considering a simple situation in which an individual is faced with choosing from among a set of fully defined, and attainable, alternatives. He may be offered the choice of a cake from a plate with six different cakes on it. We may consider seven alternative actions open to him, to take any particular cake or no cake. As a result of his action he obtains the cake of his choice, or no cake if he decides to take none. If we ignore any *implications* of the action he takes (such as suffering because the cream was sour in the cake he chose, or feeling hungry because he declined to take a cake) we have illustrated the simplest type of situation involving choice. Certain alternatives are presented, they are clearly defined and any one of them is freely available. The alternatives constitute a set of **rewards**, the action taken yields a particular reward (the chocolate eclair, or no cake, for example).

How is a choice to be made among such rewards? A schoolboy offered the choice of a 5p coin or a 10p coin by a kindly uncle may have no hesitation in choosing the latter; he *prefers* 10p to 5p; it is worth more to him, he can purchase more sweets with it. Another schoolboy may accept the 5p coin on grounds of politeness. But again it is chosen because it is *preferred*, in the respect that in acting in what he believes to be a proper way he is maximizing his sense of well-being. (At least this is what is assumed in the *utility theory* model!)

In general an individual faced with a variety of attainable rewards will be assumed to find some preferable to others, perhaps one overwhelmingly so, and will choose accordingly. The situation is seldom entirely clear cut, however, even in the case of the choice facing the schoolboy, and needs to be qualified in various respects.

(i) *The rewards need not all be ameliorative.* Some, indeed all, rewards may be unattractive. On hearing of the death of a relative we may be faced with the onerous tasks of writing a letter of sympathy or expressing our condolences in person. To most people both 'rewards' are undesirable, but usually a choice is made which represents what we *prefer.* Questions of 'duty' are part and parcel of the framework in which we work out our preferences. To consider doing nothing is merely to extend the set of rewards among which we must choose, as in the cake example above.

(ii) *The rewards may contain many components.* The government may have to choose between increasing social service charges *and* facing inevitable public dissatisfaction, or keeping them at a current inadequate level *and* subsidizing them from other areas of the economy. The presence of two (if not substantially more) components in the alternative rewards does not change the problem in essence. Admittedly it may make the expression of preferences more difficult; but a choice *will* be made and this will implicitly express the preferences that are adopted.

(iii) *Preferences will often be personal.* As far as individual action is concerned the implied preferences will reflect the current *personal* attitudes of the individual who must make the choice. What will be the better of two rewards for one person will not necessarily be the same for another. This is further illustrated when we note that:

(iv) *Preferences are conditioned by the environment.* The schoolboy would probably prefer 10p to 5p. If his current yearning is for mint humbugs there is a sense in which he may feel that 10p is about *twice as valuable* to him as 5p. On the other hand, if his burning desire is to go to the cinema and a ticket costs 15p, neither 10p nor 5p will satisfy his yearning *if he has no other resources.* 10p may still be better than 5p; but how much better? Contrast this situation with a similarly motivated schoolboy who already possesses 5p!

(v) *It may be possible to quantify preferences.* Again consider the 'mint humbug' situation. Here it may be reasonable to attach actual *numerical values* to the rewards. These values (which are commonly called **utilities**) are not necessarily the absolute *monetary* values and do not even need to be thought of as being in monetary units. If the utilities are intended to measure the strengths of preference for the two rewards, relative values will suffice—to represent the fact that 10p is twice as good as 5p measured in the scale of humbugs! But utilities need not even be proportional to the monetary values. A small (but intelligent) schoolboy may recognize that his capacity is limited to what can be purchased with 5p. If his horizon is limited to his immediate comfort 10p may be negligibly more valuable to him than 5p. At the other extreme, 10p is vastly more valuable than 5p to the schoolboy with 5p who wants to buy a 15p cinema ticket.

These observations, and qualifications, present a picture of one way in which a choice may be made between *attainable fully specified rewards*. Most of the comments relate to rather trivial examples, but they exemplify the considerations which arise in almost all situations of choice. The doctor, industrialist, politician is faced with choosing one of a set of rewards time and again. If we observe his behaviour we find the same features. A choice is inevitably made (if only the choice of suspending action for the moment). This choice is justified as yielding the *preferred* reward. The choice may in some problems vary from one individual to another—in other cases there are obvious tangible criteria which would lead most people to make essentially the same choice. In picking out a preferred reward it will be apparent that all rewards have been ranked—or at least partially ranked—in concluding that the chosen one is preferred to all others. Sometimes the rewards have obviously been more than ranked—they have been evaluated (on the basis of some combination of subjective and objective information) and the preference scheme constructed from this evaluation.

Of all types of reward, monetary ones might be thought to yield the most direct evaluation. Even the simple example of the schoolboy, however, has shown this to be a delusion, and we shall later discuss the problems of assigning a utility to money (§4.7).

Such characterizations of the behaviour of people when faced with making a choice among different rewards form part of the basis for the construction of a formal utility theory—the stimulus for discussing why and how we should assign values, or utilities, to the different rewards.

But we have not yet gone far enough to justify such a theory. As we have described the problem no question of *uncertainty* arises. It is assumed that a free choice is available from a set of fully determined, realizable, rewards. If this is the situation, our examples illustrate quite clearly that utilities (as *numbers*) are really a luxury from the point of view of representing the different rewards. Such preferences need involve no explicit assignment of numerical values. All that matters is the preference *ranking* of the rewards.

But life is not really so co-operative as to present us with a choice of distinct *determined* rewards. The individual, either in his everyday routine behaviour or as a statistician or decision-maker, is faced with the need to act in situations where a variety of rewards *may* occur, but no particular one is guaranteed to do so. What actually materializes will depend on fortuitous effects. It is the appearance of this element of uncertainty which promotes the need for *numerical* values to be assigned to rewards—that is, the need for *utilities*.

Suppose, for example, I am invited to play a game where, with probability p I win £1, and with probability $(1 - p)$ I lose £1. If I do not play, my current fortune is unchanged. The rewards are clear-cut and monetary: $-£1, £0$ or $£1$. Of these three I presumably prefer £1, but I am not offered a free choice, merely a choice of whether or not to play the game. Should I accept the game? It depends partly on the value of p. The situation is quite different when $p = 0.8$ to when $p = 0.2$! But the value of p is not all that matters as we readily see if we translate the problem to the case of the schoolboy who wants to visit the cinema.

Consider three cases:

(a) *He has 15p and is offered the game with stakes of 10p.*

(b) *He has 10p and is offered the game with stakes of 10p.*

(c) *He has 5p and is offered the game with stakes of 10p.*

The following reactions are likely:

In (a). He has the money he needs and there is little incentive to play the game. If he does he will certainly need p very close to 1 in value to offset the enormity of his discomfort if he loses.

In (b). He has *inadequate* funds for his purposes. The game offers a heaven-sent opportunity! There is little to lose in playing even if p is rather small.

In (c). The same applies as in (b) as far as winning the game is concerned. But if he loses he cannot pay his debt. This is a real moral dilemma. Your guess is as good as mine as to how he might resolve it!

So the value of p is only part of the story. The further vital ingredient is the *value* of each reward *if it happens to occur*; this depends in turn also on the current circumstances (how much he already has, what he might do if he cannot go to the cinema, and so on).

As a further illustration suppose I want to cross a busy road. I am unlikely to do so merely because there is a higher probability of doing so safely than of being run down, unless there is some compelling reason for doing so (that is, a potential reward of great value to me) to discount my natural caution. Even so, my decision may be tempered by previous experiences—suppose I have only recently come out of hospital from road accident injuries!

The fact that in real life rewards are not fully determined but relate to fortuitous events points the need for a wider concept than that of a reward as defined above. In practice we will need to choose some **action**, which has different possible *implications*. It is the implications of the actions which are the real 'rewards'; what we need to assess are the combinations of action and the fortuitous events that this action encounters. We will use the term **consequence** to describe this wider concept: the conjunction of an action and the circumstances it encounters. Thus in the traffic example *consequences* might be that (I cross the road but get knocked down) or (I wait awhile but miss my bus). What action I take will depend on three things: what values I place on these consequences, what I know of the chance mechanism (the *probability* of safely crossing the road), and what is the *conditioning* effect of my personal circumstances (present needs and previous experience).

But even in this wider context, why is there any need to assign specific *numerical values* to the consequences? Why are not relative preferences again sufficient, implying at most a utility *ordering*? Consider one more example: that of the supply of a batch of components by **TWEECO** to its associate **THECO** in the example of Chapter 2. Suppose the (unknown) proportion of defectives in the batch is θ. The possible actions might be to supply, or scrap, the batch. In either case the benefit (or disadvantage) of action depends strongly on what the value of θ happens to be. We have already seen in §2.4 that, for a particular cost scheme for production (and for the trading arrangements between the two companies), the choice of an appropriate action was crucially dependent on a *numerical* assessment of the consequences. Except in trivial circumstances (where uncertainty about θ was limited) a mere ordering of consequences was not enough!

In the TWEECO example, utilities were monetary and utility was measured *as money*. In other examples above it is clear that even if rewards are monetary it may not suffice to regard utility as equivalent to money (see §4.7). Then again, these examples were often highly *personal*, in the sense that the individual in a situation of choice chooses what *he feels* to be the best action. Others in the same situation might act quite differently. Also it is patently obvious from the examples that the individual does not always (if indeed often) consciously assign numbers to the consequences to aid his choice. He seems merely to act 'on impulse'.

Recognizing such facets in overt human behaviour, *utility theory* offers a mathematical model of the manner in which individuals take action in situations of uncertainty. To pave the way for a more detailed study of this theory, let us summarize its basic attitudes and assumptions.

It argues as follows. People do in fact take *some* action. In doing so they are responding to preferences for different consequences. How can we represent and explain such behaviour? Firstly it seems desirable to assume that certain logical rules should be adopted by the individual for handling his preferences. If these rules are obeyed it is seen to follow that (implicitly at least) he is assigning numerical values (**utilities**) to the different consequences; that these utilities are essentially unique and can in principle be determined; and that in expressing a preference for one consequence over another he is implicitly declaring that the utility of the one is greater than that of the other. Furthermore, it follows that if faced with a variety of consequences which may severally occur with different probabilities a value can be assigned to this *mixed consequence* as a natural basis for choice between different mixed consequences. That is, a utility exists for *probability distributions over the set of consequences* (a *lottery*, or *gamble*). The theory goes on to suggest that the general decision-making process may be so expressed. Each *potential* action is represented as a lottery or gamble and the appropriate action to take is the one which has highest utility in this 'mixed consequence' sense. Since the actual consequence that arises in such a scheme is not known but merely *prospective* we shall use the term **prospect** (rather than 'lottery' or 'gamble') to describe a 'mixed consequence'.

It is important to recognize the status of this theory! It is *normative*. That is to say it describes how a person *should* react, and provides a model for such 'sensible' behaviour. Not all of us are 'sensible' in our actions in this respect—utility theory aims at revealing such 'inadequacies' in the hope that we may be persuaded to be more 'sensible'. So it is not a prescription for the way people actually do behave, merely for how they *should* behave if convinced of the logical bases of the theory. Neither is it an idealized theory—it promotes no 'ideal rational man', never realized, only imperfectly emulated. On the contrary, it claims to represent the enlightened 'mere mortal'.

We shall go on to discuss informally and without proofs the essential features of utility theory, and some of its implications.

4.2 THE RATIONAL EXPRESSION OF PREFERENCES

In a decision-making situation, whether personal or objective, choice must be exercised in relation to some set of **consequences**. Any consequence, C, has the form $(a, \theta|H)$ where a represents a potential **action** or **decision**†, θ the possible fortuitous events which pertain and affect this action, and H represents the local environment which conditions our reaction to the conjunction of a and θ. Generally there will be a set, $\mathbb{C}$, of consequences in any decision-making problem; C denotes a typical member of this set. **Utility theory** commences by prescribing rules for the way in which preferences are expressed for different consequences.

Rule 1. Given two consequences C_1 and C_2 in $\mathbb{C}$, either C_1 is preferred to C_2, C_2 is preferred to C_1, or C_1 and C_2 are held in equal esteem.

This rule implies that *any* two consequences in C can be ordered in terms of preference, if only to the extent that both appear equally attractive. It precludes the possibility of consequences being incomparable. We may feel that this is not

---

† Although in real-life the *decision* and *action* are distinct, the former preceding the latter, this distinction can be ignored when we consider constructing a formal theory. The attitude adopted is that the decision *compels* the action and that the two are essentially equivalent.

always true in real life: that consequences may exist between which we are unable to express a preference, and at the same time are unwilling to declare them equally preferable. Even so it will be observed that we do take actions in the face of such consequences, so that implicitly we are ranking them, even if 'in cold blood' we claim to be unable to do so. In a normative theory of utility this implicit reaction is made explicit. It is claimed that with sufficient consideration of the nature of the consequences Rule 1 will (or should) apply.

Rule 2. If C_1, C_2 and C_3 are three consequences in $\mathbb{C}$ and C_1 is not preferred to C_2, and C_2 is not preferred to C_3, then C_1 cannot be preferred to C_3.

Again this is prescribed as a rule for *sensible* behaviour in relation to consequences. The fact that we can detect in human behaviour circularity of preferences that violates this rule does not render it inappropriate. Rule 2 speaks for itself as a basis for logical reaction to consequences, and sets the standard for the way people should behave if their action in the face of uncertainty is to be well ordered.

Rule 3. There are at least two consequences in $\mathbb{C}$ which are not held in equal esteem.

Otherwise no problem exists in the choice of action under uncertainty; it makes no difference what consequence materializes and the action we take is arbitrary. This is patently untrue except in trivial circumstances.

In prescribing the Rules 1, 2 and 3 we are simply declaring that a non-trivial, transitive, ordering exists on the preferences for the various consequences in the set $\mathbb{C}$. Such a system is adopted as the basis for what is variously described as **rational**, **consistent** or **coherent** behaviour. (See the discussion of subjective probability in §3.5.)

4.3   PREFERENCES FOR PROSPECTS AND MIXTURES OF PROSPECTS

The arguments of §4.1 make it clear that we cannot proceed far merely by considering preferences for the different consequences. In practice we are not given a free choice of consequences; for example, the element θ is beyond our control, it just happens! The true situation is that we are faced with a choice of different probability distributions over the set of consequences. We see this if we fix the action a that we are contemplating. A range of possible consequences $(a, \theta|H)$, each with the same a but different θ, may arise. They individually occur with different probabilities reflecting the random nature of θ. If we knew the probability distribution for such a prospect, the choice of a would amount to a choice of a probability distribution from among those representing the corresponding prospects. Note that the set of probabilities remains fixed, it is the set of 'outcomes' $\{(a, \theta|H)\}$, for any a, which varies.

Thus we can identify a **prospect** with a probability distribution, P, and we need to consider the expression of preferences not in terms of individual consequences but in terms of prospects (or probability distributions over the set of consequences). We assume that the choice facing the decision-maker is to be made from a set $\mathbb{P}$ of prospects, P, and that again certain rules must apply to the expression of preferences for the elements of $\mathbb{P}$. Once more the evidence for assuming that such preferences exist, and for the way in which they interrelate, is sought in the observed behaviour of individuals when they act in situations involving uncertainty. For their behaviour to be 'rational' the following model rules are prescribed, echoing and extending those of §4.2.

Rule I. Given two prospects P_1 and P_2 in $\mathbb{P}$, P_1 is preferred to P_2, or P_2 is prefered to P_1, or P_1 and P_2 are held in equal esteem.

Rule II. If P_1, P_2 and P_3 are three prospects in $\mathbb{P}$ and P_1 is not preferred to P_2, and P_2 is not preferred to P_3, then P_1 cannot be preferred to P_3.

Rule III. There are at least two prospects in $\mathbb{P}$ which are not held in equal esteem.

Although the idea of a prospect was set up by considering probability distributions over θ for a fixed action a, this is an unnecessarily limited viewpoint. Any probability distribution on $\mathbb{C}$ defines a prospect. In particular the individual **consequences** may be regarded as **prospects**, defined in terms of degenerate distributions over $\mathbb{C}$ which assign probability one to the consequence of interest and probability zero to all others. Furthermore, from any two mixed consequences P_1 and P_2 we can construct many others merely by a further probabilistic mixing of these two. Thus we might define P to be either P_1 or P_2 depending on which of two outcomes arises in a simple experiment, where the first outcome has probability p, the second $1 - p$; for some value of p in the range $(0, 1)$. Two further rules are proposed for the expression of preferences for such *mixtures of prospects*.

Rule IV. If P_1 is preferred to P_2 which is preferred to P_3, then there is a mixture of P_1 and P_3 which is preferred to P_2, and there is a mixture of P_1 and P_3 in relation to which P_2 is preferred.

Rule V. If P_1 is preferred to P_2, and P_3 is another prospect, then a particular mixture of P_1 and P_3 will be preferred to the same mixture of P_2 and P_3.

It is pertinent to consider just how realistic these rules are as a prescription for personal action. There are certainly situations where observed human behaviour breaches the rules. In particular, it is by no means unknown for someone to find a prospect P_1 preferable to a prospect P_2, which is in turn preferred to a further prospect P_3, and yet P_3 may be declared preferable to P_1. Or an individual may express preferences among some subset of prospects without even contemplating other, none-the-less viable, prospects. Then again, difficulties can arise in any attempt at a simple frequency interpretation of the probability concept in respect of a prospect P. As Fine (1973, p. 219) remarks:

> On a given day we may prefer a good meal to $1, but prefer a good meal and $5 to six good meals on the same day.

The resolution of all such difficulties is *not* found in the *normative* role of the theory: that it describes how people *should* behave not how they necessarily *do* behave.

Rules IV and V are the most restrictive and questionable. For example, they imply that there is no prospect (or consequence) that is *incomparably* more preferable, or more odious, than any other. We need these rules if we are to conclude that utilities are *bounded* (an attitude adopted in some approaches to utility theory). At first sight it may seem reasonable to believe that there are prospects that are too terrible to comprehend. What about our own death, loss of family and so on. *Utility theory* argues that however awesome these are, they are surely not immeasurable, otherwise so many simple actions would not be undertaken (like crossing the road, or sending the family off on holiday on their own). Yet not all are convinced that such actions imply (even a subconscious) valuation of 'disaster' on the same scale as more mundane eventualities. (See comment by Pearson in §1.6 above.) Cox and Hinkley (1974, p. 423) suggest that it might help in such problems to consider jointly different aspects of the utility of a prospect, i.e. to set up several preference orderings relating to the different aspects of interest. Becker, De Groot and Marschak[2] have proposed and examined models of this type.

The insistence on utility functions being *bounded* is not a unanimous one, and it also requires further assumptions on the ordering of preferences which are not made specific in the above simple statement of the rules. A formal examination of this matter, from the axiomatic standpoint, is to be found in Fishburn[3,4]. We shall see in the development of *decision theory* in Chapter 7 that it is sometimes useful, indeed, to adopt non-bounded utility functions in the study of practical problems as a pragmatic aid to the simple processing of such problems.

4.4   THE NUMERICAL ASSESSMENT OF PROSPECTS

Rules I to V for the rational expression of preferences among prospects serve as an axiomatic basis for the development of **utility theory**. As a direct logical consequence of these rules it can be shown that it is meaningful to assign to any consequence C a number $U(C)$, the **utility** of C. That is, there exists a mapping from the set, C, of consequences, to the real line. Furthermore, this mapping, or **utility function**, is to all intents and purposes unique. The utility concept is readily extended to a prospect, rather than a consequence. Since a prospect is simply a probability distribution, P, over the set of consequences we can consider the expected value of the utility function with respect to this distribution, i.e. $E(U|P)$. We define this to be the **utility of the prospect** P. The following crucial property may be deduced from the basic rules.

Property I. If P_1 and P_2 are two prospects then P_1 is preferred to P_2 if, and only if,

$$E(U|P_1) > E(U|P_2).$$

For simplicity we will denote $E(U|P)$ by $U(P)$ henceforth.

Thus the expression of preferences between prospects is seen to be equivalent to a comparison of the expected utilities—the higher the expected utility the more preferable the prospect.

The purpose of constructing a utility concept is to provide a formal model to explain what is meant by the expression of preferences between prospects. In this respect Property I implies that the *expected utility* with respect to the distribution P is the natural numerical assessment of a *prospect*.

Inevitably the same property holds for consequences, as for prospects, since the former may be viewed as *degenerate* prospects. If P_C is the distribution that assigns probability 1 to C, and probability 0 to each other consequence, then

$$C \equiv P_C.$$

By Property I, P_{C_1} is preferred to P_{C_2} if, and only if, the expected utility is higher for P_{C_1} that for P_{C_2}. But these expected utilities are merely $U(C_1)$ and $U(C_2)$, respectively; and we conclude that:

C_1 is preferred to C_2 if, and only if, $U(C_1) > U(C_2)$.

A further property of utility functions is easily demonstrated, arising directly from Property I.

Property II. If U is a utility function on C; then any function $V = \alpha U + \beta$ (where α and β are constants, with $\alpha > 0$) is also a utility function on C. Any two utility functions U and V, on C, are so related.

This is the key to the essential uniqueness of the utility function, which was referred to above. For it declares that *any two utility functions U and V, corresponding in the sense of Property I to a particular rational scheme of preferences over the set C, must be related in a linear way; so that the utility function is unique up to such linear transformations.*

4.5   THE MEASUREMENT OF UTILITIES

It is one thing to know that Rules I to V imply that a utility function exists on C, representing formally through the *expected utility property* (Property I) the nature of preferences for different prospects. Yet this is of little value if we do not know how to construct, or measure, the utility function in a particular situation. We shall divide discussion of this matter into two parts—what the basic theory provides for *formally* measuring utilities in rational preference schemes, and how this works out in practice when we try to determine an individual's utility function in a specific situation.

4.5.1   *Formal Construction of Utilities*

Suppose we consider a prospect P which is a mixture of prospects P_1 and P_2 with probabilities p and $1 - p$, respectively. Then P, being identified with this mixture, which we shall denote $\{P_1, P_2:p\}$, is held in equal esteem with it and in view of Property I it must be that

$$U(P) = U(\{P_1, P_2:p\}).$$

But by the expected-utility definition for the utility of a prospect this means that

$$U(P) = pU(P_1) + (1 - p)U(P_2).$$

Similarly, if P is a mixture of *consequences* C_1 and C_2 with probabilities p and $1 - p$, respectively; then

$$U(P) = pU(C_1) + (1 - p)U(C_2). \tag{4.5.1}$$

Consider any two consequences C_1 and C_2 in C, where C_2 is preferred to C_1. Then

$$U(C_2) > U(C_1).$$

Since U is unique only up to (increasing) linear transformations, the specific utilities of C_1 and C_2 may be assigned at will and we may arbitrarily, but legitimately, assume that $U(C_1) = 0$, $U(C_2) = 1$, say. But having made this assignment the utilities of all other consequences are uniquely defined. The arbitrary nature of these two fixed points is unimportant, they merely define the scale on which we are working. Chernoff and Moses (1959, p. 85) draw a useful parallel with the construction of temperature scales (Centigrade, Fahrenheit, etc.) in relation to the fixed points of the freezing and boiling of water.

Consider any other consequence C in C. Suppose first of all that C_2 is preferred to C, which in turn is preferred to C_1. Then Rule IV, in conjunction with Property I, implies that there is a unique mixture $\{C_1, C_2: 1 - p\}$ of C_1 and C_2 which will be held in equal esteem with C.

Thus we have immediately from (4.5.1) that

$$U(C) = (1 - p)U(C_1) + pU(C_2) = p.$$

In the same way we can assign a utility to *any* consequence intermediate to C_1 and C_2 in preference, and indeed the utility of any such C can be expressed, by (4.5.1), as a linear combination of the utilities of other such consequences.

The same approach can be used if C is preferred to C_2, or if C_1 is preferred to C.

(i) *C preferred to C_2.* There now exists a unique prospect $(C_1, C: 1 - p')$ equivalent in preference to C_2. So

$$U(C_2) = p' U(C),$$

or

$$U(C) = 1/p'.$$

(ii) *C_1 preferred to C.* For an appropriate p'' we have

$$U(C_1) = (1 - p'')U(C) + p''U(C_2).$$

So that

$$U(C) = -p''/(1 - p'').$$

Hence we have only to declare the utilities of any two particular consequences in C, and the *utility function* over C is fully determined from appropriate mixtures of these two basic consequences.

We can illustrate this in terms of the example of the schoolboy who wishes to visit the cinema. The price of the cinema ticket (in pence) is 15; suppose he currently has 5. We might quite arbitrarily assign a utility of 1 to 15, and 0 to 0. What is his utility for his current fortune of 5? The theory prescribes that there is some unique mixture of 0 and 15 (which yields 0 with probability $1 - p$ and 15 with probability p) which is equally as attractive as his current resources of 5. His utility for 5 is then simply p, *but the problem is how to determine p.* The customary response is that this can be determined by *introspection*, in terms of his subjective reaction to different possible mixing probabilities, p. It is usual to try to extract the relevant value of p by placing the problem in a betting situation. What are the minimum odds that would be accepted in a bet where if he wins he gains 10 (making his capital 15) of if he loses he must pay 5 (reducing his capital to 0). His current capital and immediate interests are crucial, they condition the choice of odds. To win means a visit to the cinema which is most attractive, to lose is not greatly more disastrous than his present situation. It may well be that he would play with as low a chance of winning as 1 in 4. This means, in consequence terms, that 5 is equivalent to $(0, 15: 0.8)$, or

$$U(5) = 0 \times 0.8 + 1 \times 0.2 = 0.2.$$

Such a conclusion means that the utility for money is not necessarily linear—it is not proportional to the monetary value in this situation. We return to fuller consideration of the utility function for money in §4.7.

This example leads naturally into the more material aspect of measuring utilities.

4.5.2   Personal Expression of Utilities

We have seen that as long as preferences are rationally expressed in the terms of Rules I to V a unique utility exists for any consequence, and this may be determined in terms of equi-preferable mixtures of other consequences of known utility. A method proposed for determining the utility function over C is to offer a variety of bets for comparative assessment, as in the example we have just discussed. It can be interesting to conduct such a programme to examine the form of the utility function which different people possess for a range of consequences. Experiments have been conducted along these lines. They are necessarily artificial both in the types of consequences considered and in terms of the environment in which the individual is asked to express his preferences. Nonetheless the results are illuminating, if hardly surprising!

Some of the observed effects are as follows. For a fixed set of consequences, vast differences are revealed from one person to another in the utility functions which arise. It is obvious from the above construction that for any individual the utility for a *particular* consequence can be extracted in terms of a variety of different bets—that is, by offering mixtures of *different* pairs of consequences for comparison with the one of interest. In this way it can be shown that even one individual's utility for that consequence is often by no means unique. Serious inconsistencies can even be revealed, where for example we might show that in one and the same situation his utility for £2 (in an assessment of monetary utility) exceeds that for £1, and is also less than that for £1! It is interesting to set up a simple classroom experiment to reveal this effect.

Does this mean that our whole elegant theory of utility is valueless? If we accept the *normative* basis of the theory; not at all! Such inconsistencies may be interpreted in two ways. They either imply that the use of hypothetical bets is an imperfect (inadequate) vehicle for extracting the true utility function that the individual possesses. Or more fundamentally that the individual is violating the rational preference Rules. But utility theory prescribes no specific technique for determining utilities—bets are only proposed as a possible aid; if they work out! And violation of the Rules does not castigate the theory; it embarrasses the individual. The view taken is that the 'obvious' propriety of the theory will point out to the individual his irrationality and encourage him to modify his preferences to accord with the Rules.

This is all very well, but a lingering disquiet remains! We still need some adequate means of measuring the utility function relevant to the individual in question, if we are to use the theory for 'rational decision-making'. We have concentrated so far on the *personal* nature of utility theory, inevitably highlighting idiosyncratic aspects. Undoubtedly many situations exist where an assessment of utilities is far less subjective; where individual differences and inconsistencies are unlikely and at least in terms of the immediate problem we might expect the measurement of utilities to be reasonably straightforward. In such cases utility theory and its derived principles for decision-making become particularly relevant, and largely uncontroversial.

For discussions of different views of utility theory and of more sophisticated models, the following additional references are relevant: Fishburn (1970), Luce (1959), and Luce and Suppes (1965). On the specific theme of the experimental measurement of utility readers should refer to Mosteller and Nogee[5]; Davidson, Suppes and Siegel[6]; Suppes and Walsh[7]; Savage[8]; and Hull, Moore and Thomas[9].

There are various other practical considerations which might (or should) be taken into account in attempting to construct utility functions. Often there is the need for a group of individuals (a committee) to express an overall utility evaluation. Hull, Moore and Thomas[9] consider this matter and give references to other work on the same theme. Utility may also, in some problems, be expected to change over time. Cox and Hinkley (1974, Chapter 11) comment on how account might be taken of some as yet unobserved relevant random variable (see also Aitchison[10]) and on the need to apply a discount function to the utility of prospects materializing at different times in the future.

4.6 DECISION-MAKING

Utility theory has immediate implications for the making of decisions. The rather restricted way in which the idea of a *consequence* was first introduced demonstrates this. We represented a consequence as a conditioned pair $(a, \theta | H)$, where a was a potential action from a set of possible actions, and θ described the environment in which the action was taken, assumed to have arisen by chance. If we imagine that θ arises from a probability distribution of possible values, then the action, a (alone), takes on the nature of a *prospect*. This happens in the following way. If we take action a, we face a set of possible consequences $(a, \theta | H)$ for varying θ, each with associated probability, $\pi(\theta)$ say. Property I of §4.4 now gives an answer to the appropriate (or best) choice of action: namely that *we should take that action, a, which has maximum expected utility* $E_\theta \{U(a, \theta | H)\}$, since this yields the most preferable result.

This makes optimum decision-making appear a simple matter! All we need do is to declare the set, C, of possible consequences, evaluate the utility function over the consequences $(a, \theta | H)$, determine the relevant probability distribution over possible values of θ, and choose a to maximize $E_\theta \{U(a, \theta | H)\}$. In principle this approach seems most attractive, but unfortunately there are serious practical difficulties. We have already touched on the problems of determining $U(a, \theta | H)$. No less are those of specifying all the possible consequences and expressing, or indeed interpreting, $\pi(\theta)$. If $\pi(\theta)$ is completely unknown, what do we do? One possibility advanced is that we then choose that a which gives the best general safeguard; in the sense that *the worst that can happen is as good as it can be*. This means choosing a to yield

$$\max_{a} \min_{\theta} \{U(a, \theta|H)\},$$

although we shall see that this is often 'too pessimistic for words'.

Then again, it may be that through subsidiary experimentation we can get more information about θ than that already available, perhaps in the form of sample data. How will this affect the situation? This will be the subject matter of Chapter 7, which discusses *decision theory* as the third major division of our study of statistical inference and decision-making.

4.7   THE UTILITY OF MONEY

It might seem that of all forms of consequence or prospect those that are purely monetary should be the easiest for which to construct a utility function. There are some problems, particularly in the commercial or industrial spheres, where as a simple first approximation we might regard the relevant consequences as expressible in direct monetary terms. Occasionally it may even suffice to accept the monetary values of the consequences *as their utilities*. But such situations are rare. Firstly, mere monetary return is often only a component of the consequences under consideration. A firm which offers a guarantee on its products might know fairly precisely what it costs them financially to honour that guarantee, but it is important also to assess the less tangible loss of goodwill which is engendered when a customer complains. Secondly, with purely monetary consequences even the large industrial concern cannot escape adopting a non-linear utility function if the sums of money at stake are large enough. A company capitalized at £1 000 000 facing a prospect of ruin or of the doubling of its capital is in essentially no different condition to the schoolboy whose pocket money does not run to the packet of mint humbugs which he desires, or the trip to the cinema. The various problems we have observed in such humble assessments of the utility of money have their counterpart at all levels. It is for this reason that utility theory has figured so widely in the study of Economics, from investment on the stockmarket to the control of the national economy.

The utility of money is a fascinating aspect of utility theory. It is worth considering one or two features of it. Some further considerations, and interesting illustrations, are given by Lindley[11].

Consider what, in probability terms, might be called a fair bet: one where the *expected monetary return* is zero. This could take the form of a game where there is probability $\frac{1}{2}$ of winning a certain sum of money (the stake) and probability $\frac{1}{2}$ of losing that same sum of money. As we go on and on playing this game an *indefinitely* large number of times we stand to neither win nor lose in the long run. So there might seem to be no reason why we should not play the game. But by the same token why should we play, expending unlimited effort for no net gain. If the probability of winning is in fact larger than $\frac{1}{2}$, say $\frac{2}{3}$, does this change the situation and render the expected return now an appropriate criterion for choice? Not really, since the prospect of an indefinite sequence of games is in any case not a realizable one—life is of limited duration. The only *practical* possibility is of a finite sequence of such games, and what we need to assess is the value of such a finite sequence or, in its simplest form, of a *single* game. The concept of mathematical expectation seems to have little relevance to this. We need to know the utility for the limited prospect that we are facing. Earlier examples have suggested how our reaction to the game will vary widely with the actual level of the stakes, in relation to our present fortunes. The millionaire *might* be *blasé* in his attitude to a game with stakes of £10 and play for the fun of it (if he enjoys gambling) even with most unfavourable odds. But even the inveterate gambler is likely to think carefully about playing a favourable game if he stands to lose all (or more than) he has. The implication is of a *non-linear utility function* for money.

This is well illustrated in an intriguing problem which has been often quoted over the years, being attributed to Daniel Bernoulli over two centuries ago and known as the **St Petersburg paradox**. A fair coin is tossed over and over again until a head first occurs. If this happens on the nth occasion we win 2^n pence. Obviously we must win something with this game. How much should we be prepared to pay for the privilege of playing the game? In terms of expectation, the long-term gain is $\sum_{n=1}^{\infty} (\frac{1}{2})^n 2^n$, which is of course infinite. We *should* be prepared to pay an unlimited entry fee if we accept expected return as an appropriate measure of utility, or if our utility function for money is linear. And yet people are most reluctant to pay any reasonable amount of money to play this game *once*. Chernoff and Moses (1959) regarded \$5 as a realistic personal limit when the monetary unit is in dollars.

The St Petersburg paradox has also been used to demonstrate the desirability of requiring that the utility function has the property of being *bounded*.We have touched on this before; it will require Rule IV for preferences among prospects *to hold for all possible prospects* in the situation under consideration. It is interesting to examine whether boundedness of the utility function is consistent with the way people actually behave. Suppose we start with the assertion that no individual prospect, P, has infinite utility. That is $U(P) < \infty$, for all P in $\mathbb{P}$. It may still happen that the utility function is unbounded from above. This means that there is no number ψ for which $U(P) < \psi$ for all P in $\mathbb{P}$. Suppose no such number did exist. Then there are prospects $P_1, P_2, P_3, \ldots$ with utilities which are respectively greater than $2, 2^2, 2^3, \ldots$. But the mixture which yields P_i with probability $(\frac{1}{2})^i$ has utility in excess of

$$2(\tfrac{1}{2}) + 2^2 (\tfrac{1}{2})^2 + 2^3 (\tfrac{1}{2})^3 \ldots,$$

which is infinite. This mixture, however, is itself a *prospect*, *P*, which we have shown to have infinite utility, contradicting our assertion that no such prospect exists. Thus we must conclude on this argument that *the utility function is bounded from above*. (Similarly we find it is bounded from below.)

This argument hinges on our accepting that no prospect has infinite utility. For a range of 'objective' practical problems it seems difficult to dispute this. The assertion is less obvious, however, in the realm of *subjective* (personal) behaviour. Arguments which show, through considering different betting situations, that the assignment of infinite utility to a prospect implies irrationality of preferences are not particularly compelling. After all, if a person really believes there to be a 'hell' too awful to contemplate we can hardly expect him to assess the whole situation of which this is one of the prospects by analysing his reactions to hypothetical betting games. Whether such a 'hell' does exist for some people seems to be more a question for psychological study; it may be that some prospect is so awful that any alternative appears immeasurably better in the prevailing state of mind of an individual. It would appear naïve to dismiss this possibility on the grounds that it violates the axioms of utility theory!

But with purely monetary consequences the assumption of a bounded utility function seems more plausible, and is usually adopted. Indeed it is common to find an even stronger constraint placed on the typical form of the utility function for money: namely that it must be *concave*, in the sense that the rate of increase diminishes with the monetary value. See Figure 4.1.

Certain features of such a utility function for money call for comment. The assumption of boundedness coupled with the obvious practical requirement that the utility $U(x)$ should be a non-decreasing function of the amount of money x, means that $U(x)$ is asymptotically parallel to the x-axis. The asymptote value is arbitrary, in view of Property II; it might as well be taken as unity. The utility for zero money is taken to be zero, and negative money is not envisaged. This presupposes firstly that we are measuring the utility function for an individual's

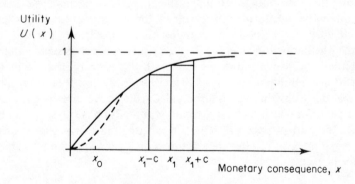

Figure 4.1   Utility function for money

(or company's) total financial state allowing for all potential or real assets, and not the resources relative to some current position. Secondly, negative resources are regarded as irrelevant as a fact of life. There seems to be no concept of *debt*, in spite of the fact that both borrowing, and unplanned debt, figure widely in contemporary society.

It seems no answer to declare that the zero on the money axis in Figure 4.1 represents some maximum envisaged debt. Observation of personal behaviour makes it most doubtful whether individuals can be claimed to possess any such (negative) money threshold, apart from any tangible constraints such as bank overdraft limits, or building society mortgage limits. Outside such areas it is a moot point whether an individual with resources x_0, say (relative to any threshold), really acts on this threshold constraint, in the sense that in no circumstances does he entertain a prospect which may potentially lose him an amount in excess of x_0. This asks a lot (perhaps too much) of human nature, even in a *normative* theory!

But the major effect of a concave form for $U(x)$ as represented by Figure 4.1— recognized for 200 years or so and first described by Daniel Bernoulli—is what is described in economics applications as the **law of diminishing marginal utility**. This says that in relation to *any* current sum of money x_1 an increase of c yields a lesser change (increase) *in utility* than does a corresponding reduction of c. See Figure 4.1. An immediate implication is that no fair game with equal stakes would be acceptable to the individual since he would have *equal* probability of his utility increasing by $U(x_1 + c) - U(x_1)$ or decreasing by the *larger* amount $U(x_1) - U(x_1 - c)$. He is said to be **risk aversive**. As Lindley (1971a) explains, he needs a 'probability [utility] premium' to make the game acceptable, a characteristic commonly observed in actual behaviour with respect to monetary gain or loss.

But some writers on utility feel this to be unrealistic as a universal concept, either for all individuals, or in particular over the whole range of values of x for a given individual. In this latter respect it is sometimes suggested as reasonable that, at least for small values of x, $U(x)$ may become *convex*, and in this region the individual abandons his caution and becomes, in an analogous sense, **risk prone**. (See dotted section of Figure 4.1.) Such an attitude is a denial of the law of diminishing marginal utility at least in relation to small amounts of money. Savage (1954) argues in these terms in his critique, and historical summary, of utility theory. In contrast, Lindley[11] prefers to maintain the concave form of $U(x)$ throughout as representative of a reasonable attitude towards the utility of money. See also Hull, Moore and Thomas[9]; and Pratt[12].

4.8   COMMENT: MATHEMATICAL REFINEMENTS: DISTINCTIONS OF ATTITUDE

The previous sections have outlined the main ideas of utility: its basis as a normative model for behaviour in the face of uncertainty, its function as a formal

expression of rational preference schemes and its implications for the construction of rules for decision-making. We have already touched on some criticisms
which might be raised. These include the denials of the normative role (that
individuals really would *always* act rationally or coherently in the sense of utility
theory, if only enlightened by the appeal of the Rules for expressing preferences);
the rejection of its personal or subjective basis as inappropriate to a formal theory
of decision-making; the difficulty of quantifying utilities in all but the simplest
situations, so that its recommendations for the taking of decisions are felt to be
insecurely based.

Precisely the opposite views are expressed by those who see utility theory as the
only basis for (or at least an important ingredient in) the design of a rational
theory of decision-making. They argue that people make decisions; in doing so
they act on implicit or explicit measures of preference expressed through utilities.
In so far as they appear incoherent in their actions this merely reflects their lack of
real appreciation of the theory, and is able to be remedied by instruction in the
principles of utility theory. Costs and consequences are vital to the taking of
decisions, and any formal expression of these (however incomplete, or subjective,
on occasions) as a basis for action is to be preferred to a 'head in sand' denial of
their form or relevance.

Fuller discussion of the rival points of view appears in various other parts of the
book (particularly Chapter 7) where more extended practical applications of the
theory, such as in *decision theory*, are presented.

Controversy apart, distinctions also exist in the detailed development of utility
theory by different workers in the field. Whilst beyond our aims to consider the
rigorous mathematical structure of the theory, it is interesting to summarize how
the emphasis varies in different treatments. The first detailed derivation of utility
theory for situations involving uncertainty† is attributed to von Neumann and
Morgenstern (1953; first published 1944) as an almost peripheral element in their
celebrated *Theory of Games and Economic Behaviour*. This attribution makes
sense from the purely mathematical standpoint, but one must acknowledge the
much earlier pioneer development by Ramsey referred to previously (§3.5).
Undoubtedly this set the scene for probabilistic utility theory, although Ramsey's
work was somewhat informally expressed, and has only recently been 're-
discovered'. See Lindley (1971*b*).

The mathematical derivation of von Neumann and Morgenstern is essentially
the one adopted informally here and in several other treatments, such as Chernoff
and Moses (1959) or the more advanced exposition by Savage (1954). One
implication of such a mathematical treatment is the boundedness property. In an
alternative approach, De Groot (1970) explains that the boundedness of the
utility function stems essentially from restrictions placed on the class of prospects

---

† Economists were greatly concerned, around the turn of the century, with a concept of
probability-less utility, now regarded as of little interest. See Savage (1954).

satisfying the preference rules. He demonstrates that a wider axiomatic system, admitting a larger class of prospects, is viable and leads to a theory of utility that does not have the boundedness condition. This is seen as an advantage in terms of application, in that in decision theory it is common to consider unbounded utilities such as **quadratic loss functions** (see Chapter 7).

In presenting in §4.5.1 what is the common basis for evaluating utilities numerically it was seen that these had a natural probability interpretation. This arose through assessing the utility of an arbitrary prospect, P, as equivalent to that of a mixture $\{P_1, P_2 : p\}$ of two pivotal prospects P_1 and P_2. The mixing probability p becomes identified with $U(P)$. This is not to say that $U(P)$ is directly interpretable as a normed probability distribution by this process. However, reversing the emphasis of De Groot, by postulating two extreme prospects $\underline{P}$ and $\overline{P}$ of *least* and *greatest* preference, Lindley[1] develops a direct probability interpretation in which $U(P)$ is a normed probability distribution over $\mathbb{P}$. That is, where $0 \leqslant U(P) \leqslant 1$ for all $P \in \mathbb{P}$.

In conclusion, it should be noted that comments on the *historical development* of the utility concept and associated theory appear in Savage (1954), Arrow[13] and Stigler[14] (non-probabilistic) and an extensive *bibliography* of developments up to about 15 years ago is given by Fishburn[15].

References

† 1. Lindley, D. V. (1971). 'A numerical measure for consequences', Chapter 4 of Lindley, D. V. (1971a), *Making Decisions*. London: Wiley.

† 2. *Becker, G. M., De Groot, M. H. and Marschak, J. (1963). 'Stochastic models of choice behaviour', *Behav. Sci.*, **8**, 41–55.

3. Fishburn, P. C. (1967). 'Bounded expected utility', *Ann. Math. Statist.*, **38**, 1054–1060.

4. Fishburn, P. C. (1975). 'Unbounded expected utility', *Ann. Statist.*, **3**, 884–896.

† 5. *Mosteller, F. and Nogee, P. (1951) 'An experimental measure of utility', *J. Political Econ.*, **59**, 371–404.

† 6. *Davidson, D., Suppes, P. and Siegel, S. (1957). 'Decision-making: an experimental approach', Chapter 6 of Edwards, W. and Tversky, A. (eds) (1967) *Decision Making*. Harmondsworth: Penguin. This has been extracted by the editors from Davidson, Suppes and Siegel (1957) (see *Bibliography*).

7. Suppes, P. and Walsh, K. (1959). 'A non-linear model for the experimental measurement of utility', *Behav. Sci.*, **4**, 204–211.

† 8. Savage, L. J. (1971). 'Elicitation of personal probabilities and expectations', *J. Amer. Statist. Assn*, **66**, 783–801.

† 9. Hull, J., Moore, P. G. and Thomas, H..(1973). 'Utility and its measurement', *J. Roy. Statist. Soc. A*, **136**, 226–247.

10. Aitchison, J. (1970). 'Statistical problems of treatment allocation' (with Discussion), *J. Roy. Statist. Soc. A*, **133**, 206–238.

† 11. Lindley, D. V. (1971). 'The utility of money', Chapter 5 of Lindley, D. V. (1971a), *Making Decisions*. London, Wiley.

*References 2, 5 and 6 above all appear in Edwards, W. and Tversky, A. (eds) (1967) *Decision Making*. Harmondsworth Penguin.

12. Pratt, J. W. (1964). 'Risk aversion in the small and in the large', *Econometrica*, **32**, 122–136.
13. Arrow, K. J. (1951). 'Alternative approaches ro the theory of choice in risk-taking situations', *Econometrica*, **19**, 404–437.
14. Stigler, G. J. (1950). 'The development of utility theory', Paris I and II, *J. Political Econ.*, **58**, 307–327 and 373–396.
15. Fishburn, P. C. (1968). 'Utility theory', *Management Sci.*, **14**, 335–378.

CHAPTER 5

Classical Inference

What we have termed the **classical** approach to statistics is distinguished by two features. It regards as the sole quantifiable form of relevant information that provided in the form of **sample data**, and adopts, as a basis for the assessment and construction of statistical procedures, long-term behaviour under assumed essentially similar circumstances. For this reason it is natural that the **frequency view of probability** serves as the only probabilistic framework within which to construct, interpret and measure the success of classical statistical procedures.

Statistics, as an organized and formal scientific discipline, has a relatively short history in comparison with the traditional deterministic sciences of chemistry, physics (even biology in its qualitative role) or their 'handmaiden', mathematics. It is only really in the present century that the major advances have been made, but after even so short a time we have an impressive heritage of achievements both in terms of principles and applications. Indeed, there has been time enough for traditions not only to be built up, but for them to encounter the customary (and healthy) opposition of alternative ideas and attitudes. In this process it is unquestionable that it is the tenets and methods of **classical inference** that play the 'traditional' role; the major rivals, particularly active and vocal over the last 35 years, are *Bayesian inference* and *decision theory*. These latter approaches, often in conjunction, as well as a variety of more individual attitudes (some of which are outlined in Chapter 8), are presenting an increasing challenge to what have for long been accepted by many as standard theory and practice.

Although with hindsight it is possible to attribute precedence for some *non-classical* ideas to isolated works in the eighteenth and nineteenth centuries, this attribution sometimes places more in the minds of early workers than some might feel was their intention. In contrast, it is certainly possible to trace the development of various important *classical* attitudes to the early 1800s. Vital probabilistic tools such as the laws of large numbers, and the Central Limit Theorem, made their appearance. Likewise the crucial concept of a sampling distribution, and the ubiquitous principle of least squares. This latter idea reflected in its origins the lack of any formalized approach to statistics; we find in the work of Laplace, Gauss and others of the period a dual concern for concepts of loss, risk or 'games against nature' anticipating the modern developments of

119

decision theory, alongside methods of assessment based on aggregate measures of long-term behaviour in the spirit of *classical* statistics.

Classical statistics developed apace, in response to practical need: in a desire to answer important questions, and develop reliable bases for statistical enquiry and the associated analysis of data, in the fields of biology, genetics, agriculture and industry. As indicated in Chapter 1, this burst of activity arose at about the turn of the century and continues to the present time. Much of the fundamental theory of estimation and hypothesis testing was developed in the period 1920–35 on two particular axes. Neyman, and E. S. Pearson, stimulated initially by problems in biology or industry, concentrated largely on the construction of principles for testing hypotheses. Fisher, with an alternative stimulus from agriculture through his work at the Rothamsted Experimental Station, gave major attention to estimation procedures. A great deal of the basis of classical inference was forged in this period by these two 'schools'. Their work was not entirely distinct either in emphasis or application. An interesting example of this is found in the contrast between *hypothesis testing* and *significance testing* (q.v.). Nor was it free from internal controversy; Fisher was particularly vocal in his criticisms of Neyman, Pearson and their associates on various issues, notably that of *interval estimation*. The literature of the 1930s well reflects this dialogue, and its repercussions have not really abated even to the present time. Fisher's concept of **fiducial probability** (q.v.) is the crucial element in the debate.

The spirit and content of the work of this period may be found in various sources. These include the published lectures of Neyman (1952), the collected papers of Fisher (1950; see also Bennett, 1971–74), Neyman (1967), Pearson (1966) and Neyman and Pearson (1967), and books by Fisher (latest editions 1959, 1966, 1970); as well as in the published discussions in the *Royal Statistical Society Journal* in the mid-1930s. Personal commentary on the seminal contributions of Neyman, Fisher and Pearson can be found in Jeffreys[1], Le Cam and Lehmann[2], Pearson[3], Savage[4], Stigler[5]; in the detailed biography of Fisher by Box (1978); and in the extended intercomparison and evaluation of the Fisherian and Neyman/Pearson approaches by Seidenfeld (1979).

In the present day it is unquestionably true that the great majority of practical statistical work uses the classical approach, being based almost entirely on the concepts, criteria and methods first advanced in the Neyman, Pearson and Fisher era.

We shall return (§5.7) to matters of intercomparison and critique, but it is interesting at this stage to note the summary comment by Cox[6]:

> . . . the great attractions of the sampling theory approach are the direct appeal of criteria based on physical behaviour, even if under idealized conditions, and the ability of the approach to give some answer to a very wide range of problems. Indeed the notion of calibrating statistical procedures via their performance under various circumstances seems of such direct appeal that it is difficult to see how some idea of the sort can be avoided. . . .

The ideas have, of course, been greatly extended and variously applied over the years, and continue so to be. As a result a vast array of complex methodology has been developed as a basis for statistical practice, and extensive associated 'case history' exists. It would be difficult to claim any comparable catalogue of specific applications or experience for the ideas of Bayesian inference or decision theory. Indeed it is disappointing that, in spite of so much discussion of the basic principles and criteria of these alternative approaches, relatively little practical application in the form of real-life case studies has been published, although a few relevant references are given later (in Chapters 6 and 7).

5.1 BASIC AIMS AND CONCEPTS

To set classical statistics in the perspective of the various different approaches to inference or decision-making, it is necessary to go beyond the brief discussion of Chapter 2. We must make the following enquiries of this approach:

(a) What is regarded as 'relevant information' to be processed in a statistical enquiry, and how is this information represented?

(b) What types of question does it seek to answer on the basis of this information, and what principles and procedures are advanced to this end?

(c) What criteria are set up to assess these principles and procedures, and how do they stand up to such scrutiny?

We will need to study the basic elements of classical statistics in fair depth, although it is no part of our aim to develop the complex methodology of the subject. This is ably dealt with at all levels in the wealth of existing textbooks and research publications—in particular the three volumes by Kendall and Stuart (1977, 1979, 1976) give a clear oversight of (predominantly) classical statistics at a fairly advanced level. See also Zacks (1971; principally estimation) and Lehmann (1959; hypothesis testing).

To keep the discussion within bounds, we shall concentrate on situations where it is assumed that a reasonable probability model exists in the form of a fully specified parametric family of distributions. Thus **sample data**, x, are assumed to arise from observing a random variable X defined on a **sample space** $\mathcal{X}$. The random variable X has a probability distribution $p_\theta(x)$ which is assumed known except for the value of the parameter θ. The parameter θ is some member of a specified **parameter space** Ω; x (and the random variable X) and θ may have one or many components.

The general form of the probability model, recognizing the lack of knowledge of the appropriate value of θ, is therefore a family of probability distributions, $\mathcal{P} = \{p_\theta(x); x \in \mathcal{X}, \theta \in \Omega\}$. The expression $p_\theta(x)$ will be used as a generic form for a probability distribution, to be interpreted as a *probability* if the data arise from a discrete sample space, or a *probability density* if $\mathscr{X}$ is continuous. Similarly, integrals or sums over the probability distribution will be liberally expressed in the particular notation described in the Preface, which avoids the need to distinguish between continuous or discrete distributions, or overtly to consider dimensionality. Thus

$$\int g(x)p_\theta(x) = \begin{cases} \sum_{x_i \in \mathscr{X}} g(x_i)p_\theta(x_i) & \text{if } \mathscr{X} \text{ discrete,} \\ \int_{\mathscr{X}} g(x)p_\theta(x)\,dx & \text{if } \mathscr{X} \text{ continuous.} \end{cases} \tag{5.1.1}$$

In some applications a more specific form is assumed for x and X, where the sample data take the form of n observations $x_1, x_2, \ldots, x_n$, of independent, and identically distributed, random variables $X_1, X_2, \ldots, X_n$. This may be alternatively described by saying that $x_1, x_2, \ldots, x_n$ is a random sample from a particular specified distribution, which will usually be a univariate one.

Where nothing is lost by doing so, we tend to ignore the question of the dimensionality of the parameter θ. This is predominantly the case throughout §§5.1–5.5, where much of the discussion of estimation and testing is simplified if θ has a single component, or if only a single component of a vector parameter is unknown, or of interest. When the dimensionality of θ is relevant to a discussion of concepts or principles it will be specifically declared in the text, although we shall stop short of developing multi-parameter extensions which involve only fairly straightforward modifications of the corresponding technique for a scalar parameter. But not all multi-parameter problems involve just an immediate extension of concept or principle. This is particularly true of the notion of *sufficiency* and of the crucial ideas of *ancillarity* and *conditionality* applied to problems where a multi-dimensional parameter can be partitioned into components; one involving parameter elements of direct interest; the other, *nuisance parameters*. Such matters are discussed in §5.6; the use of modified likelihood methods in this area is also considered in Chapter 8, as are the associated ideas of *pivotal inference*.

In assuming a prescribed parametric model, and a given set of sample data x, we implicitly rule out discussion of the large areas of statistical method known as *non-parametric statistics*, and the *design of experiments*. Little intercomparative comment is possible on the former topic. Whilst we shall need to consider some of the basic implications of the manner in which the data have arisen, we stop short of any formal detail on experimental design (although it should be noted that Fisher's notion of *randomization* is relevant to the topic of *conditional* inference). Brief comment is made in Chapter 7 on the decision-theoretic aspect of the design of experiments.

We also postpone discussion of those aspects of (predominantly) classical inference where the data play an *interventionist* role in model formulation, validation or modification, or to provide protection against model inadequacy. Chapter 9 briefly reviews the upsurge of interest in descriptive statistics under the label of *data analysis*; *adaptive methods* where data influence the form as well as the *value* of estimators or test statistics; *cross-validation* for simultaneous model scrutiny and inference; the *prediction* of future sample behaviour, and the field of *model-robust* inference procedures. Also relevant here are the discussions of *empirical Bayes methods*, and *prediction in Bayesian inference*: see Chapter 6.

Returning to the points (a), (b) and (c) at the outset of this section, some brief answers can be provided as the basis for fuller discussion in the subsequent parts of the chapter.

5.1.1 Information and its Representation

We have already remarked that the only quantitative information explicitly handled within the classical approach is *sample data*. Consequential cost considerations and prior information about the parameter θ have no *formal* role to play, but may be expected to influence the choice of what statistical procedure will be used in a particular situation and of what we should ask of its performance characteritics; for example, in the specification of a working hypothesis and significance level in hypothesis testing.

The attitude to these other facets of relevant information is well illustrated in the early papers of Neyman and Pearson on hypothesis testing. These show a persistent concern with the importance of both prior information and costs, but an increasing conviction that these factors will seldom be sufficiently well known for them to form a *quantitative basis* for statistical analysis. Fisher adopts a similar view in the dual area of estimation, but is characteristically more extreme in rejecting as logically unsound any use of the 'inverse probability' methods (of Bayesian inference) for handling prior information.

The prevailing attitude of the time to prior probabilities and costs was expressed by Neyman and Pearson[7, 8] as follows:

> We are reminded of the old problem . . . of the number of votes in a court of judges that should be needed to convict a prisoner. Is it more serious to convict an innocent man or to acquit a guilty? That will depend upon the consequences of the error; Is the punishment death or fine?; What is the danger to the community of released criminals?; What are the current ethical views on punishment? From the point of view of mathematical theory all that we can do is to show how the risk of the errors may be controlled and minimized. The use of these statistical tools in any given case, in determining just how the balance should be struck, must be left to the investigator.

> (Neyman and Pearson[7])

> It is clear that considerations of *a priori* probability may . . . need to be taken into account . . .. Occasionally it happens that *a priori* probabilities can be expressed in exact numerical form . . .. But in general we are doubtful of the

value of attempts to combine measures of the probability of an event if a hypothesis be true, with measures of the *a priori* probability of that hypothesis. . . . The vague *a priori* grounds on which we are intuitively more confident in some alternatives than in others must be taken into account in the final judgment, but cannot be introduced into the test to give a single probability measure.

(Neyman and Pearson[8])

In a subsequent paper this attitude is defended by examining the extent to which hypothesis tests 'are independent of probabilities *a priori*'. The conclusions are somewhat circumspect. Fisher[9, 10] attempted a similar exercise with regard to estimation, and introduced the controversial idea of *fiducial probability* (see Chapter 8).

(Neyman[11] tempers his caution on the use of prior information, 30 years later, in an interesting commentary on Bayesian inference in which he praises the 'breakthrough' of Robbins in the topics of *empirical Bayes methods*, and *compound decision problems*.)

So from the *classical* viewpoint we are to use only sample data in setting up procedures for statistical enquiries. How do we represent this information? The aim is to use the data x to throw light on the unknown value of the parameter θ. To this end we might consider a particular function $\tilde{\theta}(X)$ of the basic random variable X. Such a function is called a **statistic** or **estimator**; the actual value it takes, $\tilde{\theta}(x)$, is the corresponding *estimate*.

Of course, the particular data x that we obtain are only one possible realization of the experiment under study. On another occasion we might obtain x', with associated value $\tilde{\theta}(x')$. A whole range of possible values of $\tilde{\theta}(x)$ can arise, depending on the actual outcome x, each with a corresponding value for its probability (density) given by $p_\theta(x)$. The resulting probability distribution is called the **sampling distribution** of the statistic. More formally, $\tilde{\theta}(x)$ is viewed as an observation of the transformed random variable $\tilde{\theta}(X)$, where X is a random variable with probability (density) function $p_\theta(x)$. *The sampling distribution is crucial to any assessment of the behaviour of the statistic $\tilde{\theta}(X)$ as a means of drawing inferences about θ.* The particular value $\tilde{\theta}(x)$ is seen as a typical value that might arise in repeated sampling in the same situation. The properties of $\tilde{\theta}(X)$ are aggregate (long-run) ones in a frequency sense; any probability assessments are soley interpretable in terms of the frequency concept of probability. We shall consider below what this means in relation to any particular conclusion about θ which is based on a single realized value x, through the *estimate*, $\tilde{\theta}(x)$.

$\tilde{\theta}(x)$ usually effects a reduction in the data for economy of expression. For example, we might consider using the sample mean of a set of n independent univariate observations to estimate the distribution mean. Here n items are compressed into, or summarized by, a single item of information. But this is not necessary; $\tilde{\theta}(x)$ may, at the other extreme, be the original sample data in their entirety.

Example 5.1. Suppose our data consist of a random sample $x_1, x_2, \ldots, x_n$ from a Poisson distribution with mean m. We might wish to use the data to throw light on the value of the parameter m. A possible statistic of interest is $\theta(X) = \Sigma_1^n X_i$. This has a sampling distribution which is also Poisson, but with mean nm.

In particular, if nm is large, $\Sigma_i^{n-} X_i$ is approximately $N(nm, nm)$ and $\overline{X} = (1/n)\Sigma_1^n X_i$ is approximately $N(m, m/n)$. If we were to use the sample mean $\overline{X} = (1/n)\Sigma_1^n X_i$ as an estimator of m, the sampling distribution $N(m, m/n)$ serves as the basis for measuring how good this procedure is.

The expected value of $\overline{X}$ is m, so that on average *we will be correct in our inference.*

We can also determine the probability that $\overline{X}$ departs from m by a prescribed amount, and we see that this becomes smaller, the larger the sample. All these aggregate (long-run) properties are attributed to the estimator to describe its overall behaviour, or to compare it with some competitor.

But what about the actual value, $\bar{x}$, which is our estimate of m? It is simply regarded as inheriting any desirable properties represented through the sampling distribution, in the sense of being a typical value of $\overline{X}$!

Before going any further we need to be aware of some of the implications of the frequency-based probability concept underlying classical inference. The first resides in the *deceptive* simplicity of the final sentence of *Example* 5.1 above. The sampling distribution is defined over the set of prospective samples which might arise when we repeat the experiment time and again. Over such a *reference set* (or 'collective') we can say that $\overline{X}$ has expected value *m*, or that, with probability approximately 0·95, $\overline{X}$ will lie between $m - 1·96 \sqrt{(m/n)}$ and $m + 1·96 \sqrt{(m/n)}$. We can discuss such *prospective* properties of our *estimator*: make statements of **initial precision**. But suppose that we calculate $\bar{x}$ for a particular sample of size $n = 16$ and its value is 16·8! We have no way of knowing how close this estimate is to the unknown value *m*: no means of assessing *realized* accuracy, or **final precision**.

Even if we knew the extent of variability in the basic distribution, the problem remains. If our sample comes from a normal distribution $N(\mu, 1)$ with unknown mean and unit variance, so that

$$P(\mu - 0·49 < \overline{X} < \mu + 0·49) = 0·95,$$

we still do not know if the estimate $\bar{x} = 16·8$ is within 0·49 of μ, even with probability 0·95.'It may be so, if we have obtained one of the 95 % of samples for which $\overline{X}$ would be within 0·49 of μ. But we may of course have obtained one of the 5 % where this is not so!

A further crucial implication arises from what we assume about the set of repetitions of the experiments (the *reference set*) over which probability is measured. In *Example* 5.1 we considered a random sample of size *n*. The reference set is made up of all such samples, and inferences accordingly relate to such *fixed-*

size random samples. The practical problem may be one where it is natural that we have repetitions and that observations arise from a Poisson distribution, but where there is no reason why the samples have to be random or of similar size. To ignore non-randomness or variability in the sample size *n* may be the pragmatic choice (we may know nothing of the form of the non-randomness or probabilities, or relative frequencies, with which different sample sizes arise, nor wish to correspondingly complicate our analysis even if we did). But we must recognize that use of the idealized fixed-size random sample model is an assumption: inferences are *conditional* on this assumption. Whilst such conditionality may seem in this example not to be a material consideration, there are cases where it can be.

In sequential analysis, for instance, we may choose to take observations from a Poisson distribution with mean *m* but to continue sampling until the sample mean first exceeds some value, *K* say. The reference set is now quite different; it yields observations of a bivariate random variable $(\overline{X}, N)$ comprising the sample mean (intuitively less variable than before) and the *random* sample size, *N*. Within the classical approach the new probability structure is material to questions of inference about *m*, and chosen methods of inference, and their interpretation, must reflect this. (An analogous example of direct and inverse binomial sampling was described in Chapter 2.)

The fact that the same sample outcome—a sample of size *n* with mean $\bar{x}$— could lead to different conclusions, depending on the sampling procedure (that is, on the reference set) is a major debating point when *different* approaches to inference are under consideration. In particular, it conflicts with the appealing *likelihood principle*. We shall return to this point in §5.6 and in Chapters 6 and 8.

A further illustration of the difficulty of choosing an appropriate reference set arises in another simple situation. Suppose that random samples of sizes n_1 and n_2 individuals from distinct populations contain, respectively, r_1 and r_2 of a particular type. We want to draw inferences about the actual proportions, p_1 and p_2, of this type of individual in the populations. Should we use a reference set in which n_1 and n_2 are fixed, and corresponding binomial distributions? If n_1 and n_2 were *not* pre-assigned, does it matter that we conduct a statistical analysis *conditional* on the assumption that they *were*? This is precisely the situation that prompted R. A. Fisher to consider the implications of *conditionality*. Conditioning on n_1 and n_2 amounts essentially to choosing to disregard some element of the data as irrelevant to inferences about parameters of interest, and introduces in support of this policy the idea of *ancillarity* and of *ancillary statistics*, which we shall consider in more detail in §5.6. [Seidenfeld (1979) provides some interesting comment and examples on the choice of the reference set, initial versus final precision, conditionality and ancillarity.]

Passing reference having been made to the concept of *likelihood* it is appropriate now to consider this more global means of representing the information provided by the sample which does not involve its compression into a particular parameter-free statistic.

Fisher introduced the **likelihood** (or **likelihood function**) of θ for the sample data x as a means of representing the information that the sample provides about the parameter. The likelihood is merely a re-interpretation of the quantity $p_\theta(x)$. Instead of regarding $p_\theta(x)$ as the probability (density) of x for the fixed (but unknown) θ, i.e. *as a function of x over $\mathscr{X}$ with index θ*; it is alternatively thought of as *a function of θ over Ω for the observed value x*.

We must be careful not to transfer the probability interpretation from x to θ. The likelihood is not the probability (density) of θ for a given x. Apart from being unnormed, no such interpretation is conceptually justified within the classical approach. The likelihood function merely expresses how the probability (density) of the particular x changes as we consider different possible values for θ. The reverse (erroneous) interpretation is a tempting one! The very term 'likelihood' encourages it, and it is a small step to the adoption of $p_\theta(x)$ as expressing different weights of support for different values of θ. Neyman and Pearson appeared almost to suggest this in their early writings on the use of the **likelihood ratio** (q.v.) in hypothesis testing. Furthermore a minority 'likelihood school' of statisticians has developed on these lines over recent years, and is finding increasing support. We shall consider this in more detail in Chapter 8.

5.2 ESTIMATION AND TESTING HYPOTHESES— THE DUAL AIMS

Having remarked that the sampling distribution is the background against which classical statistical procedures are assessed, and that the likelihood function provides the measure of the import of x on θ, we must enquire to what ends these concepts are applied. There are basically two types of problem in the classical approach, both concerned with reducing our uncertainty about θ, but one of them more specific than the other in its aims. These are **estimation** and **testing hypotheses**.

Estimation. The enquiry here is a very general one. I have observed x; what should I infer about θ? For example, a random sample of five resistors from the output of a machine have resistances 30·4, 29·8, 29·7, 30·1, 29·8 ohms. What does this imply about the mean resistance of the output?

What is required is a mapping from the sample space $\mathscr{X}$ to the parameter space Ω, so that if x is observed we conclude that the parameter has some particular value θ, or is in some particular region $\omega(x)$ of Ω.

These two cases are referred to, respectively, as **point estimation** and **region** (or **interval) estimation**.

In point estimation, the estimator $\tilde{\theta}(X)$ serves the required purpose. $\tilde{\theta}(X)$ is a mapping from $\mathscr{X}$ to Ω, and if we obtain data x we conclude that θ has the value

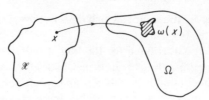

$\tilde{\theta}(x)$. In region estimation we need a wider concept than such a one-to-one transformation. In the conventional applications of region estimation the usual concept is that of a **confidence region**, which is best studied (in terms of its construction and interpretation) in the context of **tests of significance** or, for more detailed consideration of optimality properties, in terms of the behaviour of the more structured **Neyman–Pearson hypothesis tests**. We shall defer any discussion of confidence regions until later in the chapter.

In point estimation we shall need to consider what estimators $\tilde{\theta}(X)$ are appropriate in a particular situation: how we can construct them, assess their properties, compare one with another, and (if possible) ensure that we have 'made the most' of the data x. The sampling distribution of $\tilde{\theta}(X)$, and the likelihood function $p_\theta(x)$, are vital ingredients in this study, as they are when we consider parallel issues in:

Testing Hypotheses. The nature of our interest is more specific here. Consider the resistor example. It is likely that the resistors are being sold to some specification. Perhaps they are offered as 30-ohm resistors. On the same data we might now ask *if it is reasonable that the mean resistance is 30 ohms*. An inference procedure which aims to answer such a question is termed a **statistical test**. Some assumption is made: some *hypothesis* stated concerning the value of a parameter θ. Perhaps we may hypothesize that θ has some specific value θ_0, or that θ lies in some region ω which is a subset of the parameter space Ω. The data must now be processed either to provide inferential comment on the stated hypothesis, or even to support abandoning the hypothesis.

There are various types of statistical test and some confusion of terminology in the recent literature. In all cases we start by declaring the **working (basic, null) hypothesis** H to be tested, in the form $\theta = \theta_0$ or $\theta \in \omega \subset \Omega$. Distinctions between the types of test arise in respect of the following:

(a) Whether we use the data x to measure inferential evidence against H, or go further in declaring that if such contraindication of H is sufficiently strong we will judge H to be inappropriate (that is, we 'reject H').

(b) Whether or not we contemplate a specific **alternative hypothesis** $\overline{H}$ to adopt if H is rejected and construct the test, and assess its properties, in regard of $\overline{H}$ ($\overline{H}$ may be the global antithesis of H, but is often more limited).

The present tendency is for a procedure which assesses inferential evidence against H (without the prospect of rejecting H, and without consideration of an alternative hypothesis $\overline{H}$) to be termed a **pure significance test** (*pure test of significance*) or just a *significance test* (Cox and Hinkley, 1974; Cox[12]; Seidenfeld, 1979) with the measure of 'evidence against H' described as the *significance level* (*level of significance*) or, earlier, the **significance probability**. But much of the literature, including most teaching texts, uses the term **significance test** for the extended prospect where we seek a rule for 'rejecting H', with the *significance level* defined as the *maximum probability of incorrectly rejecting H* under that rule on repeated application under similar circumstances. The actual realized extent of evidence against H is sometimes termed the **critical level** (Lehmann, 1959, p. 62): it corresponds with the *significance probability*, or *significance level*, in the pure significance test.

Often the term significance test is used whether or not some specific alternative hypothesis $\overline{H}$ is contemplated; on other occasions the two-hypothesis problem, approached by the methods of Neyman and Pearson, is distinguished by the label **hypothesis test**. Neyman[13] recently remarked that he was mystified by

> the apparent distinction between tests of statistical hypotheses, on the one hand, and tests of significance on the other.

The confusion over the use of terms such as significance test and significance level is unfortunate and not readily resolved. To avoid ambiguity and the introduction of yet more terms we shall adopt the following linked terminology, which does not seem too unrepresentative of common practice.

Pure significance test; significance probability.

Significance test (rule for rejecting H); **significance level,
critical level**.

Hypothesis test (rule for rejecting H in favour of $\overline{H}$); **significance level,
critical level** and additional relevant concepts.

The development of the basic ideas, and properties, of statistical tests of one hypothesis against another is due to Neyman and Pearson, as we have remarked above. The difference in philosophy between the Neyman–Pearson hypothesis test and the pure significance test has been claimed as a crucial example of the decision-making interests of Neyman and Pearson and the inferential attitude of R. A. Fisher—with Fisher advocating the latter procedure and eschewing the need (or desirability) of a full-fledged frequency interpretation. [See Seidenfeld (1979); Kemthorne[14] and the later discussion of *fiducial inference* in Chapter 8.]

The following paragraphs summarize the main elements of the different types of statistical test within the classical approach.

Pure Significance Tests. We must now be a little more precise in defining the nature and properties of statistical tests, in particular the interpretation of such concepts as 'the inferential evidence against an hypothesis H'.

We start by declaring some hypothesis, H, of interest to serve as our working (null) hypothesis. [Cox and Hinkley (1974, p. 65) categorize ways in which H may be formulated.]

H may be **simple** or **composite**. It is simple if it fully specifies the distribution of X over $\mathscr{X}$. For example, if X constitutes a random sample of size n from $N(\mu, \sigma^2)$ the hypothesis $H:\mu = \mu_0, \sigma = \sigma_0$ (with μ_0, σ_0 having stated values) is simple: the hypothesis $H:\mu = \mu_0$ is composite since it leaves open the value of σ in $(0, \infty)$.

Suppose H is simple. We choose a statistic $t(X)$ (a **test statistic**, or **discrepancy measure**) and evaluate

$$p = P\{t(X) > t(x) | H\},$$

where $t(x)$ is the value we have observed for the test statistic. If $t(X)$ is chosen in such a way that the larger its value the more it casts doubt on H in some respect of practical importance, then p provides the basis for making inferences about H. Thus if p is *small* the data x are *highly* inconsistent with $H:p$ is the observed measure of such inconsistency. What we are really saying is that, if H is true, the data x are implausible to the extent expressed by p and in terms of discrepancies or departures reflected by our choice of test statistic $t(X)$. *Note* that, as with all classical inference procedures, we are measuring *initial precision*: p is the proportion of identical situations in which, in the long run, we would observe inconsistency at least to the extent represented by $t(x)$. *Clearly no reverse probability statement is legitimate—that p is the probability that H is true!*

It is clear that if $t(x)$ provides, in these terms, some degree of inconsistency with H, other data x' (generated on the same basis as x) for which $t(x') > t(x)$ provide at least the same degree of inconsistency as x.

The process of declaring H, choosing $t(X)$ and determining p is called a **pure significance test** and p is the **significance probability** ('significance level').

Obviously p is the observed value of same statistic. Different x yield different values of p, from a uniform distribution over $(0, 1)$. As with all inference procedures we are observing the value of some (relevant) statistic as *part* of the inferential import of the data on our model $\{p_\theta(X):H\}$. It is seldom likely that p provides *all* the information we will obtain, or would wish to obtain. Note also that whilst contemplated departures from H may influence our choice of $t(X)$ we only consider distributional behaviour *when H is true*, and it is essential that the distribution of $t(X)$, under H, is known.

Several crucial points remain to be considered. What do we do if H is composite? How, in practice, should we choose the test statistic, or discrepancy measure, $t(X)$? To what extent do prospective departures from H affect this choice? The interesting survey by Cox[12] includes discussion of these various matters.

Tests of Significance (Significance Tests). The pure significance test entertains no prospect of rejecting H. But if p is small either a rare event has occurred or perhaps H is not true, and we might decide that if p were small enough (say, less than some stated value α) we should adopt the latter possibility and *reject H*. A **level-α test of significance** formulates this idea. We choose a **significance level** α, observe x and *reject H at level α* if $P\{t(X) > t(x)|H\} \leqslant \alpha$. Equivalently, we are choosing some critical value $t_0(\alpha)$ for $t(X)$ and rejecting H if $t(x) > t_0(\alpha)$. So our sample space is being partitioned into two regions $\{S_0:S_1\}$, a **critical region** S_0 where if $x \in S_0$ *we reject H* and a **non-critical region** S_1 where if $x \in S_1$ we have '*no reason to reject H*' on the basis of the level-α test.

If H has the form $H:\theta \in \omega \subset \Omega$ our conclusions about the value of θ are as represented diagrammatically below.

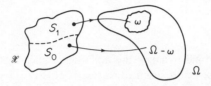

If H were true we could still obtain data in S_0 with probability

$$P(X \in S_0) = P\{t(X) \geqslant t_0 | H\} \leqslant \alpha,$$

so the *significance level provides* (an upper bound to) *the maximum probability of incorrectly rejecting H*.

But the significance level α no longer measures the observed level of inconsistency of the data with respect to H. This is given by $P\{t(X) \geqslant t(x)|H\}$ which could be quite different from (*much lower than*) α when we are lead to *reject H*. To merely report rejection of H at level α underplays the full extent of the data import. We must quote α (as a statement of the maximum risk we are prepared to tolerate of wrongly rejecting H), but good practice requires us also to state

$$p = P\{t(X) \geqslant t(x)|H\}$$

—called the **critical level** of the test—as the observed extent of the inconsistency of the data x with respect to H. (This is also the *minimum* significance level at which x would have lead to rejection of H.)

Hypothesis Tests. If we go further and explicitly declare an alternative hypothesis $\overline{H}$, which we will adopt if we have reason to reject H, wider prospects arise for test construction and determination of the behavioural characteristics of a test. The **Neyman–Pearson hypothesis test** is similar in structure to a significance test. A level-α test leads us to reject H in favour of $\overline{H}$ *or* to accept H, at level α. Problems now arise in the choice of $\overline{H}$ and composite hypotheses are almost inevitable. If H is *simple*, use of the full complement $\overline{H}$ will often be too wide-ranging to express practical interest but it is unlikely that a *simple* alternative hypothesis will be adequate. Also $\bar{H}$ now has more direct influence on the choice of the test statistic, $t(X)$. For example, if X is a random sample from $N(\mu, 1)$ and H declares that $\mu = \mu_0$, then $\bar{H} : \mu \neq \mu_0$ makes $t(X) = |\bar{X} - \mu_0|$ an intuitively plausible test statistic. But with the alternative hypothesis $\bar{H} : \mu > \mu_0$; $t'(X) = \bar{X} - \mu_0$ is surely more appealing!

Note that the hypothesis test is *asymmetric* in structure: in placing the stimulus for choice between H and $\bar{H}$ on the sample behaviour *when H is true*. It limits the probability of *incorrect rejection* of H. But the alternative prospect is also important—*the risk of incorrectly accepting H*—and with $\bar{H}$ declared it is possible to consider such a prospect. We can accordingly introduce further measures of test performance (for example, the *power* of the test), employ principles for test construction and consider questions of the optimality of performance of tests in different circumstances.

It is interesting to ask what is the *function* of such tests. Is it *inferential* or *decision-making*? The pure significance test is clearly inferential. It provides a means of delimiting our knowledge about θ, in terms of the inconsistency of the data x if θ lies in that part of Ω defined by H. The significance test, and hypothesis test, is less easily categorized. Formally we take a 'decision'—accept H or reject it (perhaps in favour of $\bar{H}$). Kempthorne[14] comments:

> The work of Neyman and Pearson was obviously strongly influenced by the view that statistical tests are elements of decision theory.

But is the situation so clear cut? Decisions should be preludes to actions. Choice of H and $\bar{H}$ *may* involve utilitarian considerations which imply corresponding actions. If so, we must ask where we have taken account of relevant costs and consequences. On the other hand, the aim may be purely inferential; a wish to choose between two descriptive statements about θ expressed by H and $\bar{H}$, with no immediate concern for future action. So are we making decisions or drawing inferences? Are estimation and testing really different in function? Is this reflected in the procedures and principles that are employed? We shall be better placed to consider these controversial points after we have studied more details of estimation and testing: the dual aspects of the classical approach.

5.3 POINT ESTIMATION

We now consider the criteria by which point estimators are assessed, and the principles on which they are constructed. As mentioned previously, the sampling distribution of the estimator is the sole basis for measuring its performance, and resulting properties inevitably relate to long-term behaviour (that is, to what happens as we repeatedly apply some principle, or use a particular estimator, in essentially similar circumstances).

Our enquiries divide naturally into three parts.

1. *What criteria may be set up to measure how good $\tilde{\theta}(X)$ is as an estimator of θ?*

2. *Measured in terms of such criteria can we delimit how good an estimator of θ we may expect to obtain, and determine conditions for obtaining best estimators?*

3. *What general methods are available for constructing estimators, and how good are the resulting estimators?*

We shall consider these points in sequence, omitting detailed proofs since these add little to the interpretative, or comparative, aspects of the different approaches to statistical analysis.

5.3.1   Criteria for Point Estimators

As a point estimate of θ based on data x, we consider an estimate $\tilde{\theta}(x)$. The quantity $\tilde{\theta}(x)$ is a realized value of the estimator $\tilde{\theta}(X)$, where X is the random variable of which our data x constitute an observation. We shall refer to the estimator as $\tilde{\theta}$ henceforth, without specific reference to the random variable X.

Unbiasedness. A simple property that $\tilde{\theta}$ may have is that it is **unbiased** for θ. This is so if the mean of the sampling distribution is equal to θ, or

$$E(\tilde{\theta}) = \theta, \tag{5.3.1}$$

for all θ.

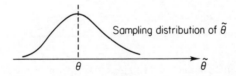

Sampling distribution of $\tilde{\theta}$

In practical terms this implies that $\tilde{\theta}$ takes *on average* the value θ, the quantity it is designed to estimate. This seems a desirable property for the estimator to have, and much of classical point estimation theory is directed to unbiased estimators. But this is not to say that only unbiased estimators are to be entertained. It may be that other desirable properties override the disadvantage of bias in an estimator, as we shall see in a moment. Then again, in alternative approaches to inference the very concept of unbiasedness may be regarded as irrelevant. This is so in some expressions of Bayesian inference (q.v.), being part of a wider dissatisfaction with the general concept of aggregate assessments. Yet in the context of the classical ('sampling-theory') approach it seems a reasonable and modest first aim for an estimator.

Certain features of unbiasedness are easily demonstrated. Many unbiased estimators may exist for a parameter θ, in a particular situation. If $\tilde{\theta}$ is unbiased for θ, then $f(\tilde{\theta})$ is generally not unbiased for $f(\theta)$. A biased estimator with *known* bias (not depending on θ) is equivalent to an unbiased estimator since we can readily compensate for the bias. Within the classical approach unbiasedness is often introduced as a practical requirement to limit the class of estimators within which an optimum one is being sought. This is notably so in the widely applied *least squares* study of linear models, and in the use of *order statistics* (see §5.3.3).

Consistency. A further criterion for assessing the behaviour of estimators concerns their behaviour as the extent of the data increases. Suppose the sample data comprise n independent observations $x_1, x_2, \ldots, x_n$ of a univariate random variable, and an estimator of θ based on this sample is $\tilde{\theta}_n$. As n becomes larger we might reasonably hope that $\tilde{\theta}_n$ improves as an estimator of θ. One way of expressing this is to require $\tilde{\theta}_n$ to get closer to θ (in some aggregate sense). We say that $\tilde{\theta}_n$ is *weakly* **consistent** for θ if $\tilde{\theta}_n \to \theta$ *in probability*, or *strongly* **consistent** if $\tilde{\theta}_n \to \theta$ *with probability* 1.

In practical terms, for a scalar (single component) parameter θ, we are essentially asking for the sampling distributions of $\tilde{\theta}_n$ to become less disperse as n increases.

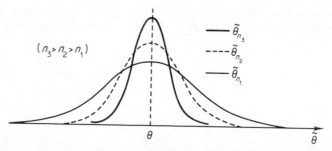

By Chebychev's inequality, a sufficient condition for weak consistency of unbiased estimators $\tilde{\theta}_n$, in this case, is that $\mathrm{Var}(\tilde{\theta}_n) \to 0$, as $n \to \infty$.

Again this seems a sensible requirement. It is quite different in spirit to unbiasedness, and we find that the two properties are largely unrelated. For example unbiased, inconsistent, estimators exist as well as biased, consistent, estimators.

Example 5.2. Suppose $x_1, x_2, \ldots, x_n$ is a random sample of observations from a distribution with mean θ, and variance σ^2. It is easily shown that $\overline{X} = (1/n) \sum_1^n X_i$ is unbiased for θ, and has sampling variance, σ^2/n. Hence it is weakly consistent. (The Central Limit Theorem enables us to assume further that, except for very small samples, *the sampling distribution of* $\overline{X}$ *is well approximated by a normal distribution, irrespective of the distribution from which the sample is chosen.)*

Consider the estimator $\tilde{\theta}_{1,n} = [n/(n+1)] \overline{X}$. *This is obviously biased, but is consistent.*

On the other hand the estimator $\tilde{\theta}_{2,n} = X_1$ (*i.e. we estimate θ by the first observation in the sample ignoring the rest*) *is unbiased,*

$$E(X_1) = \theta$$

but inconsistent, since

$$\mathrm{Var}(X_1) = \sigma^2$$

irrespective of n. Much less trivial examples are readily found, where the lack of consistency of an unbiased estimator has serious practical implications.

But there is a limiting relationship between unbiasedness and consistency. It is apparent that a consistent estimator is asymptotically unbiased.

Consistency is generally regarded as an essential property of a reasonable estimator.

Efficiency. The criteria of unbiasedness and consistency act as a primary filter in assessing possible estimators of θ. If possible we would want an estimator to be unbiased, almost certainly we should want it to be consistent. But suppose many such candidates present themselves. How should we choose between them? Again consider a scalar parameter, θ. The ubiquity of the normal distribution (via the Central Limit Theorem) makes it plausible to contrast unbiased estimators of θ in terms of their *sampling variances*. If *on the same data basis* two possible *unbiased* estimators, $\tilde{\theta}_1$ and $\tilde{\theta}_2$, are available it seems appropriate to *regard the one with smaller variance as the better one*. A concept of **relative efficiency** is developed on these terms. If

$$\mathrm{Var}(\tilde{\theta}_1) < \mathrm{Var}(\tilde{\theta}_2) \tag{5.3.2}$$

we say that $\tilde{\theta}_1$ *is more efficient than* $\tilde{\theta}_2$, in that on average it is closer (in a mean square sense) to θ. See Figure 5.1.

With biased estimators (5.3.2) is obviously not an appropriate criterion; consider $\tilde{\theta}_3$ in Figure 5.1. However, an equivalent basis for comparison can be set up in terms of *mean square error*. A general comparison is achieved by considering how small is $E[(\tilde{\theta} - \theta)^2]$, irrespective of any bias in $\tilde{\theta}$. This reduces to the above condition for relating efficiencies when the estimators are unbiased.

Here we see why unbiasedness may not necessarily be of overriding importance. Consider again $\tilde{\theta}_3$ in Figure 5.1. It may be that

$$E[(\tilde{\theta}_3 - \theta)^2] < \mathrm{Var}\,\tilde{\theta}_1,$$

possibly substantially so. Then *on average $\tilde{\theta}_3$ will be closer to θ than $\tilde{\theta}_1$* and might

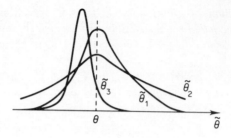

Figure 5.1

therefore be regarded as the better estimator even though it is biased. The 'local colour' of the real-life problem we are studying will, of course, influence the extent to which we are prepared to trade bias for precision in such a manner.

There is a further matter that must be considered in relation to efficiency. Suppose we conclude that of a particular set of unbiased estimators, one of them, $\tilde{\theta}$, is the most efficient. We would still wish to ask if $\tilde{\theta}$ is not only the best within the set we have considered, but is most efficient within some wider class–perhaps the set of *all* unbiased estimators; or indeed, if it is the best estimator overall irrespective of bias considerations. To answer these questions we will need some concept of *absolute* efficiency, and will return to this matter shortly.

Sufficiency. Finally we consider a criterion with less immediate intuitive appeal, but which turns out to be absolutely crucial to the later discussion of the circumstances under which best estimators may be obtained; both in terms of what we mean by 'best' and of identifying the conditions under which we attain optimality.

$\tilde{\theta}$ *is said to be* **sufficient** *for* θ *within the family* $\mathscr{P}$ *if the conditional distribution of* X, *given* $\tilde{\theta}$, *does not depend on* θ.

If $\tilde{\theta}$ is *sufficient* for θ, this means that all the information about θ contained in the data is obtained from consideration of $\tilde{\theta}$ alone, i.e. from its sampling distribution. Usually $\tilde{\theta}$ will be of lower dimension than X, so that an effective reduction in the data is achieved without loss of information about θ. This reduction can be dramatic in extent. Consider $\overline{X}$ in random sampling from $N(\mu, \sigma^2)$, where σ^2 is known. Here $X = (X_1, X_2, \ldots, X_n)$, with each X_i being $N(\mu, \sigma^2)$. The distribution of $(X_1, X_2, \ldots, X_{n-1})$, given $\overline{X} = \bar{x}$, has probability density function proportional to $\exp\{-(1/2\sigma^2)[\Sigma_1^n (x_i - \bar{x})^2]\}$, independent of μ. Thus only $\overline{X}$, and no other elements of the data, provides information about μ. An n-dimensional quantity is reduced to dimension one, with no loss of information and much greater ease of handling and interpretation.

It would be tedious, however, if, in order to show that $\tilde{\theta}$ is sufficient for θ, we should need to derive the conditional distribution of X, given $\tilde{\theta}$. Fortunately this is not necessary in view of a simpler criterion due to Fisher and Neyman.

Fisher–Neyman Factorization Criterion. $\tilde{\theta}$ *is sufficient for* θ *(in $\mathscr{P}$) if and only if*

$$p_\theta(x) = g_\theta(\tilde{\theta})h(x). \qquad (5.3.3)$$

All that is needed is that the likelihood factorizes into the product of two functions, one a function of θ and $\tilde{\theta}$ alone, the other a function of x not involving θ. This is a simple criterion to operate and yields an immediate assessment of the sufficiency, or otherwise, of $\tilde{\theta}$. One cautionary word is necessary. The function $g_\theta(\tilde{\theta})$ *need not be* (even proportional to) the marginal probability (density) of $\tilde{\theta}$. For example, consider the statistic $t = \sum_1^n x_i$ where $x_1, \ldots, x_n$ is a random sample from an exponential distribution with parameter λ. We have

$$p_\lambda(x) = \lambda^n e^{-\lambda t} = g_\lambda(t).$$

So t is sufficient for λ ($h(x) \equiv 1$), but $g_\lambda(t)$ is not proportional to the marginal density of t which is $\lambda(\lambda t)^{n-1} e^{-\lambda t}/(n-1)!$

On a point of terminology we must remark on the term *minimal set of sufficient statistics*. In a trivial sense, X is sufficient for θ. However, we will be concerned in practice with sufficient statistics of lower dimension than X, if such exist. If $\tilde{\theta}$ is sufficient and of lowest possible dimension it is said to be **minimal sufficient**, its elements constituting the *minimal set*. The *minimal sufficient statistic* $\tilde{\theta}$ must, of course, be a function of all other statistics which are sufficient for θ. A minimal sufficient statistic *of dimension one* is said to be *singly sufficient*. Clearly a singly sufficient statistic might exist for a scalar parameter: $\bar{X}$ is singly sufficient for μ in random sampling from $N(\mu, \sigma^2)$, when σ^2 is known.

But even with a scalar parameter the minimal sufficient statistic need not be singly sufficient: it may be just the whole data set X, or perhaps of intermediate dimensionality.

Example 5.3. X consists of a random sample of size n from a Cauchy distribution, with probability density function

$$\frac{1}{\pi}\{1 + (x - \theta)^2\}^{-1}. \qquad (5.3.4)$$

The factorization criterion shows that no set of statistics of dimension less than n is sufficient. We have to use the whole sample, or equivalently the ordered sample $x_{(1)}, x_{(2)}, \ldots, x_{(n)}$ *(where* $x_{(i)} < x_{(j)}$, *if* $i < j$).

Example 5.4. We have a random sample of size n from a uniform distribution on the interval $(\theta - 1, \theta + 1)$. *Here the minimal sufficient statistic for the scalar parameter* θ *is the pair* $(x_{(1)}, x_{(n)})$: *that is, the smallest and the largest observations.*

Kendall and Stuart (1979, §23.15) even quote a strange situation where the minimal sufficient statistic has dimension *less* than that of the parameter. They consider $X = (X_1, X_2, \ldots, X_n)$ with a joint normal distribution in which the mean vector is $(n\mu, 0, 0, \ldots, 0)$ and the variance–covariance matrix is

$$\Sigma = \begin{bmatrix} n-1+\theta^2 & -1 & -1 & \cdots & -1 \\ -1 & 1 & 0 & \cdots & 0 \\ -1 & 0 & 1 & \cdots & 0 \\ \vdots & \vdots & \vdots & & \vdots \\ -1 & 0 & 0 & \cdots & 1 \end{bmatrix}$$

The factorization criterion shows $\overline{X}$ to be singly sufficient for (μ, θ)! But note that here the X_i are, of course, not independent nor are they identically distributed.

Further aspects of sufficiency and related concepts will be discussed in §5.6. These involve partitioning the parameter, or sufficient statistic, into separate components. One point should be made here, however. Suppose a (vector) parameter θ is partitioned into two components (θ_1, θ_2) and $(\tilde{\theta}_1, \tilde{\theta}_2)$ is a pair of (vector) statistics. If $(\tilde{\theta}_1, \tilde{\theta}_2)$ is sufficient for $(\theta_1, \theta_2) = \theta$ it does not follow that $\tilde{\theta}_1$ is sufficient for θ_1 alone (in the sense that the conditional distribution of X given $\tilde{\theta}_1$ does not depend on θ_1) even if $\tilde{\theta}_1$ is sufficient for θ_1 when θ_2 is known. The following example gives a simple illustration of this.

Example 5.5. The data comprise a random sample of size n from $\mathbf{N}(\mu, \sigma^2)$. *Suppose* $\overline{X}, S^2$ *are the mean and variance of the prospective sample.*
  If μ *is known,*

$$\frac{1}{n}\sum_1^n (X_i - \mu)^2$$

is sufficient for σ^2. *If* σ^2 *is known,* $\overline{X}$ *is sufficient for* μ. *If* μ *and* σ^2 *are unknown,* $(\overline{X}, S^2)$ *is jointly minimal sufficient for* (μ, σ^2) *but* $\overline{X}$ *is not sufficient for* μ, *neither is* S^2 *sufficient for* σ^2.

The existence of a minimal sufficient statistic whose dimensionality is the same as (or close to) that of the parameter θ is by no means guaranteed. We have seen that it does happen (and will see later that if $\mathscr{P}$ is of a particular general form it must do so). But quite often no effective reduction of the data is possible: this is so outside a rather small set of 'standard distributions'. For example, $(\overline{X}, S^2)$ is sufficient for (μ, σ^2) in $\mathbf{N}(\mu, \sigma^2)$, but if the data came from a mixture of two distinct normal distributions, in proportions γ and $1 - \gamma$, the minimal sufficient statistic *reverts to the whole data set, however small is the value of* γ!

The major importance of sufficiency is not really its computational convenience in retaining full information about θ in a reduced form of the data. It lies more in the important *implications* the concept has for the existence and form of optimum estimators (and indeed, optimum hypothesis tests). It is vital to our further studies of the classical approach.

5.3.2 Optimum Estimators

To illustrate the general criteria just described, and to introduce the study of optimum estimators, let us consider estimating the mean, μ, of a univariate random variable by means of the sample mean $\bar{x}$ of a random sample of size n. Suppose the random variable has variance σ^2. We know that, whatever the distribution from which the sample is chosen, $\bar{X}$ has mean μ and variance σ^2/n. Thus the sample mean is *unbiased* and *consistent*, and this holds irrespective of the specific form of the distribution. Can we say more about how $\bar{X}$ compares with other possible estimators?

An immediate property of $\bar{X}$ is that, *of all linear unbiased estimators, $\bar{X}$ is the one with smallest variance.* Thus if $\tilde{\mu} = \sum_1^n \alpha_i X_i$, subject to $\sum \alpha_i = 1$, $\mathrm{Var}(\tilde{\mu})$ is minimized if we choose $\alpha_1 = \alpha_2 \ldots = \alpha_n = 1/n$. In this sense $\bar{X}$ is *optimum*.

But it is optimum in another respect! Suppose we seek, an estimator μ^* with the property that it is closest to the actual data $x_1, x_2, \ldots, x_n$; in the sense that $\sum_1^n (x_i - \mu^*)^2$ is as small as possible. This is achieved for $\mu^* = \bar{x}$, and we say that $\bar{X}$ is the **least squares** estimator of μ. Much of classical estimation theory is concerned with such **minimum variance linear unbiased estimators**, or with *least squares estimators*. In the widely applied area of linear models, where the basic observations came from distributions whose means are *linear* functions of the components of a (vector) parameter θ, these principles are paramount and lead to the multiplicity of methods in analysis of variance, regression, and so on.

How important are these properties that we have demonstrated for $\bar{X}$? In the first place, minimizing the variance of linear estimators is only a limited aim. Why restrict attention to *linear* estimators, or *unbiased* ones. We might do much better for non-linear, or biased, estimators. In certain cases this is true, as we shall see in a moment.

Then again, the least squares principle (*per se*) is even more restricted. It has no concern even for the long term sampling behaviour of the estimator—it is *entirely specific to the actual set of data we have obtained*. It *inherits* some optimum properties, however, in particular situations and these give it a transferred importance. For example, for the linear model it turns out that the least squares estimator is identical to the minimum variance linear unbiased estimator. Then again, if we are prepared to be more specific about the form of the distribution being sampled, it acquires an even greater importance, as we shall see in §5.3.3.

Thus minimum variance linear unbiased estimators represent a limited form of optimality, and the least squares principle produces estimators with this property under certain circumstances. How useful a concept of optimality this is must be considered in the light of how much better we can expect to do if we relax the unbiasedness, or linearity, requirement; also in terms of the effort involved in constructing better non-linear estimators.

One appealing feature of the minimum variance linear unbiased estimator is its universality, is being essentially independent of the probability model. We saw this above in respect of estimating μ by $\bar{x}$. But it is instructive to consider through some examples how satisfactory $\bar{X}$ is as an estimator of μ for different particular distributions.

Example 5.6. $X \sim N\ (\mu, \sigma^2)$. *Here* $\tilde{X} \sim N(\mu, \sigma^2/n)$, *so its exact sampling distribution is known. In this situation we shall see that* $\bar{X}$ *is truly optimum among all possible unbiased estimators, not merely linear ones. It has the absolute minimum variance (and expected mean square error) that can be achieved. In contrast, the sample median m has larger sampling variance. In large samples* var *(m) is approximately* $\pi\sigma^2/2n$, *so that its asymptotic efficiency relative to* $\bar{X}$ *is 0.637. In other words it requires a sample about 57 per cent larger to estimate the mean* μ *by the median with equal precision to that obtained using the sample mean.*

Example 5.7. Consider a binomial distribution $B(n, p)$. *Again* $\bar{X}$ *is the globally optimum estimator of the mean* $\mu = np$.

Example 5.8. The data arise as independent observations of random variables X_1; *each* X_i *has the form* $X_i = \exp Y_i$ *where* $Y_i \sim N(\theta, 2\xi)$. *That is, we have a random sample from a* **lognormal** *distribution. In this situation* $\bar{X}$ *is unbiased for the mean* $\mu = e^{\theta + \xi}$. *However, an alternative estimator*

$$\hat{\mu} = \exp\left(\bar{Y} + \tfrac{1}{2}S^2\right),$$

where $\bar{Y}$ *and* S^2 *are the sample mean and variance of* $Y_i = \log_e X_i$, *is asymptotically unbiased and has somewhat smaller variance than* $\bar{X}$. ($\hat{\mu}$ *is the* **maximum likelihood estimator**, *see* §5.3.3 *and* §5.3.4.) *It is easy to modify* $\hat{\mu}$ *to produce an unbiased estimator of* μ, *which for all sample sizes has smaller variance than* $\bar{X}$ *and in this sense is a better estimator.*

Example 5.9. A random sample is chosen from an arbitrary distribution with mean μ, *variance* σ^2. *The Central Limit Theorem ensures that asymptotically* $\bar{X} \sim N(\mu, \sigma^2/n)$. *Whatever the distribution sampled there is no other estimator of* μ *with smaller asymptotic variance. We say that* $\bar{X}$ *is the* **most efficient** *estimator of* μ, *efficiency being defined in terms of asymptotic behaviour.*

Increasing disadvantage in using $\bar{X}$ to estimate μ is illustrated by the following two examples.

Example 5.10. Consider a uniform distribution on the interval $(\mu - \tfrac{1}{2},\ \mu + \tfrac{1}{2})$. *Here* $\bar{X}$ *is unbiased, consistent, and has sampling variance* $1/12n$. *However, if* $X_{(1)}$

and $X_{(n)}$ are respectively the smallest and largest potential sample values, the estimator $\frac{1}{2}[X_{(1)} + X_{(n)}]$, the **mid-range,** *is also unbiased and consistent, with variance $\{2(n + 1)(n + 2)\}^{-1}$. Thus the mid-range is more efficient than $\overline{X}$ for all n, by an order of magnitude. Asymptotically $\overline{X}$ has zero efficiency relative to $\frac{1}{2}[X_{(1)} + X_{(n)}]$.*

And even more extreme, we have:

Example 5.11. Consider the Cauchy *distribution with probability density function*

$$\frac{1}{\pi}\{1 + (x - \theta)^2\}^{-1}$$

[*as* (5.3.4) *in Example 5.3*].

The parameter θ measures central tendency and we might again consider $\overline{X}$ as a possible estimator. But now we encounter real difficulties. No moments exist for the Cauchy distribution, so that θ is not the mean. Furthermore $\overline{X}$ possesses no moments, and is inconsistent as an estimator of θ. In fact, whatever the sample size n, $\overline{X}$ has the distribution (5.3.4) and the Central Limit Theorem does not apply. Having no mean and being inconsistent, $\overline{X}$ is now useless. On the other hand, the sample median, m, is unbiased, and consistent, for θ, with asymptotic variance $\pi^2/4n$. This compares strangely with Example 5.6, for distributions of apparently so similar a shape.

The results in Example 5.11 do not contradict the previously declared general unbiasedness, consistency and asymptotic normality of $\overline{X}$. These results hold in general *provided X possesses a mean and variance,* which is not so for the Cauchy distribution.

Minimum Variance Bound (MVB) Estimators

The above examples demonstrate clearly that the criterion of minimizing the variance among linear unbiased estimators provides a somewhat restricted view of optimality. A far more satisfactory yardstick is provided by what is known as the Cramér-Rao inequality. This provides an absolute standard against which to measure estimators in a wide range of situations. It tells us what is the best we can hope for in terms of the accuracy of an estimator, under what circumstances we can achieve this, and what is the actual form of the best estimator if it exists. The results depend on the form of the *likelihood function,* and on the existence of *sufficient statistics.*

Suppose our sample data x have likelihood function $p_\theta(x)$ and, for the moment, that θ is a scalar parameter. We make certain assumptions (quite generous ones) about the behaviour of $p_\theta(x)$. Termed the **regularity conditions,** these declare that the first two derivatives of $p_\theta(x)$ with respect to θ exist for all θ, and that certain operations of integration and differentiation may be interchanged. This involves some uniform convergence requirements as well as the fact that *the range of variation of X does not depend on the parameter θ.*

We will consider estimating some function $\gamma(\theta)$ by an unbiased estimator $\tilde{\gamma}$. If $\gamma(\theta) \equiv \theta$, we have the special case of unbiased estimation of θ alone. Some preliminary results are necessary. Define the **log-likelihood** $L_\theta(x)$ by

$$L_\theta(x) = \log_e[p_\theta(x)].$$

Then it is readily confirmed that

$$E\left\{\frac{\partial L_\theta(X)}{\partial \theta}\right\} = 0, \tag{5.3.5}$$

$$E\left[\left\{\frac{\partial L_\theta(X)}{\partial \theta}\right\}^2\right] = -E\left[\frac{\partial^2 L_\theta(X)}{\partial \theta^2}\right].$$

The quantity $\partial L/\partial \theta$ is called the **score** by Fisher; so that (5.3.5) says that the score has zero expectation. It is now easy to show that

$$\mathrm{Var}(\tilde{\gamma}) \geqslant -\{\gamma'(\theta)\}^2 / E\left(\frac{\partial^2 L}{\partial \theta^2}\right), \tag{5.3.6}$$

where $\gamma'(\theta)$ is the derivative of $\gamma(\theta)$ with respect to θ. The right-hand side of (5.3.6), which in general is a function of θ, thus represents the lowest value that might arise for the variance of any unbiased estimator of $\gamma(\theta)$. It is called the **Cramér–Rao lower bound**, and it is commonly believed that it was discovered only relatively recently (in the mid 1940s). Often attributed to Cramér and Rao[15] (for work published in 1946 and in 1945, respectively), it appeared earlier (in 1942) in a paper by Aitken and Silverstone[16]. Indeed a similar result and method of proof was given by Dugué (in 1937). An even earlier attribution is to R.A. Fisher (in work published in 1922).

If we can actually obtain an estimator $\tilde{\gamma}$ which is unbiased for $\gamma(\theta)$ and has variance

$$-\{\gamma'(\theta)\}^2 / E\left(\frac{\partial^2 L}{\partial \theta^2}\right) = \{\gamma'(\theta)\}^2 / E\left[\left\{\frac{\partial L}{\partial \theta}\right\}^2\right] = \{\gamma'(\theta)\}^2 / \mathrm{Var}\left(\frac{\partial L}{\partial \theta}\right),$$

we know that this is truly the best we can achieve, in terms of having smallest variance among *all* unbiased estimators. Such an estimator is called the **Minimum Variance Bound (MVB) estimator**.

Two special cases deserve mention. Firstly, if $\gamma(\theta) \equiv \theta$, we have $(E[\{\partial L/\partial \theta\}^2])^{-1}$ as the lower bound to the variance of unbiased estimators of θ itself. Secondly, if the data consist of independent observations $x_1, x_2, \ldots x_n$ of random variables $X_1, X_2, \ldots, X_n$ each with common probability (density) $f_\theta(x)$, say, then

$$p_\theta(x) = f_\theta(x_1) f_\theta(x_2) \ldots f_\theta(x_n).$$

Thus,

$$E\left[\left\{\frac{\partial L}{\partial \theta}\right\}^2\right] = nE\left[\left\{\frac{\partial \log_e f_\theta(X)}{\partial \theta}\right\}^2\right]$$

and (5.3.6) becomes

$$\mathrm{Var}\,(\hat{\gamma}) \geqslant \{\gamma'(\theta)\}^2/nE\left[\left\{\frac{\partial \log_e f_\theta(X)}{\partial \theta}\right\}^2\right] = \{\gamma'(\theta)\}^2/nI(\theta).$$

Fisher referred to $I_s(\theta) = E[\{\partial L/\partial \theta\}^2]$ as the *amount of* **information** in the sample; by analogy, in the case just described, $I(\theta) = I_s(\theta)/n$ is the amount of information in a single observation. The idea of 'information' here has intuitive appeal, since for fixed n the larger $I_s(\theta)$, or $I(\theta)$, the better the estimator we might potentially obtain through the MVB estimator *if it exists*. We consider other concepts of 'information' in Chapter 8.

For unbiased estimation of θ itself, we have (rewriting $\hat{\gamma}$ as $\tilde{\theta}$) that

$$\mathrm{Var}(\tilde{\theta}) \geqslant (I_s)^{-1} = (nI)^{-1}.$$

Notice how in the case of independent observations the best we can hope for is unbiased estimators with variance of order n^{-1}.

Example 5.12. If we have a random sample from $N(\mu, \sigma^2)$ *with* σ^2 *known, then* $I = 1/\sigma^2$. *So the Cramér-Rao lower bound for estimation of* μ *is* σ^2/n. *But* $\mathrm{Var}(\overline{X}) = \sigma^2/n$, *so that* $\overline{X}$ *is the MVB estimator of* μ.

This example shows that the MVB estimator can exist. The important question to ask is under what circumstances this will be so. These are in fact easily determined, and we find that $\hat{\gamma}$ *is the MVB estimator of* $\gamma(\theta)$ if and only if

$$\frac{\partial L}{\partial \theta} = k(\theta)\{\hat{\gamma} - \gamma(\theta)\}, \tag{5.3.7}$$

where $k(\theta)$ does not involve the data x. This result has many implications.

(i) $E[(\partial L/\partial \theta)] = 0$, so that $E(\hat{\gamma}) = \gamma(\theta)$ confirming that the MVB estimator of $\gamma(\theta)$ is unbiased. If we carry out a parallel study of biased estimators we find that there is a similar lower bound for the *mean square error*, but that the conditions for this to be attained again imply that the estimator *is unbiased*. So we return to the MVB estimator as the optimum one.

(ii) From (5.3.6) and (5.3.7) the variance of the MVB estimator is easily seen to take the simple form $\gamma'(\theta)/k(\theta)$.

(iii) (5.3.7) provides the means of constructing the MVB estimator if it exists.

Considering the form of $\partial L/\partial\theta$, we obtain by inspection the estimator $\tilde{\gamma}$ and from (ii) its variance.

(iv) (5.3.7) implies that $\tilde{\gamma}$ is *singly sufficient* for θ. Thus the existence of a singly sufficient statistic is necessary if an MVB estimator is to be found. On the other hand the relation (5.3.7) requires more than just sufficiency of $\tilde{\gamma}$.

Example 5.13. $x_1, x_2, \ldots, x_n$ is a random sample from $N(0, \sigma^2)$.

$$\frac{\partial L}{\partial(\sigma^2)} = -\frac{n}{2\sigma^2} + \frac{\Sigma x_i^2}{2\sigma^4} = \frac{n}{2\sigma^4}\left\{\frac{1}{n}\Sigma x_i^2 - \sigma^2\right\}.$$

So $(1/n)\Sigma x_i^2$ is the MVB estimator of σ^2, with variance $2\sigma^4/n$. But

$$\frac{\partial L}{\partial\sigma} = \frac{n}{\sigma^3}\left(\frac{1}{n}\Sigma x_i^2 - \sigma^2\right),$$

cannot be put in the form (5.3.7). So no MVB estimator exists for σ: that is, for the standard deviation rather than for the variance.

This shows how an MVB estimator may exist for one function $\gamma(\theta)$ of θ, but not for another function $\psi(\theta)$. In fact we can be much more precise about this.

(v) If $\hat{\theta}$ is singly sufficient for θ, there is a *unique* function $\gamma(\theta)$ for which an MVB estimator exists, viz. $\gamma(\theta)$ satisfying (5.3.7). So the existence of a singly sufficient statistic ensures that an MVB estimator exists for a unique function $\gamma(\theta)$ of θ.

The complete situation is assessed when we recognize the following fact.

(vi) When θ is one-dimensional the only situations for which a singly sufficient statistic exists for θ are those where the data arise as realizations of independent random variables $X_1, X_2, \ldots, X_n$ each having a common distribution whose probability (density) function has the form

$$\exp\{A(\theta)B(x) + C(\theta) + D(x)\}. \tag{5.3.8}$$

This is the so-called **exponential family** of distributions, and includes common distributions such as the *normal, binomial, Poisson and gamma. Only in such situations will we encounter MVB estimators and singly sufficient statistics for a scalar parameter θ.*

Thus we have encountered the promised conditions for single sufficiency. Outside the exponential family sufficiency can still lead to *some* reduction in the dimensionality of the set of statistics we need to consider, as we saw in Example 5.4. See also §5.3.4.

For a discussion of minimum variance bound estimation in the presence of several parameters, and in particular the MVB of a *single* function of a set of parameters, see Kendall and Stuart (1979, Chapter 17).

Recognizing the somewhat limited range of situations where [that is, range of functions $\gamma(\theta)$ for which] MVB estimators can be found, we are forced to ask a further question. If a MVB estimator cannot be found for $\gamma(\theta)$, is there nonetheless an unbiased estimator $\tilde{\gamma}$ with variance uniformly (in θ) lower than that of any other estimator γ^*. See Figure 5.2.

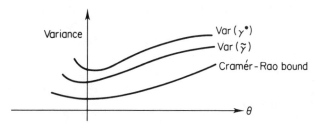

Figure 5.2

This is possible, although again the existence of a singly sufficient statistic is necessary to obtain such **uniformly minimum variance** (UMV) estimators.

In addition, if $\tilde{\gamma}$ is to be the UMV estimator of $\gamma(\theta)$, we need it to satisfy a further condition: namely to be **complete**.

Completeness is a useful concept in estimation and testing (but with surprisingly little immediate intuitive appeal).

Definition. $\tilde{\gamma}$ is complete if for any statistic $h(\tilde{\gamma})$, independent of θ, the statement $E[h(\tilde{\gamma})] = 0$ implies that $h(\tilde{\gamma})$ is identically zero. This says *there is no non-trivial unbiased estimator of zero based on $\tilde{\gamma}$.*

If $\tilde{\gamma}$ is sufficient and complete then it turns out that there is a *unique* function $h(\tilde{\gamma})$ of $\tilde{\gamma}$ which is unbiased for any $\gamma(\theta)$ and that this $h(\tilde{\gamma})$ is the UMV estimator of $\gamma(\theta)$. The estimator $h(\tilde{\gamma})$ is readily justified and constructed, in principle, from the *Rao-Blackwell theorem* which says that if we have an unbiased estimator, and a sufficient statistic, then the expected value of the unbiased estimator *conditional on the sufficient statistic* is unbiased, *is a function of the sufficient statistic* and has variance no larger than that of the unbiased estimator.

Alternatively we may seek the unique $h(\tilde{\gamma})$ directly.

Example 5.14. A univariate random variable X has a binomial distribution $\mathbf{B}(n, p)$. Our data consists of a single observation of X. We see that X is sufficient for p, and it is easily confirmed that X/n is the MVB estimator of p. We want to estimate p^2. No MVB estimator exists, but we can find the UMV estimator. X is not only sufficient, but complete, Also

$$E[X(X-1)] = n(n-1)p^2,$$

so that $X(X-1)/[n(n-1)]$ is unbiased for p^2 and a function of a sufficient and complete statistic. It is therefore the UMV estimator of p^2.

Two further points are worth stressing.

Suppose the regularity conditions are not satisfied. Will we necessarily obtain inferior estimators in such situations? On the contrary, Example 5.10 demonstrates a situation where the range of the random variable depends on the parameter and an estimator may be found which is outstandingly better than an MVB estimator, in having variance of order n^{-2} rather than n^{-1}.

The actual effort involved in obtaining an estimator has not figured in our study of optimum estimators. But this may be important. For example, in estimating σ^2 in $N(0, \sigma^2)$ the sample variance $S^2 = (1/n)\Sigma x_i^2$ is optimal. However, if $X_{(1)}, X_{(2)}, \ldots, X_{(n)}$ are the *ordered* potential sample members (the **order statistics**), the estimator

$$V = 1.7725 \sum_1^n (2i - n - 1) X_{(i)}/[n(n-1)] \tag{5.3.9}$$

is more easily calculated and is unbiased with efficiency about 98 per cent relative to S^2 for all sample sizes. (See Downton[17].) Far more extreme examples of economy of effort for similar efficiency may be found.

We now understand the conditions under which optimum point estimators exist for a scalar (single) parameter θ and we have examined their form. Whilst this aspect of the subject is important we must face up to what can be done in practice if no optimum estimator exists, or if the effort in obtaining such an estimator is prohibitive. A range of practical procedures exist for actually constructing estimators. Some we have briefly mentioned. The following section reviews these and comments on the properties of the estimators they produce.

5.3.3 Methods of Constructing Estimators

There are many practical methods which are used for constructing estimators. Several of these methods are briefly described below.

Maximum Likelihood

This is perhaps the most widely used method and the resulting estimators have some (mainly asymptotic) optimum properties. First detailed by Fisher in the early 1920s[†], it proposes the estimation of θ by that value $\hat{\theta}$ for which the

---

† The principle was used by Lambert around 1760 and D. Bernoulli about 13 years later—though with no attempt at justification.

likelihood function $p_\theta(x)$ is a maximum, for the given data x. The estimator $\hat{\theta}$ is called the **maximum likelihood estimator** (m.l.e.).

Essentially, the principle is that the value of the parameter under which the obtained data would have had highest probability (density) of arising must be intuitively our best estimator of θ. We must avoid the temptation to reverse this probability interpretation and regard the likelihood function $p_\theta(x)$, for fixed x, as providing (subjective) probability measures for different possible values of θ. The very term 'likelihood' encourages this in a way; 'likelihood' sounds akin to 'probability'. It also endorses the principle based on it—if we have a measure of the 'likelihood' of θ, what has more immediate appeal than to choose that value whose 'likelihood' is a maximum! Although other approaches (see §8.2) do interpret $p_\theta(x)$ as providing at least relative measures of credence for different values of θ, no such 'θ-probabilities' are advanced in the classical approach. Probability remains attached to X, not θ; it simply reflects inferentially on θ. In this spirit, it is difficult to formally justify 'maximizing the likelihood'. It rests largely on any intuitive appeal, but is strength lies in the fact that resulting estimators often process what the classical approach advances as desirable properties.

Again we concentrate on a scalar (one-component) parameter θ, although here (as throughout this chapter) immediate extensions to more general θ are possible, and some of the relevant considerations and results are discussed later (in §5.3.4 and in §5.6). It is usual to consider the *log-likelihood* $L_\theta(x)$, rather than $p_\theta(x)$, because of the added ease this introduces particularly for data in the form of random samples from a fixed distribution.

In many situations $L_\theta(x)$ is particularly well behaved, in being continuous with a single maximum away from the extremes of the range of variation of θ. Then $\hat{\theta}$ is obtained simply as the solution of

$$\frac{\partial L}{\partial \theta} = 0, \tag{5.3.10}$$

subject to

$$\left.\frac{\partial^2 L}{\partial \theta}\right|_{\hat{\theta}} < 0.$$

Example 5.15. $X_1, X_2, \ldots, X_n$ *are independent Poisson random variables, each with mean* θ.

$$L_\theta(x) = -n\theta + \sum_1^n x_i \log \theta - \sum_1^n \log(x_i!).$$

$\partial L/\partial \theta = 0$ *yields* $\hat{\theta} = (1/n)\Sigma x_i = \bar{x}$ *as the maximum likelihood estimate. The estimator* $\bar{X}$ *is also the MVB estimator in this situation.*

But in other situations computational difficulties can arise.

(i) *No explicit solution may exist for* (5.3.10). Then iterative numerical methods must be used and care is needed to ensure proper convergence. This is particularly serious when we have a multi-component θ, or when there may be several *relative* maxima of L from which we must choose the absolute maximum. For example, in estimating the location parameter θ for the Cauchy distribution (5.3.4) on the basis of a random sample of size n, many maxima may occur and the derivative of the log-likelihood function is typically of the form shown in Figure 5.3. The Newton-Raphson, or fixed-slope Newton, methods may fail completely due to points of inflection in L sending the iterate off to infinity, or due to the iterate cycling indefinitely as a result of a very steep slope in $\partial L/\partial\theta$ in the region of a root of (5.3.10). The method of the false positions, or some similar method, may be needed to avoid these difficulties. (See Barnett[18].)

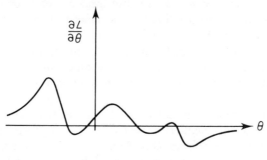

Figure 5.3

(ii) *L may be discontinuous, or have a discontinuous first derivative, or a maximum at an extremal point.* Again special precautions will be needed. An interesting case arises when we have independent observations of a random variable with probability density proportional to $\exp\{-|x-\theta|^{1/2}\}$; L may look like this:

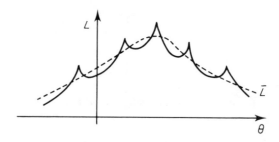

Figure 5.4

The cusps make it impossible to employ (5.3.10) and again numerical methods will be needed. One appealing off-shoot of this example (first proposed by Daniels[19]) is to consider maximizing a smoothed version of L, rather than L itself, perhaps

$$\bar{L} = \frac{1}{2\varepsilon} \int_{\theta-\varepsilon}^{\theta+\varepsilon} L_\theta(x)\, d\theta$$

for different ε. Little seems to be known of this modification. Obviously if ε is too large we lose resolution and must expect poor estimators. But it is easily shown that this approach can lead on occasions, for suitable ε, to estimators which are more efficient than $\hat{\theta}$ (Barnett[18]).

Let us consider what can be said in general about the properties of maximum likelihood estimators. In summary, the properties of $\hat{\theta}$ are as follows.

(a) Under fairly general conditions $\hat{\theta}$ is *consistent*. Thus it is *asymptotically unbiased*, but usually *biased in small samples*.

(b) If there is a singly sufficient statistic, $\hat{\theta}$ *is a function of it*. If furthermore an MVB estimator exists for θ, then *this is the same* as $\hat{\theta}$. In this case $\hat{\theta}$ is, of course, optimum. If no MVB estimator exists *little can be said of the efficiency of $\hat{\theta}$ in small samples*. Its variance is often difficult to determine, although in view of (a) and (d) (below) it may be approximated by

$$\left(-E\left[\left\{ \frac{\partial^2 L}{\partial \theta^2} \right\} \right] \right)^{-1}, \qquad \text{or even} \qquad \left(-\frac{\partial^2 L}{\partial \theta^2}\bigg|_{\hat{\theta}} \right)^{-1},$$

in large samples.

(c) If $\hat{\theta}$ is the maximum likelihood estimator of θ, $\gamma(\hat{\theta})$ *is the maximum likelihood estimator of* $\gamma(\theta)$.

(d) $\hat{\theta}$ *is asymptotically fully efficient*; that is,

$$\text{Var}(\hat{\theta}) \to [I_s(\theta)]^{-1},$$
$$= [nI(\theta)]^{-1},$$

for independent observations from a common distribution.

(e) $\hat{\theta}$ *is asymptotically normally distributed*; that is,

$$\hat{\theta} \sim N(\theta, [I_s(\theta)]^{-1}).$$

Whilst we must take care about the possibly poor behaviour of $\hat{\theta}$ from small samples, in cases where no MVB estimator exists, the above catalogue of desirable asymptotic properties ensures the maximum likelihood method of an important role in classical point estimation.

Attention to asymptotic behaviour has lead to the study of a wider class of **best asymptotically normal** (BAN) estimators, which have asymptotic normal distributions and also satisfy other desirable criteria, such as consistency and (asymptotic) efficiency but without some of the computational complexity of the maximum likelihood estimator. The m.l.e. is, under appropriate regularity conditions, a BAN estimator (see Zacks, 1971, pp. 244 *et seq.*). The method of estimation we next consider also yields BAN estimators.

Minimum Chi-squared

This is another general method of constructing point estimators, applicable when fairly extensive data are available. It finds less application than the other methods discussed here, and again cannot be claimed to produce universally good, let alone optimum, estimators in small samples.

We shall illustrate its form in a special case. Suppose our data, as n independent observations of a random variable X with probability (density) $f_\theta(x)$, has been reduced to the form of a frequency distribution. That is, the frequencies of the observations in k different categories are $n_1, n_2, \ldots, n_k$; where $\Sigma_1^k n_i = n$. The probability of X falling in the ith category may be expressed in terms of an integral (or sum) of $f_\theta(x)$ over an appropriate range of values of x. Denote this probability by $\pi_i(\theta)$. Then the **method of minimum chi-squared** prescribes that we estimate θ by minimizing the quantity

$$U = \sum_1^k \left\{ \frac{(\tilde{\pi}_i - \pi_i(\theta))^2}{\pi_i(\theta)} \right\},$$

where $\tilde{\pi}_i$ is the simple empirical estimate of $\pi_i(\theta)$ given by the relative frequency n_i/n.

In achieving this minimization, techniques and difficulties are involved which are akin to those encountered in the maximum likelihood approach. The similarity does not rest at this procedural level. We find that *minimum chi-squared estimators* are asymptotically equivalent to maximum likelihood estimators. However, in small samples, no assurances are possible. The minimum chi-squared estimator need not even be a function of a singly sufficient statistic if one exists. It is usually biased, and may have low efficiency relative to alternative estimators. The principal appeal is an intuitive one.

So-called *modified minimum chi-squared* estimators are obtained if we replace the probabilities $\pi_i(\theta)$ in the denominator of U by their relative frequency estimates $\tilde{\pi}_i$.

Estimating Equations

The methods of maximum likelihood and of minimum chi-square may both involve obtaining the estimator $\tilde{\theta}$ as the solution of an equation of the form

$$g(x, \tilde{\theta}) = 0. \tag{5.3.11}$$

For example, the m.l.e. is obtained from (5.3.10) and the minimum chi-squared estimator from $(\partial U/\partial \theta) = 0$. The particular equations to be solved in these methods were determined by the method being used. A more general approach to point estimation consists of estimating θ by $\tilde{\theta}$: the solution of an equation of the form (5.3.11), subject to some regularity conditions, and the requirement that

$$E_\theta\{g(x, \theta)\} = 0 \tag{5.3.12}$$

for all $\theta \in \Omega$. Any equation (5.3.11) satisfying the regularity conditions and (5.3.12) is termed an unbiased **estimating equation** for θ.

The estimating equation is said to be *optimum* if it yields the minimum possible value for

$$E_\theta\{g^2(x, \theta)\}/E_\theta[\{\partial g(x, \theta)/\partial \theta\}^2]$$

uniformly in θ. See Godambe[20].

Method of Moments

This was one of the earliest methods of point estimation to be proposed. It has immediate intuitive appeal and is very simple to employ. Suppose the data consist of independent observations from some distribution. The moments of the distribution will typically be functions of the components of θ, and we have only to equate enough of the sample moments to the corresponding population moments to obtain equations for estimating θ. *The resulting estimators will normally be biased, and little can be said of their efficiency properties.* The real appeal is simplicity of application.

Example 5.16. $X_1, X_2, \ldots, X_n$ *are independent random variables with a common gamma distribution, having probability density function*

$$f_{r,\lambda}(x) = \lambda(\lambda x)^{r-1} e^{-\lambda x}/\Gamma(r) \qquad (r > 0, \lambda > 0).$$

We have $E(X) = r/\lambda$, *and* $E(X^2) = r(r+1)/\lambda^2$. *So If* $\bar{X} = (1/n)\Sigma_1^1 X_i$, *and* $S^2 = (1/n)\Sigma X_i^2$, *we obtain moment estimators of* r *and* λ *as*

$$\tilde{r} = \bar{X}^2/(S^2 - \bar{X}^2), \qquad \tilde{\lambda} = \bar{X}/(S^2 - \bar{X}^2).$$

We obtain useful quick estimates which might serve, at least, as starting values for an iterative maximum likelihood study.

Least Squares

Here we meet one of the most frequently applied estimation principles. Its origins go back to the work of Laplace and Gauss. Earlier we have remarked on its *sample specific* nature, and on the fact that for linear models it yields linear unbiased estimators of minimum variance. Let us consider this in more detail. It is assumed that our data x arise as an observation of a vector random variable X (of dimension n) whose expected value depends on known linear combinations of the p components of the vector parameter θ. It is further assumed that the variance matrix of X is known up to a scale factor. Thus

$$x = A\theta + \varepsilon, \tag{5.3.13}$$

where A is a known $(n \times p)$ matrix of full rank (usually, but not necessarily), and the random vector ε has zero mean and variance matrix

$$\text{Var}(\varepsilon) = E(\varepsilon\varepsilon') = V\sigma^2,$$

where the $(n \times n)$ symmetric matrix V is known precisely.

In this situation the **extended least squares principle** states that we should estimate θ by minimizing the quadratic form

$$(x - A\theta)'\, V^{-1}(x - A\theta), \tag{5.3.14}$$

and this produces an estimator

$$\tilde{\theta} = (A'V^{-1}A)^{-1}\, A'V^{-1}X \tag{5.3.15}$$

with variance matrix

$$(A'V^{-1}A)^{-1}\sigma^2. \tag{5.3.16}$$

The estimator (5.3.15) with variance (5.3.16) is seen to have smallest variance (on a component to component basis) among all unbiased estimators of θ which are linear combinations of the elements of X.

We must ask how good is $\tilde{\theta}$ outside the class of linear estimators. With no assumptions about the form of the *distribution* of ε little can be said. However, in the special case where ε has a multivariate normal distribution the criterion of minimizing (5.3.14) is precisely what is involved in the *maximum likelihood* approach to this problem. Thus the various properties of maximum likelihood now transfer to the least squares procedure. In particular $\tilde{\theta}$ is fully efficient in an asymptotic sense, and its large sample properties are immediately determined.

In many applications the data x, whilst still of the form (5.3.13), arise as *uncorrelated* observations with common variance. Then V is the $(n \times n)$ identity matrix, and in large samples the (unbiased) least squares estimator $\tilde{\theta} = (A'A)^{-1}A'X$ is essentially normally distributed, in view of the Central Limit Theorem, with variance matrix $(A'A)^{-1}\sigma^2$. Thus probability statements may be made about $\tilde{\theta}$, formally equivalent to those which are appropriate through the maximum likelihood approach when the error structure is described by independent normal distributions.

Either as a principle in its own right with its endowed *linear* optimality properties, or with the wider optimality that arises from a *normal* error structure, the principle of least squares is applied to a vast range of statistical problems based on a linear model. The majority of classical statistical techniques (regression, analysis of variance and so on, in their variety of form and complexity) are based on such linear models, which are advanced as reasonable first approximations for the study of practical problems across the whole range of disciplines in which statistics is applied. In this sense, least squares appears as a very widely applied principle for the construction of point estimators. We shall see later that corresponding procedures for testing hypotheses concerning linear models assume an equal, if not greater, practical importance.

Estimation by Order Statistics

Suppose a set of data arises as independent observations $x_1, x_2, \ldots, x_n$ of some common univariate random variable. We might place these observations in ascending order, obtaining $x_{(1)}, x_{(2)}, \ldots, x_{(n)}$ which can be regarded as observed values of (correlated) random variables $X_{(1)}, X_{(2)}, \ldots X_{(n)}$. These random variables are called **order statistics,** and it is possible for useful estimators to be constructed in terms of $X_{(1)}, X_{(2)}, \ldots, X_{(n)}$.

The ordered observations have a natural function in describing the properties of the sample. For example $x_{(1)}$ and $x_{(n)}$ are the *extreme* members of the sample and are often of interest in their own right, for instance in meteorological problems. Then again, $x_{(n)} - x_{(1)}$ is the **sample range,** $\frac{1}{2}(x_{(1)} + x_{(n)})$ is the **sample mid-range,** and $m = x_{(k)}$ or $m = \frac{1}{2}[x_{(k)} + x_{(k+1)}]$ (depending on whether $n = 2k - 1$, or $n = 2k$) is the **sample median:** all with interpretative importance. Intuitively we might expect the sample range to reflect the *dispersion* of the distribution being sampled, and the sample mid-range or sample median to reflect its *location or central tendency.* Thus if μ, and σ, are parameters representing measures of location, and dispersion, we might seek estimators of μ, or σ, based on $\frac{1}{2}[X_{(1)} + X_{(n)}]$ say, or $X_{(n)} - X_{(1)}$.

We have already seen that this can be a sensible approach. In Example 5.10 the mid-range was seen to be a most desirable estimator of the mean of a uniform distribution on $(\mu - \frac{1}{2}, \mu + \frac{1}{2})$, whilst in Example 5.11 the median appeared as a useful estimator of the location parameter of the Cauchy distribution. We shall also consider later (Chapter 9) the use of order statistics in *robust* inference about location or dispersion parameters.

The importance of the order statistics is highlighted by their appearance as maximum likelihood estimators, or minimal sufficient statistics, in certain situations.

Example 5.17. Suppose X has a uniform distribution on the interval $(\mu - \sigma, \mu + \sigma)$. Then $(X_{(1)}, X_{(n)})$ are minimal jointly sufficient for (μ, σ). Also the maximum likelihood estimators of μ and σ are

$$\hat{\mu} = \tfrac{1}{2}(X_{(1)} + X_{(n)}) \quad and \quad \hat{\sigma} = \tfrac{1}{2}(X_{(n)} - X_{(1)}).$$

If not of interest as a natural expression of some property of the data, or not arising from more general criteria, we might still consider using the order statistics for an alternative expression of the probability model. In this guise it may happen that some convenient linear combination of $X_{(1)}, X_{(2)}, \ldots, X_{(n)}$ produces a reasonable estimator. This again has been demonstrated. When X is $N(0, \sigma^2)$, we saw that the linear form V given as (5.3.9) lacks little in efficiency as an estimator of σ^2.

The major use of order statistics is in a rather special class of problems, where we are sampling from a distribution assumed to have a distribution function of the form $F[(x-\mu)/\sigma]$. Estimators of the form

$$\mu^* = \sum_1^n \alpha_i X_{(i)} \quad \text{and} \quad \sigma^* = \sum_1^n \beta_i X_{(i)}$$

are considered, and the extended least squares principle can be used to determine the minimum variance unbiased estimators of this form. The **'best linear unbiased estimators'** (BLUEs) are expressed in terms of the mean vector, and variance matrix, of the parameter-free reduced order statistics, $U_{(i)} = [X_{(i)} - \mu]/\sigma$. The BLUEs are asymptotically fully efficient, and although little can be proved in general about their small sample behaviour they seem to behave surprisingly well in many cases. Their prime disadvantage, however, is the difficulty in calculating their variances. This can involve extensive numerical integration, or summation, or the availability of detailed tabulated material. Efforts have been directed to more tractable approximations. See David (1981).

5.3.4 Estimating Several Parameters

Passing reference has already been made to multi-parameter situations. In §5.3.1 we had cause to comment on the concept of *sufficiency* and, in §5.3.2, to refer to the *minimum variance bound criterion*, in the presence of several parameters. The particular methods of estimation described in §5.3.3 frequently admit immediate extension to the multi-parameter case; indeed, the *least squares method* was specifically developed for such a situation and the *method of moments* and *use of order statistics* were illustrated for two-parameter problems.

One method of constructing estimators of several parameters merits more detailed comment and that is the *method of maximum likelihood*. Suppose our data consist of a set of independent observations $x_1, x_2, \ldots, x_n$ of a random variable x with probability (density) function $f_\theta(x)$, where θ is a vector parameter with components $\theta_1, \theta_2, \ldots, \theta_p$. The principle remains the same as in the one-parameter case: namely, to estimate θ by that value $\hat{\theta}$ (with components $\hat{\theta}_1, \hat{\theta}_2, \ldots, \hat{\theta}_p$) in the parameter space Ω, which makes the likelihood as large as possible. Under suitable regularity conditions the likelihood function [and equivalently the log-likelihood $L \equiv L_\theta(x) = \sum_{j=1}^{n} \log f_\theta(x_j)]$ will have relative maxima which are obtained as solutions of the set of equations

$$\frac{\partial L}{\partial \theta_i} = 0 \qquad (i = 1, 2, \ldots, p) \tag{5.3.17}$$

subject to the matrix $\{\partial^2 L/\partial \theta_r \partial \theta_s\}$ being negative definite.

In general there may be several relative maxima identified by this procedure and we then have the task of picking that one which is the absolute maximum, and whose position $\hat{\theta} \in \Omega$ is then the *maximum likelihood estimate*.

Note that the difficulties described in the one-parameter case, of non-explicit solutions of the likelihood equation and ill-conditioning of numerical methods, are even more likely to be encountered (and in more acute form) when we have several parameters. Furthermore, when the range of X depends on θ, added complications arise and relatively little is known of the behaviour of $\hat{\theta}$.

One case, however, is straightforward and yields a *unique* maximum likelihood estimator with properties akin to those obtained in the corresponding one-parameter situations.

Suppose there exists a set of statistics $\tilde{\theta}_1, \tilde{\theta}_2, \ldots, \tilde{\theta}_t$ which are jointly sufficient for $\theta_1, \theta_2, \ldots, \theta_p$. It is easily established that the maximum likelihood estimator, $\hat{\theta}$, must be a function of $\tilde{\theta}_1, \tilde{\theta}_2, \ldots, \tilde{\theta}_t$. But more important, if $t = p$ [which is so only for the multi-parameter exponential family: (5.3.8), but where θ is a vector and $A(\theta)B(x)$ is replaced by $\sum_{j=1}^{p} A_j(\theta)B_j(x)$] the maximum likelihood estimator is unique. Thus uniqueness of $\hat{\theta}$ is ensured by the minimal sufficient statistic having the same dimensionality as Ω. For now (5.3.17) has a unique solution $\hat{\theta}$ and $\{\partial^2 L/\partial \theta_r \partial \theta_s\}_{\theta=\hat{\theta}}$ is negative definite.

As for the one-parameter case, $\hat{\theta}$ is consistent under very wide conditions (even if the minimal sufficient statistic has dimension in excess of p). Furthermore, if the regularity conditions hold, $\hat{\theta}$ can be shown to be efficient and asymptotically normal; specifically, $\hat{\theta}$ is asymptotically $N(\theta, V)$, where

$$V^{-1} = -E\left\{\frac{\partial^2 L}{\partial \theta_r \partial \theta_s}\right\} = E\left\{\frac{\partial L}{\partial \theta_r}\frac{\partial L}{\partial \theta_s}\right\} \tag{5.3.18}$$

is known as the *information matrix*.

When the minimal sufficient statistic has dimension p, some economy of effort arises from noting that asymptotically

$$V^{-1} = -\left\{\frac{\partial^2 L}{\partial \theta_r \partial \theta_s}\right\}_{\theta=\hat{\theta}} = \left\{\frac{\partial L}{\partial \theta_r}\frac{\partial L}{\partial \theta_s}\right\}_{\theta=\hat{\theta}},$$

so that we can approximate the variance–covariance matrix of $\hat{\theta}$ from the inverse of the matrix of second derivatives of L (or of products of first derivatives) evaluated at $\hat{\theta}$, without the need to obtain expected values.

As for the one-parameter case, little can be said in general about the behaviour of $\hat{\theta}$ in small samples.

If we estimate a parameter $\psi = \psi(\theta)$, where ψ is a one-to-one transformation, we have simply that $\hat{\psi} = \psi(\hat{\theta})$. If, furthermore, $\psi(\theta)$ is differentiable (at the true parameter value θ) the information matrix for ψ is readily obtained as $u'Vu$, where u is the vector of values $(\partial\psi/\partial\theta_r)$ $(r = 1, 2, \ldots, p)$, and V is the information matrix for θ.

Example 5.18. For a random sample $x_1, x_2, \ldots, x_n$ from the lognormal distribution discussed in Example 5.8, we have

$$L = -\frac{n}{2}\ln(2\pi\xi) - \frac{1}{4\xi}\sum_{i=1}^{n}(y_i - \theta)^2,$$

where $y_i = \ln x_i$. Thus $\hat{\theta} = \bar{y}$ and $\hat{\xi} = s^2/2$, where $s^2 = \left[\sum_{i=1}^{n}(y_i - \bar{y})^2\right]/n$.

Asymptotically $\hat{\theta}$ and $\hat{\xi}$ are independent with variances $(2\xi)/n$ and $(2 - \xi^2)/n$, respectively.

The lognormal distribution has mean $\mu = \exp(\theta + \xi)$ and variance $\sigma^2 = [\exp(2\xi) - 1]\exp[2(\theta + \xi)]$. Thus we have as maximum likelihood estimate of μ,

$$\hat{\mu} = \exp(\hat{\theta} + \hat{\xi}) = \exp(\bar{y} + s^2/2).$$

It can be shown that the asymptotic variance of $\hat{\mu}$ is

$\{2(\xi + \xi^2)\exp[2(\theta + \xi)]\}/n$, which is less than $\operatorname{var}(\bar{x}) = \sigma^2/n$.

The properties of $\hat{\theta}$ sketched above apply only in a range of standard situations where appropriate conditions are satisfied. It must be stressed that they do not inevitably carry over to other situations; indeed the maximum likelihood estimator need not necessarily be unique or consistent, and asymptotic normality is not inevitable. Particular cases of importance include *estimating a subset* of the components of θ, the occurrence of relative maxima of L at the *boundary of Ω*, *failure* of the regularity conditions, correlated or *non-identically distributed observations* and the prospect that the dimensionality of θ *depends on the sample size*. In this last case $\hat{\theta}$ might not be consistent; when $\hat{\theta}$ is consistent, its asymptotic variance—covariance matrix *need not be given by the inverse of the information matrix.* [See Cox and Hinkley (1974, Chapter 9) for a straightforward review of some of these situations.]

Motivated by the idea of maximum likelihood ratio tests (q.v.), an interesting approach (Nelder and Wedderburn[21]) to the fitting of successively more complicated models to a set of data proposes maximizing the likelihood of the current model at each stage and assessing quantitatively the extent to which the model fits the data by calculating $-2\{$maximized log-likelihood$\}$, which is termed the *deviance* in this context. Applied to the study of *generalized linear models*, informal methods are proposed for determining when the appropriate degree of fit has been achieved.

Some further remarks, on the effect of nuisance parameters and incidental parameters, and on *conditional* maximum likelihood estimation, appear in §5.6.

5.4 TESTING STATISTICAL HYPOTHESES

We must now review the dual concept of testing hypotheses. It is convenient to do this in a manner directly parallel to that employed in the previous section in the discussion of point estimation. That is to say we will consider what criteria are set up to represent and assess the performance of a statistical test, discuss in terms of such criteria what may be meant by an optimum test and determine the conditions under which such a test will arise. Finally we examine any general methods of test construction and the performance of the tests they generate.

The distinctions were drawn, in §5.2, between the ideas of a *pure significance test* (with no alternative hypothesis specified and no mechanism for rejecting the basic hypothesis, H), a *significance test* (as a rule for rejecting H, without specific regard for any alternative hypothesis) and an *hypothesis test* (as a formal mechanism for rejecting or accepting H in direct comparison with a specified alternative hypothesis $\overline{H}$). Many of the crucial conceptual points were outlined in the earlier section (§5.2). In moving on to such topics as performance criteria, optimality properties and the construction of tests, the more substantial matters relate to the most highly structured form (that of the hypothesis test) and we shall proceed to consider this in some detail.

5.4.1 *Criteria for Hypothesis Tests*

In the classical approach, the basis for accepting or rejecting some hypothesis, H, about θ is the so-called **test of significance**. Before discussing its general form as a Neyman–Pearson **hypothesis test**, with a specific alternative hypothesis $\overline{H}$ specified, it is useful to illustrate some particular features through a simple example.

Suppose our data consist of observations $x_1, x_2, \ldots, x_n$ of independent random variables $X_1, X_2, \ldots, X_n$, each of which has a normal distribution, $N(\mu, \sigma^2)$, with known variance, σ^2. We want to test a *simple* (or *point*) hypothesis $H: \mu = \mu_0$. We shall do this by evaluating some function $t(x)$ of the data x, *accepting H* if $t(x)$ is within a particular range of values, and *rejecting H* otherwise. The performance characteristics of such a procedure are again to be expressed in *aggregate* terms, so that we are concerned with the probability distribution of $t(X)$, the so-called **test statistic**.

We might, for example, decide to reject or accept H on the basis of the value of the sample mean, $\bar{x}$. This is intuitively sensible in that we know that $\bar{X}$ is a good estimator of μ. Accordingly, we might decide to accept H if $\bar{x}$ is sufficiently close to μ_0, and to reject H if $\bar{x}$ is not close to μ_0. Thus the basis for rejecting H could be that $|\bar{x} - \mu_0| > c$, for some prescribed value c. How should we choose c? We obviously want to avoid rejecting H when it is true. This cannot be guaranteed other than by *always* accepting H, which would not be sensible since we would then never reject H when it is false. As a compromise we could ask that the probability of rejecting H *when it is true* should be small enough 'to suit our practical needs', perhaps 0·1, say. This implies that we are prepared to accept a risk of 10 per cent of *incorrectly rejecting* H. This is expressed in terms of the *test statistic* $\bar{X}$ by the fact that when H is true

$$P(|\bar{X} - \mu_0| > c) = 0{\cdot}1,$$

which implies that $c = 1{\cdot}645\sigma/\sqrt{n}$ and yields a *critical region* for rejecting H of the form

$$S_0 = \{x : |\bar{x} - \mu_0| > 1{\cdot}645\sigma/\sqrt{n}\}.$$

So if $x \in S_0$ we reject H, otherwise we accept H. Such a procedure is called a 10% *test of significance*, the **significance level** (here 10 per cent, but this depends on the circumstances of the problem) being the risk we are prepared to bear of being wrong in rejecting H. The test of significance operates on the principle that we will accept H unless we witness some event which has sufficiently small probability of arising when H is true. This is what Barnard calls 'the principle of the disbelief in tall stories'.

But how should we choose the significance level in a particular situation, why should we use the particular test statistic $\bar{X}$, and what is the justification for the specific rejection criterion based on the absolute value of $\bar{x} - \mu_0$?

No *formal* guidance is offered on the choice of the significance level. If different actions are envisaged depending on whether we accept or reject H, the attitude adopted is that the significance level should be chosen in the light of how serious it is (what it would cost) if *we reject H when it is true*. This type of error, termed **type I error**, is regarded as of prime importance. The significance level chosen expresses the risk we are prepared to accept of committing such an error. In some cases we may need to ask for this to be severely restricted, say to 0·1 per cent.; in other cases we may even accept a potential risk of 10 per cent. Thus consequences of different actions are relevant to the choice of test, but are used only subjectively and informally. They are not quantified, nor processed as part of the basic information.

This attitude is justified on the grounds of the 'intangibility' of costs and consequences, as we have already remarked. An alternative is to declare no significance level at the outset; but to evaluate the test statistic and to determine the probability of getting such a value or *a more extreme one* (in the spirit of the rejection criterion) if H is true. This probability is called the **critical level** of the test. In the example it is

$$P\{|\overline{X} - \mu_0| > |\overline{x} - \mu_0| \,|\, \mu = \mu_0\}.$$

The critical level is often interpreted as a measure of how significant our actual test result has turned out to be. Thus suppose we conduct a 10% test of $H : \mu = \mu_0$, and find that $\sqrt{n}(\overline{x} - \mu_0)/\sigma = 2\cdot58$. We obviously reject H for such a test. The test was designed to limit the risk of type I error to at amost 10 per cent. But the critical level is only 1 per cent, so we would have rejected even when the maximum risk of type I error that we can tolerate is only 1 per cent. To quote the critical level is to provide, in an aggregate sense, a more precise statement of our findings about the propriety of H.

As described, the test of significance is concerned merely with limiting the risk of type I error. But for any significance level a vast assortment of tests can be constructed with varying rejection criteria, and based on a multitude of different test statistics. All that is required is that P (reject $H \,|\, H$ true) $= \alpha$, where α is the significance level. Even restricting attention to the statistic $\overline{X}$ various possibilities exist with no way (so far) of distinguishing between them. Consider the need for a 10 per cent test. The following three *rejection criteria* all meet this need:

(i) $|\overline{x} - \mu_0| > 1\cdot645\sigma/\sqrt{n},$

(ii) $\overline{x} - \mu_0 > 1\cdot282\sigma/\sqrt{n},$

or even

(iii) $|\overline{x} - \mu_0| < 0.126\sigma/\sqrt{n}.$

In each case the probability of rejecting H, *when it is true*, is obtained from the standarized normal distribution, $N(0, 1)$. It is precisely 0.1 in each case, as illustrated by the shaded regions in Figure 5.5.

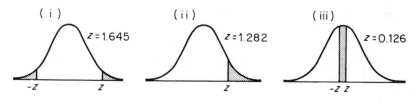

Figure 5.5

Obviously some further condition must be imposed. For we find in case (iii), for example, that the probability of rejecting H *is larger when H is true than when it is false*.

This is most unsatisfactory. We avoid such obvious anomalies by recognizing another type of error, **type II error**, which we must guard against. This is *the possibility of accepting H when it is false*. In general terms we want the probability of type II error to be as small as possible. At this stage we are *forced* to declare an alternative hypothesis, $\overline{H}$, as a statement of what is meant by H being 'false'.

Otherwise there is no probability model specified, against which to assess the implications of the falseness of H. We are now concerned with what has been termed above an **hypothesis test**; the outcome of which will be either to accept the declared hypothesis, H, or the specified alternative, $\overline{H}$.

In our example, $\overline{H}$ might take the *composite* form. $\overline{H}$: $\mu \neq \mu_0$. The probability of type II error is the probability that the test leads to acceptance of H when in fact $\overline{H}$ prevails. However, when $\overline{H}$ applies there is no unique probability in view of its *composite* nature. For *each* $\mu \neq \mu_0$ there is a probability, P (accept $H|\mu$), of type II error. Thus we introduce a function $\beta(\mu)$, called the **power function**, to assess the behaviour of the test with regard to type II error. This is defined as

$$\beta(\mu) = P\,(\text{reject } H|\mu) = 1 - P\,(\text{accept } H|\mu), \qquad (5.4.1)$$

and ideally we want to choose that test for which $\beta(\mu)$ is 'as large as possible'. Consider $\beta(\mu)$ in the three cases above. These typically appear as follows:

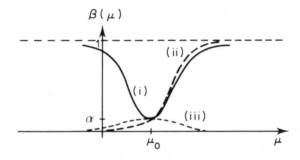

Figure 5.6

We immediately recognize a difficulty. Tests (ii) and (iii) have little appeal due to their very low power when $\mu < \mu_0$, or for all $\mu \neq \mu_0$, respectively. And yet (i) is not everywhere superior. In fact (ii) has greater power than (i) for $\mu > \mu_0$. So no clear-cut distinction can immediately be drawn. We can at least rule out (iii) as '*inadmissible*', in that it is '*dominated*' by (i), which has power greater than that of (iii) *for all* $\mu \neq \mu_0$. This idea of *admissibility* is important in hypothesis testing, and it has an equally (if not more) important counterpart in decision theory. Ideally we would wish to choose from amongst any set of tests under consideration that which has *uniformly greatest power*; indeed we would hope to go further and distinguish one test which is **uniformly most powerful** (UMP) in the complete class of tests of a given significance level. Uniformly most powerful tests represent the optimum behaviour we seek. We must consider whether it is realistic to hope to achieve this, and if so how we might derive UMP tests.

One notable feature of the hypothesis test is the asymmetric way in which H and $\overline{H}$ are regarded. In seeking to minimize the probability of type II error, with an overriding constraint placed on the probability of type I error, we are in a sense regarding H as more important than $\overline{H}$. That is, the *major* concern is to ensure that the probability of rejecting H when it is true is suitably small. As a consequence, the opposite concern for rejecting $\overline{H}$ when H is true may suffer. Consider μ close to μ_0, but not equal to μ_0, for test (i) in Figure 5.6. The power may be very poor.

Why take this asymmetric viewpoint? Simply because it is not easy to treat H and $\overline{H}$ on the same footing. If we reduce the probability of type I error we find (usually†) that the probability of type II error increases, and vice versa. It might make sense in some situation to compromise by constructing a procedure which specifies both risks at chosen points in H and $\overline{H}$, rather than limiting the type I risk to some prescribed level and tolerating what this implies about the type II risk. This is done for example in some sampling inspection and quality control problems. But the asymmetric test is the prevailing form of hypothesis test on the classical approach. This asymmetry has certain implications. We cannot interchange H and $\overline{H}$ without altering the whole nature of the test and this forces us to choose H and $\overline{H}$ quite specifically in relation to practical details of the problem under study.

The basic hypothesis H, whose incorrect rejection we wish primarily to safeguard, is usually chosen on a natural, *benefit of the doubt*, basis. A manufacturer of biscuits might claim that his packets weigh 200 g on average. To test this, as consumers, it would seem fair and reasonable to adopt this claim as the basic hypothesis H, and safeguard its incorrect rejection by only rejecting it if we have adequate grounds for doing so.

The different roles of H and $\overline{H}$ are of course distinguished in the terminology of the subject. We shall here talk of $\overline{H}$, in the usual way, as the **alternative hypothesis**. The basic hypothesis H is often called the **null hypothesis**. We shall instead refer to it as the **working hypothesis**, in view of the lack of agreement in the literature as to whether *null* necessarily implies *simple*. Such a restriction would hamper the general development of the following sections. Indeed, we should note that a simple hypothesis is of somewhat limited practical interest. In the biscuit example we may not be really concerned with the hypothesis $H : \mu = 200$. In different circumstances $H : 199 < \mu < 201$, or $H : \mu \geqslant 200$, may be more realistic. Whilst the latter type of one-sided hypothesis does figure widely in studying hypothesis tests, the former is usually abstracted as a *simple* hypothesis $H : \mu = 200$, with a resulting element of unreality.

The final illustrative point concerns the choice of the test statistic. We chose to use $\overline{X}$, because of its desirable properties as a point estimator. Is this necessarily a legitimate extrapolation? The answer is essentially, yes! For example, we find for case (i) that if we increase the sample size, for the same significance level, the test becomes more powerful. See Figure 5.7(a). This merely reflects the *consistency* of

† Some exciting work by Barnard[22] and Robbins and Siegmund[23] on tests of *power one* uniformly throughout the alternative hypothesis region calls into question whether the conventional belief that we must inevitably trade off one type of risk against another is necessarily true.

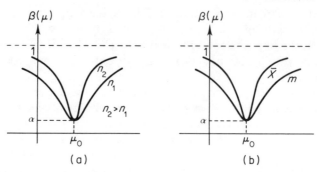

Figure 5.7

$\overline{X}$ and can be exploited in choosing the sample size n to yield a test of a *specified* power at some chosen value of $\mu \neq \mu_0$. Then again, if we construct a test of $H : \mu = \mu_0$ based on the sample median m, rather than on $\overline{X}$, we find that for common sample size and significance level the former is less powerful than the latter. See Figure 5.7(b). Here we are witnessing in another form the *greater efficiency* of $\overline{X}$.

The concept of *sufficiency* was seen to be crucial to the existence of optimum point estimators. In the area of hypothesis tests it again turns out to play a vital role in determining whether or not UMP tests exist.

5.4.2   Uniformly Most Powerful Tests

Let us consider now to what extent it is possible to achieve the ideal advanced in the previous section: that of a UMP test. Some of the central results in this area will be reviewed, but no formal proofs are presented. A detailed treatment of the material summarized below, and relevant extensions, is given by Lehmann (1962).

We start with a more formal description of the hypothesis test. The *working hypothesis H*, and the *alternative hypothesis* $\overline{H}$, serve to partition the *parameter space* Ω. Under H, θ lies in a subspace ω; under $\overline{H}$, θ lies in the complementary subspace, $\Omega - \omega$.

$$H : \theta \in \omega, \qquad \overline{H} : \theta \in \Omega - \omega.$$

The *sample space* $\mathscr{X}$ is also partitioned by the test procedure into two complementary subspaces, S_0 and S_1: the *critical region* and *non-critical region*, respectively. The test proceeds on the policy:

reject H, if $x \in S_0$;   accept H, if $x \in S_1$ ($\mathscr{X} - S_0$).

The choice of test $\{S_0, S_1\}$ is guided by the desire to safeguard the *incorrect rejection* of H, by assigning a *significance level* α, to be an upper limit to the probability of incorrect rejection of H. That is,

$$P(X \in S_0|\theta) \leqslant \alpha, \qquad \text{all } \theta \in \omega. \tag{5.4.2}$$

Among tests $\{S_0, S_1\}$ of level α, that is satisfying (5.4.2), one is judged better than another in terms of how good it is at guarding against the second type of error, *of accepting H when it is false*. Consider a particular $\theta \in \Omega - \omega$. If in one level-$\alpha$ test it happens that $P(X \in S_1|\theta)$ is less than it is in a second level-α test, the first test is adjudged better at the particular θ value; we say that at θ it has greater **power**, $P(X \in S_0|\theta)$. [$P(X \in S_0|\theta)$ is the probability of correctly rejecting H, for $\theta \in \Omega - \omega$]

The ultimate aim is to construct a test $\{S_0, S_1\}$ satisfying (5.4.2) and for which $P(X \in S_0|\theta)$ is maximized *simultaneously for all* $\theta \in \Omega - \omega$. This, if it exists, is the **UMP level-α test** of H against $\overline{H}$.

We must note that in defining a level-α test by means of (5.4.2) there is no guarantee that the probability of incorrectly rejecting H will be as large as α for any value of $\theta \in \omega$. Accordingly we introduce a further concept. A level-α test $\{S_0, S_1\}$ is said to have **size**, ξ, where

$$\xi = \sup_{\theta \in \omega} P(X \in S_0|\theta).$$

The main point at issue must be to determine whether we can reasonably expect to encounter situations where it is possible to maximize the power of a level-α test at all values of $\theta \in \Omega - \omega$ simultaneously. We shall approach this general enquiry in simple stages.

In their early writings on hypothesis testing, Neyman and Pearson considered the most basic example, of testing *one simple hypothesis against another*. In this situation Ω consists of just two members, θ_0 and θ_1. H declares that $\theta = \theta_0$ is the true probability model whilst $\overline{H}$ states the opposite, that $\theta = \theta_1$. The likelihood function, for data x, takes just two forms, $p_0(x)$ or $p_1(x)$, respectively. Neyman and Pearson suggested that it was sensible to base the acceptance or rejection of H on the relative values of $p_0(x)$ and $p_1(x)$, that is to reject H only if the **likelihood ratio** $p_0(x)/p_1(x)$ is sufficiently small. This amounts to rejecting H only if the sample data, x, are sufficiently more probable when $\theta = \theta_1$ than when $\theta = \theta_0$. Again we must take care over the interpretation of this criterion; the likelihood ratio is intended to measure the relative chances of x arising, not the relatively probabilities of θ_0 and θ_1 (however appealing this latter view might seem)!

When considering general hypotheses $H:\theta \in \omega$ and $\overline{H}:\theta \in \Omega - \omega$, Neyman and Pearson extended this intuitive criterion to the ratio

$$\max_{\theta \in \omega} p_\theta(x) / \max_{\theta \in \Omega - \omega} p_\theta(x). \tag{5.4.3}$$

They proposed the extended principle of rejecting H only if (5.4.3) is sufficiently small, and we shall consider the implications of this in some detail as a practical means of test construction when we discuss **likelihood ratio tests** in §5.4.3.

But whilst presenting this principle as a natural intuitive basis for testing H against $\overline{H}$ in general, they subsequently showed that in the particular case of testing one *simple* hypothesis against another it was in fact the *optimum* procedure. This is expressed in the celebrated Neyman–Pearson lemma.

Neyman–Pearson Lemma.† *If a partition $\{S_0, S_1\}$ of the sample space $\mathscr{X}$ is defined by*

$$S_0 = \{x\colon\ p_1(x) > kp_0(x)\} \qquad S_1 = \{x\colon\ p_1(x) < kp_0(x)\}$$

with $\int_{S_0} p_0(x) = \alpha$; then $\{S_0, S_1\}$ is the most powerful level-α (size-α) test of $H\colon \theta = \theta_0$ against $\overline{H}\colon \theta = \theta_1$.

There is, of course, no question of *uniformly* greatest power in this simple situation since $\overline{H}$ is *simple*. However, the importance of the result is that it can be used to demonstrate the existence, and form, of *uniformly most powerful tests* in certain more complicated situations, where $\overline{H}$ is composite. Not least important in this effect is the fact that it really is possible to encounter UMP tests, a point we had not previously established. Consider the following example.

Example 5.19. $x_1, x_2, \ldots, x_n$ *is a random sample from $\mathbf{N}(\mu, \sigma^2)$, where σ^2 is known. We wish to test two simple hypotheses, $H\colon \mu = \mu_0$ against $\overline{H}\colon \mu = \mu_1$ $(> \mu_0)$, on a level-α test. By the lemma the most powerful test has critical region*

$$S_0 = \{x\colon\ \exp[\overline{x}(\mu_1 - \mu_0)/\sigma^2 - (\mu_1^2 - \mu_0^2)/2\sigma^2] > k\}.$$

But this means that we reject H if $\overline{x} > c$, for a suitable constant c, i.e. when $\overline{x}$ is particularly large, which seems a sensible criterion. Notice how the resulting test statistic, $\overline{X}$, is sufficient; also $\overline{X}$ is the MVB estimator of μ.

Now c must be chosen so that $P(\overline{X} > c | \mu_0) = \alpha$, which gives $c = \mu_0 + z_{2\alpha}\sigma/\sqrt{n}$, where $z_{2\alpha}$ is the double-tailed 2α-point of $\mathbf{N}(0, 1)$. So we have the most powerful level-α (size-α) test of H against $\overline{H}$.

But the interesting point is that c does not explicitly involve μ_1. *So the same test is optimum whatever the value of μ_1 and we conclude that this level-α (size-α) test is uniformly most powerful for testing $H\colon\ \mu = \mu_0$ against the composite alternative hypothesis $\overline{H}\colon\ \mu > \mu_0$.*

Thus we have obtained a UMP test for a particular form of **one-sided** composite alternative hypothesis. If we take $\alpha = 0\cdot1$ in Example 5.19 we have precisely the test given as case (ii) in the previous section.

† The enunciation of this lemma, and similar results later, is not strictly correct in one respect. If X has a discrete component it *may* not be possible to exactly satisfy the size-α condition. This is easily remedied by introducing the idea of a **randomized test**, to cover cases where the event $p_1(x) = kp_0(x)$ has non-zero probability and we cannot achieve exactly some prescribed size α for the test. Here we *randominize* our decision for any x satisfying $p_1(x) = kp_0(x)$ by assigning such an x to S_0 with some probability ϕ ($0 < \phi < 1$): choosing ϕ to yield the required size α for the test. We shall not pursue this matter further, but see, for example, Lehmann (1959, Chapter 3).

It will be obvious that an analogous UMP *one-sided* level-α test can be obtained for H: $\mu = \mu_0$ and $\overline{H}$: $\mu < \mu_0$. Both of these tests also have level α in the **two-sided** situation with H: $\mu = \mu_0$ and $\overline{H}$: $\mu \neq \mu_0$, and each is uniformly most powerful over a particular region: viz. one, or other, side of μ_0. But the tests are not the same test, of course, and we can conclude that there is consequently *no* UMP test in the two-sided situation. So in spite of the previously discussed intuitive advantages of test (i) of the previous section (see Figure 5.5) for the two-sided case, *it cannot be optimum.*

This illustrates a general effect. When testing a *simple* working hypothesis about a *single* parameter against a *one-sided* alternative hypothesis, a UMP test may exist. In fact *it is guaranteed to do if a singly sufficient statistic exists for* θ. When the alternative hypothesis is two-sided *there is in general no UMP test*. This seems a somewhat bleak conclusion when viewed in the light of the different types of practical situation of interest (where alternative hypotheses may be two-sided, θ may have many components, $\overline{H}$ may be composite). To extend the attribution of optimum properties to tests in these wider important practical situations it is necessary to make stronger or more specific assumptions concerning the test situations, or test structures. There are basically two ways in which this can be done.

(a) We can make some particular assumptions about the probability model, and seek UMP tests for such assumed probability models. This limits the range of application of the resulting tests, but need not be a severe restriction.

(b) We can retain a perfectly general probability model, but place further conditions on the types of tests which are to be considered. Such conditions are motivated by essentially practical interests, in the sense that they constitute no serious restriction to the class of tests we should wish to consider *in terms of practical needs*. UMP tests may then exist within the restricted class of tests, and their non-existence in the wider class of all possible tests is no embarrassment.

Let us consider just some of the results available under the principles (a) and (b).

Restricted Probability Models

In certain situations the probability model, whilst not specified in any detail, does satisfy the general condition that it has **monotone likelihood ratio**. This is defined as follows for a single (component) parameter θ.

Suppose for any θ and θ', with $\theta < \theta'$, $p_\theta(x)$ and $p_{\theta'}(x)$ are not identical and $p_{\theta'}(x)/p_\theta(\theta)$ *is a non-decreasing function of a real-valued function* $\tilde{\theta}(x)$ *of the data. Then* $p_\theta(x)$ *is said to have monotone likelihood ratio in* $\tilde{\theta}(x)$.

In this situation it can be shown that a *UMP level-α test exists for testing*

$H: \theta \leqslant \theta_0$ against $\overline{H}: \theta > \theta_0$. This has critical and non-critical regions

$$S_0 = \{x: \tilde{\theta}(x) > c\}, \qquad S_1 = \{x: \tilde{\theta}(x) < c\},$$

where c is chosen to ensure $P(X \in S_0/\theta_0) = \alpha$. (Thus the test also has size α.)

So we now encounter a UMP test of a *composite one-sided* hypothesis H, against a *composite one-sided alternative* hypothesis $\overline{H}$. In fact, in Example 5.19, the probability model has monotone likelihood ratio in $\bar{x}$ and the test described in the example is UMP for the wider situation with $H: \mu \leqslant \mu_0$ and $\overline{H}: \mu > \mu_0$.

To understand the practical implications of this assumed probability structure, we need to know in what situations it arises. It is quite frequently encountered, and again the idea of sufficiency is crucial. We find that single parameter families of probability distributions with monotone likelihood ratio arise as conditional distributions given a singly sufficient statistic. In particular the single parameter *exponential family* [see (5.3.8)] with $A(\theta) \equiv \theta$ (that is, with so-called *natural parameter*, θ) has monotone likelihood ratio in $B(x)$, and UMP tests of one-sided hypotheses exist with rejection criterion based on the magnitude of $B(x)$.

But if we are prepared to make this even stronger assumption that we are sampling from the single parameter exponential family,† UMP tests exist even for some two-sided hypotheses. We have the following result.

For testing H; $\theta \leqslant \theta_1$ or $\theta \geqslant \theta_2(\theta_1 < \theta_2)$ against $\overline{H}$: $\theta_1 < \theta < \theta_2$ a UMP level-α test exists with

$$S_0 = \{x: c_1 < B(x) < c_2\}, \qquad S_1 = \{x: B(x) < c_1 \text{ or } > c_2\},$$

where c_1 and c_2 are determined by $P(X \in S_0|\theta_1) = P(X \in S_0|\theta_2) = \alpha$. (Once again this UMP level-α test actually has size α.)

But no UMP test exists if we interchange H and $\overline{H}$!

Restricted Types of Test

The alternative means of obtaining optimum tests is to place some limitations on the types of test which are to be considered.

Referring back to the previous section, we considered testing $H: \mu = \mu_0$ against $\overline{H}: \mu = \mu_0$ on the basis of a random sample of n observations from $N(\mu, \sigma^2)$, with σ^2 known. Of the three cases examined, one of them [case (iii)] was obviously unsatisfactory. The symmetric test (i) has more appeal, in having increasing power as μ departs more and more from μ_0. And yet it is not UMP, in being dominated for $\mu > \mu_0$ by the asymmetric test (ii). (See Figure 5.6.) Indeed we have now seen that (ii) represents the UMP test of $H: \mu \leqslant \mu_0$ against $\overline{H}: \mu > \mu_0$ and no UMP test can exist for the two-sided situation.

However, there is a strong reason why test (ii) should not be entertained for

---

† The term 'single parameter' needs qualification—strictly what is meant is a single *unknown* parameter. $N(\mu, \sigma^2)$, with σ^2 known, is one example.

testing $H:\mu = \mu_0$ against $\overline{H}:\mu = \mu_0$, in spite of its optimum behaviour for $\mu > \mu_0$. This is its unacceptable performance for $\mu < \mu_0$, where the probability of rejecting H becomes smaller and smaller as μ goes further from μ_0. Indeed *this probability is not even as large as the significance level of the test.* This observation suggests that in order to avoid such unsatisfactory behaviour it would be sensible to consider only those tests where

$$P(\text{reject } H|H \text{ false}) \geqslant P(\text{reject } H|H \text{ true}) \qquad \text{(for all } \theta)$$

—surely a modesty but important restriction. Such tests are said to be **unbiased**. On this restriction, test (ii) is not a contender in the above situation, and the question of whether an optimum *unbiased* test exists needs to be studied. Thus we now seek **uniformly most powerful unbiased** (UMPU) tests.

A general level-α test of $H:\theta \in \omega$ against $\overline{H}:\theta \in \Omega - \omega$, with critical and non-critical regions, S_0 and S_1, respectively, is said to be unbiased if

$$\begin{aligned} P(X \in S_0|\theta) \leqslant \alpha, \qquad \theta \in \omega, \\ P(X \in S_0|\theta) \geqslant \alpha, \qquad \theta \in \Omega - \omega. \end{aligned} \qquad (5.4.4)$$

(In Chapter 7 we shall see this as a particular form of a wider concept of *unbiased decision rules.*)

Restricting attention to unbiased tests we find that optimum tests now exist in further situations. Thus it happens that *for the single parameter exponential family* (where Ω is the *natural* parameter space) *a UMPU test, of level α, exists for testing* $H:\theta = \theta_0$ *against* $\overline{H}:\theta \neq \theta_0$. *This has*

$$S_0 = \{x:B(x) < c_1 \text{ or } > c_2(c_1 < c_2)\}, \qquad S_1 = \{x:c_1 < B(x) < c_2\},$$

where c_1 and c_2 are determined by $P(X \in S_0|\theta_0) = \alpha$ and $|(d|d\theta)\{P(X \in S_0|\theta)\}|_{\theta_0}$ $= 0$: this latter condition expressing the fact that the power function has a minimum at $\theta = \theta_0$.

Example 5.20. $x_1, x_2, \ldots, x_n$ *is a random sample from* $N(\mu, \sigma^2)$ *with σ^2 known. Here we have a member of the exponential family with $B(x) \propto \overline{x}$. The UMPU test, at level α, of $H:\mu = \mu_0$ against $\overline{H}:\mu \neq \mu_0$ has*

$$S_0 = \{x: \overline{x} < c_1 \text{ or } > c_2\}, \qquad S_1 = \{x: c_1 < \overline{x} < c_2\},$$

where $P[c_1 < \overline{x} < c_2|\mu_0] = 1 - \alpha$ *and* $c_1 + c_2 = 2\mu_0$. *This implies that* $c_1 = \mu_0 - z_{2\alpha}\sigma/\sqrt{n}$, $c_2 = \mu_0 + z_{2\alpha}\sigma/\sqrt{n}$. *Suppose $\alpha = 0\cdot10$, then $c_1 = \mu_0 - 1\cdot645\sigma/$* $\sqrt{n}$ *and $c_2 = \mu_0 + 1\cdot645\sigma/\sqrt{n}$, so that the test (i) of the previous section is in fact the UMPU test.*

An alternative means of restricting the types of tests to be considered is also based on natural practical requirements. It seems sensible to ask that certain transformations of the sample space should not essentially alter the statistical inferences that we draw. For example a change of scale or origin in the observations should not affect the conclusions of a test. Thus we might restrict attention only to those tests which are *invariant* under prescribed types of transformation. Within this restricted class we again find that optimum tests may exist, **uniformly most powerful invariant** (UMPI) tests, where they do not do so in the unrestricted environment. The restrictions of *unbiasedness* and *invariance* tend to complement each other, in that UMPU tests may exist where UMPI tests do not, and vice versa. See Lehmann (1959), Chapter 6. However, when both exist in some situation, they coincide.

Where optimum tests do not exist either in these restricted frameworks, or as a consequence of some special probability structure, more pragmatic criteria are adopted. As one example, **locally most powerful** tests may be employed, which ensure that the probability of rejection is maximized for those $\theta \in \Omega - \omega$ *which are in some sense close to the working hypothesis region ω*. Again see Lehmann (1959), Chapter 8.

In much of the discussion so far we have acted as if θ has just a single (unknown) component. This is most unrealistic; in the majority of practical situations θ has several components and we will wish to test joint hypotheses about these components. Consider $N(\mu, \sigma^2)$ for example. There are two components here, namely μ and σ^2, and it is unlikely that we shall have any precise knowledge of the value of either of them. We might wish to test $H: \mu = \mu_0; \sigma \leqslant \sigma_0$ against $\overline{H}: \mu \neq \mu_0, \sigma > \sigma_0$; or $H: \mu = \mu_0$ against $\overline{H}: \mu \neq \mu_0$, when σ^2 *is unknown*. In the second case σ^2 is termed a **nuisance parameter**; its value is not of interest to us, but being unknown it affects the form and properties of any test of H against $\overline{H}$.

In the main no global UMP tests exist in such multi-parameter problems. But again UMPU or UMPI tests may sometimes be found, and the existence of sufficient statistics again turns out to be crucial. Rather than attempt to summarize the results which are available, we shall merely report a few illustrative results.

When sampling from $N(\mu, \sigma^2)$, with σ^2 unknown, the customary tests of $H: \mu = \mu_0$ against $\overline{H}: \mu \neq \mu_0$, or of $H: \mu \leqslant \mu_0$ against $\overline{H}: \mu > \mu_0$ are the familiar *one-sided* and *two-sided t-tests*. These are in fact optimum in the sense of being both UMPU and UMPI. Likewise the usual tests of σ^2, with μ unknown, based on the χ^2 distribution are both UMPU and UMPI, although the *unbiased* two-sided test is not of course based on equal ordinates, or equal tail areas, of the χ^2 distribution. Since we are here considering a member of the exponential family of distributions, we find in fact a more universal optimality in some situations. For example, a UMP test exists for $H: \sigma \leqslant \sigma_0$ against $\overline{H}: \sigma > \sigma_0$, *but not for H and $\overline{H}$ interchanged*. For one-sided tests of μ when σ^2 is unknown it turns out that a UMP test exists for $H: \mu \leqslant \mu_0$ against $\overline{H}: \mu > \mu_0$ (or vice versa) *provided $\alpha \geqslant \frac{1}{2}$*. But this latter result is hardly of much practical relevance!

5.4.3. Construction of Tests

We come now to the question of how to construct tests in practical situations. As with optimum estimators, so the study of optimum tests has pinpointed the circumstances in which these exist and has demonstrated their form. Thus in some cases, where optimum tests exist, the detailed nature of the test can be derived from the general results. All that is involved are the (sometimes substantial) probability calculations needed to determine the specific rejection criterion for a test of particular H and $\overline{H}$, at the prescribed significance level α. We saw, for example, that the UMP level-α test of $H : \mu \leqslant \mu_0$ against $\overline{H} : \mu > \mu_0$ in the case of sampling from $N(\mu, \sigma^2)$, with σ^2 known, implies rejection of H when $\bar{x} > \mu_0 + z_{2\alpha}\sigma/\sqrt{n}$.

So where certain types of optimum test exist the specific form can often be derived from the general characterization results illustrated in the previous section. But this covers only a few situations; optimum tests are rare. What is needed is a universal procedure (cf. maximum likelihood in point estimation) for constructing tests in different situations, not tied to optimality considerations. Such a procedure, intuitively based but providing some reasonable performance characteristics in many cases, is that which yields:

Likelihood Ratio Tests

We saw, in terms of the *Neyman–Pearson lemma*, how the most powerful test of one simple hypothesis against another is based on the value of the ratio of the likelihoods under the two hypotheses. Even earlier in historical sequence, Neyman and Pearson had proposed the value of the likelihood ratio as a natural basic criterion for acceptance or rejection in any significance test. It was so proposed as a practical procedure, without specific regard to the properties of the resulting tests. As such it is still the principal method that is adopted in constructing tests, and we shall consider briefly how it operates and what may be said of the tests, **likelihood ratio tests**, that it produces.

Suppose we wish to test $H : \theta \in \omega$ against $\overline{H} : \theta \in \Omega - \omega$, on the basis of sample data x. It would seem intuitively reasonable that the data provide greater support for H in preference to $\overline{H}$ the larger is the likelihood $p_\theta(x)$ under H than under $\overline{H}$. But $p_\theta(x)$ is not unique in either case, it takes different values for different values of θ. A procedure is proposed where the ratio of the *maximized* likelihoods (over H and $\overline{H}$) is employed as a basis for accepting or rejecting H. Specifically, the level-α *likelihood ratio test* demands rejection of H if this ratio λ is sufficiently small. This is equivalent to using a critical region for rejection

$$S_0 = \{x : p_{\hat{\theta}_0}(x)/p_{\hat{\theta}}(x) < c\}, \tag{5.4.5}$$

where

$$p_{\hat{\theta}}(x) = \max_{\theta \in \Omega} p_\theta(x) \quad \text{and} \quad p_{\hat{\theta}_0}(x) = \max_{\theta \in \omega} p_\theta(x),$$

Thus $\hat{\theta}$ is the *maximum likelihood estimator of* θ, and $\hat{\theta}_0$ *the constrained maximum likelihood estimator* when θ is assumed to lie in the subspace ω. The constant c is constrained by the condition

$$P(X \in S_0|\theta) \leqslant \alpha, \qquad \theta \in \omega. \tag{5.4.6}$$

and usually we ask that equality holds for some $\theta \in \omega$, that is, that the test has *size* α.

Example 5.21. $x_1, x_2, \ldots, x_n$ *is a random sample from* $\mathbf{N}(\mu, \sigma^2)$ *with* μ, σ^2 *unknown.*

$$H: \mu = \mu_0; \qquad \overline{H}: \mu \neq \mu_0$$

Here σ^2 *is a nuisance parameter.*

Now $\hat{\mu} = \bar{x}$, $\hat{\sigma}^2 = (1/n)\sum_1^n(x_i - \bar{x})^2$, *and the constrained maximizing values under H are* $\hat{\mu}_0 = \mu_0$, $\hat{\sigma}^2 = (1/n)\sum_1^n(x_i - \mu_0)^2$.

Thus

$$\lambda = p_{\hat{\mu}_0,\, \hat{\sigma}_0^2}(x)/p_{\hat{\mu},\, \hat{\sigma}^2}(x) = (\hat{\sigma}/\hat{\sigma}_0)^n.$$

So we reject H if $(\hat{\sigma}/\hat{\sigma}_0)^n \leqslant c$.
But $\hat{\sigma}/\hat{\sigma}_0 = [1 + t^2(n-1)]^{-1/2}$, *where* $t = \sqrt{n}(\bar{x} - \mu_0)/s$ *with*

$$s^2 = \frac{1}{n-1}\sum_1^n(x_i - \bar{x})^2.$$

So $\lambda < c$ *is equivalent to* $|t| > k$, *for suitably chosen k. For a level-α test k must be chosen to satisfy*

$$P(|t| > k|\mu_0) = \alpha.$$

This yields the familiar two-sided *t-test, which we have already noted as being* UMPU.

It is interesting to see that the likelihood ratio test can have optimum properties outside the elementary case of two simple hypotheses. But this effect is not so surprising. The test is based on the maximum likelihood estimator which we have seen depends on minimal sufficient statistics, and which is in fact optimum if an optimum estimator exists. What we are witnessing is the transfer of these properties to the hypothesis testing situation. But as for maximum likelihood estimation there is no guarantee that the likelihood ratio test always behaves well for small samples. It will often be biased, and may have anomalous properties in special situations.

On the other hand it does offer some optimum properties in an asymptotic sense, again reflecting the behaviour of maximum likelihood estimators. Commonly H amounts to a reduction in the dimension of the parameter space Ω

due to placing linear constraints on the components of θ. Suppose θ has k components, and H imposes $m \leqslant k$ independent linear constraints so that (possibly with a reparameterization) we have $\theta_1, \ldots, \theta_m$ forced to assume specified values. That is,

$$H: \theta_1 = \theta_{1_0}, \ldots, \theta_m = \theta_{m_0}.$$

If the sample size is large then, as a result of the asymptotic normality of the maximum likelihood estimator, we find that the likelihood ratio statistic λ has a distribution essentially independent of that from which the sample has arisen. In fact $-2 \log \lambda$ is asymptotically distributed according to a χ^2 distribution with m degrees of freedom, and the detailed form of the rejection criterion for a level-α test immediately follows. This test is asymptotically UMPI. In the special case of the linear model (5.3.13) with normal error structure these results apply exactly for any sample size n $(n > k)$.

In Summary then, likelihood ratio tests.

(i) quite often produce reasonable, even optimum, procedures in finite sample size situations,

(ii) are asymptotically well behaved, and tractable, in being based on the χ^2 distribution in an important class of problems.

A comparison of likelihood ratio tests with other possible tests is made by Peers[24].

Monte Carlo, or Simulation, Tests

A simple empirical form of significance test has been receiving some attention in recent years. First proposed by Barnard[25], its properties have been investigated and illustrated by, among others, Hope[26], Besag and Diggle[27] and Marriott[28]. The principle of the so-called *Monte Carlo test*, or *simulation test*, is as follows. We want to test some simple hypothesis H on the basis of the observed value t of an appropriate test statistic, T. Under H we simulate a set of m samples of similar size to that under consideration and calculate the corresponding values $t_1, t_2, \ldots, t_m$ of T. We have then merely to observe the rank of t in the augmented set of values $\{t, t_1, t_2, \ldots, t_m\}$ to determine the exact significance level of the test.

The approach has obvious attractions: avoidance of the need to deal with complicated distributional behaviour of T, easy explanation and interpretation for the layman. Modern computing facilities make it far more feasible than previously, especially as m in the region of only 100 or so may suffice (Besag and Diggle[27]). The question of the power of the test against specified alternatives needs to be considered in particular cases. The test also has the initially strange property of a *variable* critical region depending on the simulation set. Thus, depending on $(t_1, t_2, \ldots, t_m)$, a specific value t may or may not lead to rejection. (See Marriott[28], Hope[26].)

5.5   REGION AND INTERVAL ESTIMATES

The third component in the classical approach is the use of the data x to construct a region within which the parameter θ may reasonably be expected to lie. Thus, instead of a one-to-one mapping from $\mathscr{X}$ to Ω as is involved in point estimation, $\mathscr{X}$ is mapped into a family of subsets of Ω, $\mathscr{S}(X)$. Just how this family is to be chosen, and what is the interpretation and application of its members, needs to be carefully considered. One thing is clear: the *function* of such a procedure. It is entirely *inferential*, in concluding on the basis of data x that θ lies in a subspace $\mathscr{S}(x) \subset \Omega$.

Ideally we should again proceed through the three stages of definitive criteria, bases for optimality and methods of construction. Instead we shall telescope these by dealing only with illustrative examples which touch on all three aspects.

The common expression of a region or interval estimate in the classical approach is in terms of **confidence regions** and **confidence intervals**. The idea is due to Neyman in the late 1930s, although some hints of an apparently similar concept appear as early as 1814 in the work of Laplace. Developing alongside this idea was a parallel concept of **fiducial regions** and **fiducial intervals** due to Fisher. In their practical expression they coincided for the single parameter problems to which they were originally applied, and it was some time before a basic distinction of attitude and end-result became apparent. This arose through the celebrated Behrens–Fisher problem of drawing inferences about the difference in the values of the means of two normal distributions with different unknown variances. There resulted from these conflicting attitudes what was perhaps the most turbulent controversy in the history of statistics, as the research literature of the time demonstrates. We shall separate the protagonists here, by considering the *confidence* concept at this stage as essentially a development (or re-emphasis) of the test of significance. At a later stage (in Chapter 8) we take up the rather more nebulous concept of *fiducial probability* and *fiducial estimation*.

We start with the idea of a one-sided confidence interval for a *scalar* parameter θ. Suppose we wish to use the data x to declare a *lower bound* for θ. We cannot do this with any certainty, unless we adopt the trivial bound $-\infty$. But we could define a value $\underline{\theta}(x)$ with the property that

$$P\{\underline{\theta}(X) \leqslant \theta | \theta\} \geqslant 1 - \alpha, \qquad \text{for all } \theta. \tag{5.5.1}$$

If α is close to zero then $\underline{\theta}(X)$ is most likely to be less than θ in an aggregate sense. Accordingly we call $\underline{\theta}$ a **lower confidence bound** for θ with **confidence level** $1 - \alpha$. Alternatively, $[\underline{\theta}, \infty)$ is a **lower one-sided** $1 - \alpha$ **confidence interval** for θ.

In practice we will substitute the observed x in $\underline{\theta}$, to obtain a numerical value for this lower confidence bound, as $\theta(x)$. But it is important to recognize the proper probability interpretation of a *confidence* statement, namely that in the long run a proportion $1 - \alpha$ of values $\underline{\theta}(x)$ will be less than θ. Whether a *particular* $\underline{\theta}(x)$ in some current situation is less, or greater, than θ is entirely uncertain. We have only any transferred assurance that arises from the long-run property (5.5.1).

Usually $\inf P\{\underline{\theta}(X) \leqslant \theta | \theta\}$ will equal $1 - \alpha$, and is called the **confidence coefficient**. In some cases $P\{\underline{\theta}(X) \leqslant \theta | \theta\}$ is independent of θ, in which case the confidence coefficient is attained at every value of θ.

Example 5.22. $x_1, x_2, \ldots, x_n$ *is an independent random sample from* $\mathbf{N}(\mu, \sigma^2)$ *with* σ^2 *known. Consider*

$$\underline{\theta}(x) = \bar{x} - z_{2\alpha}\sigma/\sqrt{n}.$$

We have

$$P\{\bar{X} - z_{2\alpha}\sigma/\sqrt{n} < \mu | \mu\} = 1 - \alpha,$$

so that $\underline{\theta}$ *is a lower* $1 - \alpha$ *confidence bound for* μ, *with confindence coefficient* $1 - \alpha$, *attained whatever is the value of* μ.

In a similar manner we can obtain an *upper* $1 - \alpha$ confidence bound $\bar{\theta}$. Then, combining $\underline{\theta}$ at confidence level $1 - \alpha_1$ with $\bar{\theta}$ at confidence level $1 - \alpha_2$ we obtain a **two-sided** $1 - \alpha_1 - \alpha_2$ **confidence interval** $(\underline{\theta}, \bar{\theta})$ satisfying

$$P\{\underline{\theta}(X) < \theta < \bar{\theta}(X) | \theta\} \geqslant 1 - \alpha_1 - \alpha_2, \qquad \text{for all } \theta. \qquad (5.5.2)$$

If both the lower and upper bounds act with probabilities independent of θ, then we can obtain a two-sided $1 - \alpha$ confidence interval with *confidence coefficient* $1 - \alpha$ attained for all θ.

Several questions arise from these general observations. Obviously a variety of confidence bounds will exist in any situation depending on the statistic on which they are based. In Example 5.22 a lower bound was constructed in terms of the *sample mean* $\bar{X}$; but we could alternatively have used the median, m. Then again if we consider a two-sided $1 - \alpha$ confidence interval, *based on a particular statistic*, there remains open the choice of α_1 and α_2 satisfying $\alpha_1 + \alpha_2 = \alpha$. So again we must ask what *criteria* exist for the choice of a particular confidence bound or interval; is there some optimum choice, what are its characteristics and how should we construct confidence bounds or intervals in specific situations?

Consider two-sided confidence intervals. One convenient principle is to choose $\alpha_1 = \alpha_2$ to produce so-called *central* confidence intervals based on *equal tail-area probabilities*. If the sample statistic on which the confidence interval is based has a distribution which is symmetric about θ, the confidence interval may be symmetric about that statistic. Compare the central confidence interval for μ in $\mathbf{N}(\mu, \sigma^2)$ based on $\bar{x}$, which is symmetrical about $\bar{x}$, with that for σ^2 based on the sample variance s^2, which is asymmetrical about s^2.

But the appeal of *central* confidence intervals is entirely intutive. More basic criteria for good confidence intervals have been proposed.

Shortest Confidence Intervals

In considering confidence intervals of the form $(\bar{x} - \alpha, \bar{x} + \beta)$, with $\alpha > 0$ and $\beta > 0$, for μ in $N(\mu, \sigma^2)$, it is easy to show that any confidence level $1 - \alpha$ the *central* interval (with $\alpha = \beta$) has the shortest length. However, in other situations, it is not always the *central* interval which has this property.

The shortness of $(\theta, \bar{\theta})$ seems a desirable practical feature for a confidence interval, and is one basis for choice between different possible confidence intervals, or for the designation of a best interval. One problem, however, is that the length of the interval may vary with the actual observation x on which it is based. If this is so we can widen the criterion, and express it in terms of the *expected* length of the interval, $E\{(\bar{\theta} - \theta)|\theta\}$.

We find that intervals based on $\partial L/\partial \theta$ have asymptotically minimum expected length; a result which ties in naturally with the optimum asymptotic properties of the maximum likelihood estimator and with the nature of MVB estimators.

Indeed the asymptotic distribution of the score $\partial L/\partial \theta$ provides a convenient means of constructing confidence intervals. We have asymptotically that

$$\psi_\theta(x) = \frac{\partial L}{\partial \theta} \bigg/ \sqrt{I_s(\theta)} \sim N(0, 1),$$

and often $\psi_\theta(x)$ is monotone in θ thus enabling probability statements about ψ to be inverted to obtain confidence intervals, or bounds, for θ.

Most Accurate Confidence Intervals

An alternative criterion for choosing between different confidence intervals has been described under a variety of different labels. It is based on minimizing the probability that the confidence interval contains *incorrect* (or *false*) values of θ. On this basis we set out to obtain what Neyman referred to as 'shortest' confidence intervals. In the present context this is an inappropriate term—it has little to do with the physical size of the interval and we have already used the word with a different and more natural interpretation. Kendall and Stuart (1979) propose 'most selective' as a better description, but Lehmann's (1959) use of the term 'most accurate' seems best to represent the basic idea.

Consider the general concept of a confidence region $\mathcal{S}(X)$ of level $1 - \alpha$. That is

$$P\{\theta \in \mathcal{S}(X)|\theta\} \geqslant 1 - \alpha, \qquad \text{all } \theta \in \Omega. \tag{5.5.3}$$

[Note that $\mathcal{S}(X)$ is a *family* of subspaces of Ω, and that it is X, *not* θ, which is the random variable.]

Suppose we now consider some value θ' of the parameter which we assume to be *incorrect* or *false*. It is not the true value in the situation being studied. There will be many $\mathscr{S}(X)$ satisfying (5.5.3), and for each we can determine

$$P\{\theta' \in \mathscr{S}(X)|\theta\}.$$

The particular one, $\mathscr{S}_1(X)$ say, for which this probability of containing the false value θ' is a minimum is said to be the level-$(1-\alpha)$ confidence region which is **most accurate** at the value θ'. This is similar *in spirit* to seeking a test with greatest power at some value θ' in the alternative hypothesis subspace.

If we can go further and find a unique $\mathscr{S}_1(X)$ which simultaneously minimizes $P\{\theta' \in \mathscr{S}_1(X)|\theta\}$ for all θ' other than the true value, we call $\mathscr{S}_1(X)$ **uniformly most accurate** (UMA), and regard it as the optimum level-$(1-\alpha)$ confidence region.

The study of best confidence regions from this viewpoint must proceed by considering under what circumstance UMA confidence regions exist, what is their form, and so on. We shall not pursue this in detail. The similarity of approach to that in hypothesis testing, remarked on above, does in fact provide the answer. For we find that there is a natural duality between the two ideas, and the question of the existence (or non-existence) of UMA confidence regions is really just a re-expression of the same question about UMP tests. It turns out that in situations where UMP tests exist so do UMA confidence regions; where they do not, but restricted forms exist (such as UMPU tests), the same is true of confidence regions on a transferred restriction to, for example, UMA *unbiased* confidence regions. Furthermore the actual form of an optimum test directly generates the form of the associated optimum confidence region. We can illustrate this in the following way.

Suppose we construct a level-α test of $H:\theta = \theta_0$ and this has non-critical region S_1. As we consider different values for θ_0 we get a *set* of non-critical regions $S_1(\theta_0)$. If we now define

$$\mathscr{S}_1(X) = \{\theta : X \in S_1(\theta)\}$$

then $\mathscr{S}_1(X)$ is a confidence region for θ, of level $1-\alpha$.
This is so since $\theta \in \mathscr{S}_1(x)$ if and only if $x \in S_1(\theta)$, and thus

$$P\{\theta \in \mathscr{S}_1(X)|\theta\} = P\{X \in S_1(\theta)|\theta\} \geqslant 1-\alpha.$$

So we obtain a confidence region for θ as the set of values of θ for which we would accept H, for given x. But we can go further and show that if $S_1(\theta_0)$ corresponds to the UMP test of H against $\overline{H} : \theta \in \Delta \subset \Omega$; then $\mathscr{S}_1(X)$ is uniformly most accurate for $\theta \in \Delta$, i.e. it uniformly minimizes

$$P\{\theta' \in \mathscr{S}_1(X)|\theta\} \qquad \text{for all } \theta' \in \Delta.$$

Example 5.23. $x_1, x_2, \ldots, x_n$ *is a random sample from* $N(\mu, \sigma^2)$ *with* σ^2 *known. For a level-α test of* $H:\mu = \mu_0$ *against* $\overline{H}:\mu > \mu_0$ *we accept* H *if*

$$\bar{x} < \mu_0 + z_{2\alpha}\sigma/\sqrt{n}.$$

Thus $\bar{x} - z_{2\alpha}\sigma/\sqrt{n}$ *constitutes a level-*$(1 - \alpha)$ *lower confidence bound for* μ, *and since the test is* UMP *the lower confidence bound is* UMA *in the sense that*

$$P\{\bar{X} - z_{2\alpha}\sigma/\sqrt{n} < \mu'|\mu\}$$

is uniformly minimized for all $\mu' > \mu$.

For multi-parameter problems (i.e. where θ is a vector), the principles of hypothesis testing lead to corresponding confidence interval, or confidence region, inferences. Distinctions arise depending on whether we seek *marginal* confidence statements about a single component θ_i of θ, when other components are unknown; or whether we require a *joint* confidence statement relating to more than one (perhaps all) of the components of θ. The former interest leads to 'composite' confidence intervals or regions, so-called because the corresponding test is of a composite hypothesis. For several components we need to determine 'simultaneous' confidence intervals or regions. In simple cases these may be 'separable', e.g.

$$\{\underline{\theta}_1 < \theta_1 < \bar{\theta}_1; \underline{\theta}_2 < \theta_2 < \bar{\theta}_2\},$$

but, more often, the components are tied together through some function $g(\theta)$, e.g.

$$\{\underline{\theta} < g(\theta) < \bar{\theta}\},$$

in a way which renders interpretation problematical.

Conceptual difficulties in the confidence interval method arise in their most acute form for multi-parameter problems (although they can also arise with a scalar parameter—see, for example, Robinson[29]). Some details, and references, on multi-parameter confidence intervals are given by Kendall and Stuart (1979), and Seidenfeld (1979, Chapter 2) considers basic principles.

> I feel a degree of amusement when reading an exchange between an authority in 'subjectivistic statistics' and a practicing statistician, more or less to this effect:
> *The Authority*: 'You must not use confidence intervals; they are discredited!'
> *Practicing Statistician*: 'I use confidence intervals because they correspond exactly to certain needs of applied work.'
>
> Neyman[30]

5.6   ANCILLARITY, CONDITIONALITY, MODIFIED FORMS OF SUFFICIENCY AND LIKELIHOOD

We have already considered some aspects of the crucial role played by *sufficiency* in the classical approach to statistical inference. We recall that $\tilde{\theta}$ based on sample data x is *sufficient for* θ if the conditional distribution of x given $\tilde{\theta}$ does not depend on θ; $\tilde{\theta}$ is *minimal sufficient* if there is no other sufficient statistic θ' of dimension *lower* than that of $\tilde{\theta}$. If θ has dimension p then, effectively, the best we can hope for is to encounter a minimal sufficient statistic which also has dimension p with the associated advantages that arise with respect to estimation or hypothesis testing.

The fundamental importance of sufficiency is reflected in what is sometimes referred to as:

The Sufficiency Principle.   If data x arise from a model $p_\theta(x)$ and $\tilde{\theta}$ is minimal sufficient for θ, then identical inferences should be drawn from any two sets of data x_1 and x_2 which happen to yield the same value for $\tilde{\theta}$.

Thus, once we know the value of $\tilde{\theta}$ any addition representation of the data (not expressible as a function of $\tilde{\theta}$) is uninformative about θ provided that we are using the correct probability model. But is such 'additional representation' of no value to us? One function of the residual information in x, *given* $\tilde{\theta}$, might be to examine the adequacy of the assumed model. This and other possible advantages lead us to define (in a sense almost complementary to that of a sufficient statistic) the notion of an **ancillary statistic—a function of the data whose marginal distribution does not depend on** θ.

Example 5.24. If $X_1, X_2, \ldots, X_n$ are independent $\mathbf{N}(\theta, 1)$ then $\overline{X}$ is singly sufficient for θ. However $X_1 - X_2$ has a distribution $\mathbf{N}(0, 2)$ which does not depend on θ—it is ancillary. Indeed $T = (X_1 - X_2, X_1 - X_3, \ldots, X_1 - X_n)$ is an $(n-1)$-dimensional ancillary statistic.

The following modification of this example illustrates one of the principle features of ancillarity.

Example 5.25. $X_1, X_2, \ldots, X_N$ are independent $\mathbf{N}(\mu, 1)$ but the number of random variables, N, is also a random variable—it can take either of two values n_1 or n_2 with equal probability. Thus the data consist of an observation of the vector random variable $(N, X_1, X_2, \ldots, X_N)$. It is easily seen that $\overline{X}$ is still singly sufficient for μ and N is an ancillary statistic carrying no information about μ. Thus nothing would seem to be lost in drawing inferences about μ if we condition our analysis on the observed value of the ancillary statistic N.

This suggests a principle (which we formalize later) of *conditional* statistical inference, where we condition on the value of an ancillary statistic. Note that given $N = n_1$, say, we have that $\overline{X} \sim \mathbf{N}(\mu, 1/n_1)$ and N provides no information about μ although (and this is often a major function of the ancillary statistic) it does provide information about the accuracy with which we can draw inferences about μ [since $\text{Var}(\overline{X}) = 1/n_1$] in the realized situation: that is, in cases when $N = n_1$.

Note that the statistic T of Example 5.24 is also ancillary in Example 5.25 for any n not exceeding the observed value of N; in both examples $\sum_1^n (X_i - \overline{X})^2$ is also ancillary with a χ^2 distribution on $(n-1)$ degrees of freedom.

A somewhat anomolous feature is illustrated in the following example.

Example 5.26. $X_1, X_2, \ldots, X_n$ are independent uniform on $(0, \theta)$. Denote by $X_{(1)}$ and $X_{(n)}$ the smallest and largest potentially observed values. Then $X_{(n)}$ is singly sufficient for θ. Also $X_{(n)}/X_{(1)}$ has a distribution that does not depend on θ and is ancillary. Thus $(X_{(1)}, X_{(n)}/X_{(1)})$ is sufficient but one component is ancillary, the other is not sufficient!

Such prospects, coupled with the lack of uniqueness of ancillary statistics, and the absence of any general method for determining them, has lead to a more restrictive form of ancillarity concept being advanced for practical usage. Since we will want whenever possible to base inferences on *minimal sufficient statistics*, we will resort to an operational definition of ancillarity in the following form. Suppose we have a minimal sufficient statistic $\tilde{\theta}$ for θ, and the dimension of $\tilde{\theta}$ *exceeds* that of θ. If we can partition $\tilde{\theta}$ in the form $\{S, T\}$ where the marginal distribution of T does not depend on θ then T is said to be an **ancillary statistic**, and S is sometimes referred to as **conditionally sufficient**, in recognition of the desirability (in the spirit indicated above) of using the statistic for drawing inferences about θ *conditional* on observing $T = t$ for the ancillary statistic. It seems sensible to choose T to have maximum possible dimension, although again uniqueness is not guaranteed.

That a minimal sufficient statistic can have an ancillary component is easily demonstrated—indeed, in Example 5.25, we see that $(\overline{X}, N)$ is minimal sufficient with N ancillary.

The notion of conducting a statistical analysis conditional on the value of an ancillary statistic is formalized in:

The Conditionality Principle . Suppose that $\tilde{\theta} = \{S, T\}$ is minimal sufficient for θ and T is ancillary. Then inferences about θ are appropriately drawn in terms of the sampling behaviour of $\tilde{\theta}$ under the assumption that T is constrained to the value t observed in the particular sample that has been obtained.

Ancillarity and conditionality have especial importance in a class of situations where the parameter θ can be partitioned into two (possibly vector) components (ψ, ϕ) with different relative status. The component ψ is that part of the parameter of principal practical importance, whilst ϕ (although crucial to the probability specification) is of no direct practical interest. The component ϕ is usually referred to a **nuisance parameter**. If all possible combinations of ψ and ϕ can occur, it can be advantageous (indeed may be essential for tractability) to introduce a modified notion of ancillarity in which the minimal sufficient statistic $\tilde{\theta} = \{S, T\}$ has the properties that the distribution of T depends only on the nuisance parameter ϕ and the conditional distribution of S, given $T = t$, depends only on ψ. The conditionality principle then prompts us to draw inferences about ψ in terms of the behaviour of S conditional on the assumption that $T = t$.

The ideas of ancillarity and conditional inference have been implicit in classical statistical methods for a long time—conditional inference figured in the early work of Fisher, and the notion of an ancillary statistic was explicit in his book on *Statistical Methods and Scientific Inference* (Fisher, 1956). The literature of the last 20 years or so shows substantial interest in these topics in the various respects of definition, implication, basic principle and application. Some of the basic ideas, and difficulties, are described by Basu[31, 32] and Cox[33]; and Cox and Hinkley (1974, Chapter 2) provide a succinct summary. (In Chapter 8 we shall consider some recent refinements of the use of ancillarity as reflected in the *pivotal approach* to inference.)

Sprott[34] extends the discussion of ancillarity by considering the roles of **marginal sufficiency** and **conditional sufficiency** in the presence of nuisance parameters. In this development, as in all of classical statistics, the likelihood function plays a crucial role,

We have seen the likelihood function used directly in estimation and hypothesis testing, and remarked (in Chapter 2) on its central role in Bayesian inference. This role is formalized in what is known as:

> **The Likelihood Principle.** We wish to draw inferences about a parameter θ in a parameter space Ω. If two sets of data x_1 and x_2 have likelihood functions which are proportional to each other then they should support identical inferential conclusions about θ.

There are really two versions of this principle—the *weak* version where x_1 and x_2 arise under a common probability model (cf. notion of sufficiency) and the *strong* version where the models differ but relate to a common parameter and parameter space.

A typical example of the latter arises if we contrast direct and inverse binomial sampling. If in each of a sequence of independent trials, an event $\mathscr{E}$ can occur with probability θ, and $\mathscr{E}$ occurs r times in n trials, then the strong likelihood principle implies that our inferences about the value of θ should be the same whether we had *decided* to conduct n trials and happened to observe $\mathscr{E}$ on r occasions or we had *needed* to conduct n trials in order to observe a (prescribed) number r of occurrences of $\mathscr{E}$

This is because the likelihoods in the two cases are proportional to each other: being

$$\binom{n}{r} \theta^r (1 - \theta)^{n-r} \quad \text{and} \quad \binom{n-1}{r-1} \theta^r (1 - \theta)^{n-r},$$

respectively. That is to say *the sampling rule is irrelevant*. However, the statistics, and their sampling distributions, differ in the two cases; we have r as an observation from $\mathbf{B}(n, \theta)$ in the first case or n as an observation from a negative binomial distribution with parameters (r, θ) in the second. This difference in the sampling distributions *does* matter in the classical approach. (A simple example on the calculation of an upper 95 per cent confidence bound for θ is given in §6.8.2.) So what does this imply about the classical approach, or about the likelihood principle?

We have seen some of the difficulties that arise in using the likelihood function for inference about θ in the classical approach. These include less than optimum performance levels in finite samples, computational and mathematical intractability, anomolous behaviour for certain types of model (see §5.3.3) and so on. The problems are compounded in multi-parameter situations especially where we have nuisance parameters. In response to such difficulties, modified forms of likelihood have been proposed and studied with the aim of eliminating unwanted parameters. Any detailed discussion is beyond our scope here, but a basic exposition of the form and use of such notions as **integrated likelihood, maximum relative likelihood, marginal likelihood** and **conditional likelihood** is given by Kalbfleisch and Sprott[35]. Cox[36] unifies some of these ideas under a more general concept of **partial likelihood**.

Some of the general principles, and specific concepts, of this section take us beyond the realm of strictly classical statistical methods. But this is no disadvantage at this stage. The general principles provide a yardstick commonly used for measuring the classical approach, and indeed other approaches. The extended concepts will serve us well when we later consider some more specific alternative approaches.

5.7   COMMENT AND CONTROVERSY

Within their terms of reference the classical principles and procedures described above seem to make good sense. The criteria proposed for estimation and hypothesis testing have great appeal in their framework of aggregate measures of assurance assessed through the idea of a sampling distribution. Indeed, the extent to which it is possible to characterize statistics which have optimum properties, and to construct these, is impressive. And even the brief review of the available practical methods for deriving estimators and tests of significance gives a clear indication of the wealth of statistical methodology which has been developed within the classical approach.

Nonetheless *classical statistics* does not lack its critics, and in the spirit of this book we must now consider the nature of this criticism. In general terms the controversy is directed to two fundamental factors.

The first is the total preoccupation of classical statistics with a frequency-based probability concept, which provides the justification for assessing the behaviour of statistical procedures in terms of their long-run behaviour. This is at the very heart of the classical approach. Criticism in this area questions the validity of attributing *aggregate* properties to *specific* inferences. In particular, the various concepts of unbiasedness, consistency and efficiency of point estimators, also the use of tail-area probabilities in hypothesis testing, are claimed by some to be largely irrelevant. The probability model must be built on a sample space, which is assumed (conceptually at least) to provide the constant basis for considering repeated experience under similar circumstances to those which prevail in the problem at hand. That such a *'collective'* exists (even conceptually) is sometimes disputed, the specification of the sample space may be claimed to be arbitrary or largely subjective, some critics go further in doubting the honesty of the classical approach in its declaration that all inferences have a solely frequency-based interpretation.

The second front of criticism concerns the limits the classical approach places on what it regards as 'relevant information'; namely the attitude that it is sample data alone for which quantification and formal analysis are appropriate. Prior information and consequential costs have claims as crucial components of the information to be processed, and criticism becomes particularly strong in cases where these are readily quantifiable. In this respect the question of the *function* of classical procedures becomes relevant too; whether, for example, the test of significance is an *inferential* or *decision-making* tool.

Let us take up some of these points in more detail.

5.7.1 Initial and Final Precision

Consider a simple problem in which we take a random sample $x_1, x_2, \ldots, x_n$ of observations from a distribution with mean μ. Our purpose is to estimate the unknown mean, μ. To this end we consider different ways of reducing the data to a single quantity $\tilde{\theta}(x)$, which is to be our estimate of μ. We might for example use the *sample mean*, $\bar{x}$, or the *sample median*, m, or perhaps the *mid-range*, $\frac{1}{2}[x_{(1)} + x_{(n)}]$. We have already considered, in terms of criteria proposed in the classical approach, which if any of these procedures is sensible, or best, in different circumstances.

Attempting to answer such questions raises a fundamental issue of how we are to assess 'accuracy' or 'precision' in estimation (indeed, the same issue arises in all aspects of inference and decision-making). A distinction can be drawn between quite different notions: of **initial precision** and of **final precision**. On the one hand, the *procedure* of using the sample mean (or some other measure) to estimate μ could be assessed in terms of *how well we expect it to behave*: that is, in the light of different possible sets of data which might be encountered. It will have some *average* characteristics which express the precision we *initially* expect, i.e. before we take our data. This involves averaging over the sample space, $\mathscr{X}$, and it is the principle which we have seen adopted in the classical approach. (It occurs in *decision theory* as well.) But it is sometimes raised as a criticism that this is the *only* aspect of assessment applied in *classical* statistical procedures, in that the sampling distribution of the estimator $\tilde{\theta}(X)$ (or in particular $\bar{X}$, etc.) is the sole basis for measuring properties of the estimator.

An alternative concept of *final precision* can be set up, in which we aim to express the precision of an inference in the specific situation we are studying. Thus if we actually take our sample and find that $\bar{x} = 29.8$, how are we to answer the question 'how close is 29.8 to μ'? This is a most pertinent question to ask—some might claim that it is the supreme consideration. Within the classical approach we must rest on any transferred properties of the long-term behaviour of the *procedure* itself. If $\bar{X}$ has high probability of being close to μ, we are encouraged to feel that 29.8 'should be close to μ'. Classical statistics regards this as the appropriate attitude to take to the question of the precision of the estimate—*that it is typical of the procedure that generates it.*

But not everyone is satisfied by this attitude. Some critics suggest that it is misconceived! Suppose from data we obtain a 95 per cent confidence interval for μ as $29.5 < \mu < 30.2$. It is suggested that it is of little value to someone needing to act in this situation to know that the '95 per cent confidence' attaches not to the specific statement made, but to what proportion of such similarly based statements will be correct in the long run. Such an interpretation may justify the statistician in his professional activities at large, (in the long run his advice is correct on 95 per cent of occasions), but what of the present situation? A statistically naive client will almost inevitably attach the 95 per cent to the statement $29.5 < \mu < 30.2$; the statistician himself can hardly avoid doing so implicitly when he puts the information to any practical use! Any probability interpretation of such an attitude must be expressed in non-frequency terms, for example as a statement of *subjective* probability. But with what justification? The Bayesian approach does not provide the justification; although supporting such a statement in principle, it may lead to a quite different numerical result for the confidence interval.

Lindley (1965b) argues this way in an example on a test of significance, where the sample mean $\bar{x}$ is being used to test the population mean θ in a $N(\theta, \sigma^2)$ distribution.

> ... the probability quoted in the example of a significance test is a frequency probability derived from random sampling from a normal distribution: if one was to keep taking samples from the distribution the histogram of values of $\bar{x}$ obtained would tend to $N(\theta, \sigma^2/n)$. But the interpretation of the probability is in terms of degree of belief, because the 5% or 1%, is a measure of how much belief is attached to the null hypothesis $[H : \theta = \theta]$. It is used as if 5% significance meant [in terms of Bayesian inference] the posterior probability that θ is near θ is 0.05. This is not so: the distortion of the meaning is quite wrong in general. It may, as with $N(\theta, \sigma^2)$, be justifiable, but this is not always so. (p. 68.)

Lindley's example really says that in order to relate the results of a statistical enquiry to some *specific* problem (that is, to assess *final precision*) we find ourselves forced to adopt a (possibly invalid) *subjective* or *degree-of-belief* interpretation: that the *frequency* basis of the *procedure* has no relevance in assessing what we really know in some situation *after we have carried out that procedure.*

Another line of criticism of the aggregate concept in classical statistics disputes its relevance in a more fundamental way: not merely saying that final precision is unmeasurable, or will be philosophically invalid. It claims that the cart is being put before the horse, and that *final precision* is all that matters and cannot be sought through principles concerned *only* with *initial precision.* In this respect, tail-area probabilities in a test of significance are claimed to be irrelevant to the rejection or acceptance of an hypothesis. The argument goes as follows. If we are testing $H : \mu = 30$ and find $\bar{x} = 31 \cdot 6$, what sense can there be in considering values of $\bar{x}$ in excess of $31 \cdot 6$ (or even, less than $28 \cdot 4$) that *may* have occurred, *but have not.* Jeffreys (1961) put this rather well:

> . . . *a hypothesis which may be true may be rejected because it has not predicted observable results which have not occurred.* This seems a remarkable procedure. (p. 385.)

This attitude needs to be contrasted with the *intuitive* appeal of the principle of drawing inferences on the basis of the set of possible results which are as extreme, or more extreme, than those which have actually occurred. De Groot[37] considers the possibility of 'interpreting a tail area as a posterior probability or as a likelihood ratio' in an attempt at reconciliation of classical and Bayesian attitudes on this topic.

The distinction between initial and final precision in the classical approach is well illustrated by the often quoted example (discussed by Welch[38], and others) on estimating the mean of a uniform distribution on the interval $(\mu - \tfrac{1}{2}, \mu + \tfrac{1}{2})$. In Example 5.10 we saw that the midrange $\tfrac{1}{2}(x_{(1)} + x_{(n)})$ provides an *extra efficient* estimate of μ, in that its sampling variance is of order $1/n^2$. Thus for reasonable sized samples we might expect to obtain very precise estimates of μ. But consider the following situation. We have a large sample, but it so happens that the sample range is small. Suppose $x_{(1)} = 2 \cdot 0$, $x_{(n)} = 2 \cdot 1$. In terms of *final precision* all we can really claim is that

$$x_{(n)} - \tfrac{1}{2} < \mu < x_{(1)} + \tfrac{1}{2}, \quad \text{or} \quad 1 \cdot 6 < \mu < 2 \cdot 5,$$

and this remains so *however large the sample.* Thus the high efficiency of the *estimator* $\tfrac{1}{2}(X_{(1)} + X_{(n)})$ provides little real comfort; the final conclusion remains most imprecise.

Admittedly, the larger n is, the more and more improbable it is that we obtain such a small sample range. But suppose we do!

The terms *initial precision* and *final precision* were used by Savage (1962) in an informal discussion of this topic.

5.7.2   Hypothesis Tests and Decisions

We have frequently stressed the distinction between the different functions of a statistical enquiry: that of inferring properties of the underlying probability structure for its better understanding, and that of decision-making where inferential knowledge is applied to the choice of appropriate action. This distinction cannot be ignored when we consider *classical* statistics. Whilst claimed by many to be an entirely inferential approach in its various expressions (point estimation, region estimation and hypothesis testing) there is a sense in which the last of these, hypothesis testing, has a somewhat ambiguous function, as we have already noted in §5·2.

On the one hand it might appear that the test of significance serves merely, and solely, to refine or clarify our knowledge of the probability structure appropriate to the problem under investigation. As such it is purely an inferential procedure, on a level with point or region estimation. But is it really correct to regard the test of significance as simply a means of delimiting the probability model with no reference, or relevance, to the taking of decisions? Its very nature, and the language used to describe it, give pause for thought. We set up hypotheses and *reject* or *accept* them. In many situations there will obviously be practical implications in rejecting or accepting some hypothesis, *H*, and incorrect conclusions (of type I or type II) will involve contingent, and possibly quite different, consequences. If we accept that certain actions may follow from conducting a test of significance, should we not attempt to quantify these and incorporate them formally in the structure of the test, thus admitting its decision-making role?

This is the basic dilemma, that tests of significance inevitably will be employed as aids to, or even as the basis for, the taking of practical action. And yet their method of construction does not allow for the costs of possible wrong actions to be taken into account in any precise way. Costs and consequences are very much part of the relevant information to be processed; and yet even when these are readily quantifiable there is no machinery for formally incorporating them in the test. This is a deliberate feature of the classical test of significance, for reasons already discussed: namely, that the common 'intangibility' of consequences makes them inappropriate as a *universal* ingredient of relevant information in an 'objective' approach to statistics.

The only extent to which consequential cost factors are considered is in relation to the choice of the working hypothesis *H*, and of the significance level of the test. We are advised to choose both *H* and the significance level with regard to how important it is (what it costs) if we incorrectly reject *H*. No specific guidance is given on how this is to be assessed, and in any case its relevance seems bound up with what we see as the *function* of the test.

If the test is believed to be purely inferential in function it is difficult to see the relevance of costs, or of a prior choice of a significance level. On this viewpoint the responsibility for acting on the result of the test falls on a second party—the statistician infers, the user acts on the joint basis of the inference and what *he* knows about costs and consequences. In this respect (if the function of the test is solely inferential) it would seem that it is only the *critical level*, not the *significance level*, which matters since this most accurately represents the *inferential* import- ance of the test. No significance level needs to be assigned; costs and consequences do not seem to have even the tangential relevance to such an assignment that they possess in a decision-making situation. This is very much the attitude implicit in the use of a *pure significance test*. Note, furthermore, how the dispute about initial and final precision must again arise here.

But all of the discussion of what constitutes a good or optimum *hypothesis test* has been based on the assumption that a particular *level* of test is needed. To this extent it seems that at least an *informal* decision-making role is to be attributed to the procedure! If so, there seems no alternative but to accept the test in the compromise spirit in which its originators presented it: namely that in order to have a universally applicable procedure covering cases where consequences are not fully quantifiable they can only be afforded informal consideration. If they are well specified, of course, we may want to incorporate them explicitly through, for example, the methods of *decision theory* (see Chapter 7), and we shall see shortly how this may be achieved. It will be interesting to look back at that stage to see what is the status of the test of significance in decision-theoretic terms.

An informative discussion of the 'role of significance tests' is given by Cox[12]. The debate about the inferential or decision-making function of the classical approach is highlighted in the co-ordinated set of papers by Birnbaum[39], Lindley[40], Pratt[41], Smith[42] and Neyman[30] (all of which are contained in a most interesting whole volume of the journal *Synthese* on the 'foundations of probability and statistics' published in memory, and appreciation of the work, of Allan Birnbaum).

5.7.3 Counter Criticism

Some substantial criticisms of the classical approach have been advanced. How are they countered? Any approach to inference or decision-making must rest on certain premises, which constitute the practical and philosophical framework within which the approach operates. Most criticism ultimately hinges on the nature of these premises. In the classical approach the two major assumptions are that probability statements must be based solely on a frequency view of probability, and that sample data are the only 'objective' form of information to be assessed. To criticize, for example, the lack of any direct measure of final precision, or an invalid expression of final precision, is to dispute the appropriate- ness of the basic premises, particularly the frequency probability standpoint. Such criticism may be countered by saying that no claims are made for any direct measure of final precision, so that the approach cannot be blamed for the fact that someone chooses to extend its aims, and that in doing so invalid interpretations are adopted. Indeed some would go further and claim that final precision is not an appropriate concept to seek to represent.

Jeffreys' criticism is sometimes answered by pointing out that the tail-area probability criterion is not advanced for any inherent properties it has in its own right, but that it is derived as a consequence of more basic considerations of power and the nature of alternative hypotheses. We have seen how optimum tests frequently involve a rejection criterion expressed in terms of extreme values of the test statistic, and hence imply the consideration of tail-area probabilities. The proof of the pudding is thus in the eating! If general principles make sense (within the approach) their practical expression has the necessary support.

This brings us to the crux of any intercomparison, or cross-criticism, of different approaches to inference or decision-making. Each approach defines its own aims and objectives which limit its operations: that is, the types of question it seeks to answer and the form in which such answers are to be expressed. Any approach is open to criticism on two counts: that its internal limitations are unrealistic, or that they are not consistently maintained in the practical application of the approach.

In this latter respect we have, for example, the criticism of classical statistics for seeming to necessitate an intuitive (degree-of-belief, say) interpretation of certain inferences, such as of the confidence interval statement '$29\cdot5 < \mu < 30\cdot2$'. The approach must be prepared to answer such criticisms, in this case presumably by saying that no such degree-of-belief interpretation is intended, or implied. The only proper interpretation of the confidence level is to regard the particular interval as typical of what happens in the long-run, and that this is as much assurance as we can rightly expect.

The former type of criticism—that the internal limitations of the approach are unrealistic—needs rather different treatment. The onus is on both the plaintiff and defendant to justify their position. We have already considered some of the reasons why the restrictions to sample-data information, and frequency probability, are maintained in classical statistics. Any claim that these restrictions are unrealistic would *itself* need to be substantiated. Arguments must be presented, for example, to demonstrate that prior information and consequences can reasonably form integral parts of the information to be assessed, or that a degree-of-belief view of probability should accompany (or replace) the frequency view, and so on. Likewise, if the absence of any measure of final precision (such as for our confidence interval, $29\cdot5 < \mu < 30\cdot2$) is to be held against classical statistics, it would surely be necessary to show that some other approach satisfactorily remedies like lack. (Bayesian inference does claim to do so.) At various points in this book such matters are argued in some detail.

Another form of criticism is based on the argument that a particular approach violates some general principle which is not specific to the approach *per se*, but is felt to be so important that for *any* approach to be reasonable it should encompass that principle.

This is exemplified in complaints that the classical approach violates the *strong likelihood principle* (see §5.6) in, for example, allowing inferences to depend on the way in which the data were obtained. Also, since the concept of *coherency* (or consistency) in the personalistic view of probability and decision-making implies use of Bayes' theorem and hence of the strong likelihood principle, such violation has attributed to it a much more substantial importance. Accordingly, some would argue that the classical approach is untenable.

Undoubtedly, such argument requires serious debate, but rather than merely adopting without question the inviolability of coherency and the strong likelihood principle, those very principles themselves require investigation. As Cox[6] remarks:

> The conflict is by itself of no importance, unless one regards the strong likelihood principle as justified by some further arguments, such as Bayesian theory. . . .

Then, in turn, it is necessary to attempt to assess whether Bayesian coherency should occupy a crucial, inevitable, central position. Whilst admitting the importance of coherency within the framework of personalistic inference, Cox[6] and Cox and Hinkley (1974) provide strong counter argument to its relevance outside the area of personal behaviour. Cox[6] concludes:

> it seems to me that at an immediate practical quantitative level there is little role for personalistic probability as a primary technique in the analysis of data.

What becomes clear is how often the debate revolves on individual attitudes and views as to what is right and proper, and that reconciliation is hardly a realistic hope. If one person believes that the frequency view of probability is the only proper basis for developing statistical principles, and another feels the same of the subjective view, it may be unnatural to expect either one to prove their case. For instance, Lindley[43] remarks:

> . . . I have recently become convinced that the most important difference between the two schools is that whereas a Bayesian uses, as a prop for his methods, the prior distribution, the prop for the other approaches is the . . . [sample space]. Bayesians are often attacked for the arbitrariness of their prior. A reply is that the choice of . . . [sample space] involves similarly arbitrary selections.

Arguments sometimes turn out to be circular, in that conclusions on one approach (based on its own premises) contradict the *premises* of another approach and *vice versa*. This is the situation in much of the cross-criticism of different approaches to statistical analysis, as we have just witnessed in the discussion of the strong likelihood principle. All this process of argument is constructive provided that attitudes are not allowed to harden: that minds remain open, and different methods are used in different situations with an honest desire to assess their value, uncluttered by philosophical preconceptions.

References

1. Jeffreys, H. (1974). 'Fisher and inverse probability', *Int. Statist. Rev.*, **42**, 1–3.
† 2. LeCam, L. and Lehmann, E. L. (1974) 'J. Neyman—On the occasion of his 80th birthday', *Ann. Statist.*, **2**, vii–xiii.
† 3. Pearson, E. S. (1974). 'Memories of the impact of Fisher's work in the 1920s', *Int. Statist. Rev.*, **42**, 5–8.
† 4. Savage, L. J. (1976). 'On rereading R. A. Fisher' (with Discussion), *Ann. Statist.*, **4**, 441–500.
† 5. Stigler, S. M. (1973). 'Studies in the History of Probability and Statistics. XXXII Laplace, Fisher and the discovery of the concept of sufficiency', *Biometrika*, **60**, 439–445.
† 6. Cox, D. R. (1978). 'Foundations of statistical inference: the case for eclecticism', *Austral. J. Statist.*, **20**, 43–59.
7. Neyman, J. and Pearson, E. S. (1933). 'On the problem of the most efficient tests of statistical hypotheses', *Phil. Trans. Roy. Soc. A*, **231**, 289–337.Reprinted in Neyman and Pearson (1967).
† 8. Neyman J. and Pearson, E. S. (1933). 'The testing of statistical hypotheses in relation to probabilities *a priori*', *Proc. Camb. Phil. Soc.*, **29**, 492–510. Reprinted in Neyman and Pearson (1967).
9. Fisher, R. A. (1930). 'Inverse probability', *Proc. Camb. Phil. Soc.*, **26**, 528–535. Reprinted in Fisher (1950).
10. Fisher, R. A. (1933). 'The concepts of inverse probability and fiducial probability referring to unknown parameters', *Proc. Roy. Soc. A*, **139**, 343–348.
† 11. Neyman, J. (1962). 'Two breakthroughs in the theory of statistical decision making', *Rev. Int. Statist. Inst.*, **30**, 11–27.
† 12. Cox, D. R. (1977). 'The role of significance tests', *Scand. J. Statist.*, **4**, 49–70.
13. Neyman, J. (1976). 'Tests of statistical hypotheses and their use in studies of natural phenomena', *Commun. Statist. –Theor. Meth.*, **A5**, 737–751.
14. Kempthorne, O. (1976). 'Of what use are tests of significance and tests of hypothesis', *Commun. Statist. –Theor. Meth.*, **A5**, 763–777.
15. Rao, C. R. (1945). 'Information and accuracy attainable in the estimation of statistical parameters', *Bull. Calcutta Math. Soc.*, **37**, 81–91.
16. Aitken, A. C. and Silverstone, H. (1942). 'On the estimation of statistical parameters', *Proc. Roy. Soc. Edinb. A*, **61**, 186–194.
17. Downton, H. F. (1966). 'Linear estimates with polynomial coefficients', *Biometrika*, **53**, 129–141.
18. Barnett, V. D. (1966). 'Evaluation of the maximum-likelihood estimator where the likelihood equation has multiple roots', *Biometrika*, **53**, 151–165.
19. Daniels, H. E. (1961). 'The asymptotic efficiency of a maximum likelihood estimator', in *Proceedings of the Fourth Berkeley Symposium on Mathematical Statistics and Probability*. Berkeley, Calif.: University of California Press, pp. 151–163.
20. Godambe, V. P. (1960). 'An optimum property of regular maximum likelihood estimation', *Ann. Math. Statist.*, **31**, 1208–1211.

21. Nelder, J. A. and Wedderburn, R. W. M. (1972). 'Generalised linear models', *J. Roy. Statist. Soc. A*, **135**, 370–384.

† 22. Barnard, G. A. (1971). 'Practical applications of tests with power one' (with Discussion), *Bull. Int. Statist. Inst.*, **43**, 389–393.

† 23. Robbins, H. A. and Siegmund, D. (1971). 'Confidence intervals and interminable tests' (with discussion), *Bull. Int. Statist. Inst.*, **43**, 379–387.

24. Peers, H. W. (1971). 'Likelihood ratio and associated test criteria', *Biometrika*, **58**, 577–587.

25. Barnard, G. A. (1963). Discussion of Bartlett, M. S. (1963)'The spectral analysis of point processes', *J. Roy. Statist. Soc. B*, **25**, 264–296.

26. Hope, A. C. A. (1968). A simplified Monte Carlo significance test procedure. *J. Roy. Statist. Soc. B*, **30**, 582–598.

27. Besag, J. and Diggle, P. J. (1977). 'Simple Monte Carlo tests for spatial patterns', *Appl. Statist.*, **26**, 327–333.

† 28. Marriott, F. H. C. (1979). 'Barnard's Monte Carlo tests: how many simulations?', *Appl. Statist.*, **28**, 75–77.

29. Robinson, G. K. (1975). 'Some counter-examples to the theory of confidence intervals', *Biometrika*, **62**, 155–161.

† 30. Neyman, J. (1977). 'Frequentist probability and frequentist statistics', *Synthese*, **36**, 97–131.

31. Basu, D. (1959). 'The family of ancillary statistics', *Sankhyā A*, **21**, 247–256.

32. Basu, D. (1964). 'Recovery of ancillary information', *Sankhyā A*, **26**, 3–16.

33. Cox, D. R. (1971). 'The choice between alternative ancillary statistics, *J. Roy. Statist. Soc. B*, **33**, 251–255.

34. Sprott, D. A. (1975). Marginal and conditional sufficiency', *Biometrika*, **62**, 599–605.

35. Kalbfleisch, J. D. and Sprott, D. A. (1970). 'Application of likelihood methods to models involving large numbers of parameters' (with Discussion), *J. Roy. Statist. Soc. B*, **32**, 175–208.

36. Cox, D. R. (1975). 'Partial likelihood', *Biometrika*, **62**, 269–276.

37. DeGroot, M. H. (1973). 'Doing what comes naturally: interpreting a tail area as a posterior probability or as a likelihood ratio', *J. Amer. Statist. Assn*, **68**, 966–969.

† 38. Welch, B. L. (1939). 'On confidence limits and sufficiency with particular reference to parameters of location', *Ann. Math. Statist.*, **10**, 58–69.

† 39. Birnbaum, A. (1977). 'The Neyman–Pearson theory as decision theory, and as inference theory; with a criticism of the Lindley–Savage argument for Bayesian theory', *Synthese*, **36**, 19–49.

† 40. Lindley, D. V. (1977). 'The distinction between inference and decision', *Synthese*, **36**, 51–58.

41. Pratt, J. W. (1977). '"Decisions" as statistical evidence "Birnbaum's and confidence concept"', *Synthese*, **36**, 59–69.

† 42. Smith, C. A. B. (1977). 'The analogy between decision and inference', *Synthese*, **36**, 71–85.

† 43. Lindley, D. V. (1971). Discussion on paper by G. A. Barnard in Godambe and Sprott (1971), p. 303.
    See also:

† 44. Pearson, E. S. (1962). 'Some thoughts on statistical inference', *Ann. Math. Statist.*, **33**, 394–403.

    and: The prepared contributions and discussion by Bartlett, Barnard, Cox, Pearson, Smith and others in Savage (1962).

CHAPTER 6

Bayesian Inference

6.1 THOMAS BAYES

The Rev. Thomas Bayes' contribution to the literature on probability theory consists of just two papers published in the *Philosophical Transactions* in 1763[1] and 1764. Of these the first, entitled 'An essay towards solving a problem in the doctrine of chances', is the one which has earned Bayes a crucial place in the development of statistical inference. Both papers were published after his death, having been communicated to the Royal Society by his friend Richard Price who added his own introduction, comment and examples on the work. The second paper was concerned with some further details in the derivation of a particular result in the 'Essay'. Bayes' principal contribution, the use of **inverse probability** was further developed later by Laplace.

There is still some disagreement over precisely what Bayes was proposing in the 'Essay'. Two ideas can be distinguished and these are described by various names in the literature. For present purposes we refer to these as **Bayes' theorem**, and his **principle of insufficient reason**, these being perhaps the most common terms used. Bayes' reasoning was informal, in the spirit of the age, and the modern expression of these two ideas has been constructed out of the mere hints he provided, and the examples given by him and Price to illustrate them. It is in connection with this current formalization that most dispute arises, notably in two respects:

(a) whether the concept of *inverse probability* (stemming from Bayes' theorem) is presented as a general inferential procedure even when it implies, or demands, a degree-of-belief view of probability;

(b) how universally Bayes intended the *principle of insufficient reason* to be applied, and whether it provides a description of the state of **prior ignorance**.

This dispute on Bayes' intentions need not concern us here. It has been amply discussed elsewhere, an individual view is given by Hogben (1957, pp. 110–132).

190

Also a more accessible, slightly edited, version of the 'Essay' and Price's 'Appendix' has been provided, with bibliographical notes by Barnard[2], from which the reader may reach his own conclusions. What seems to be widely agreed is that Bayes' work is noteworthy in three respects; in his use of continuous rather than discrete frameworks, in pioneering the idea of estimation through assessing the chances that an 'informed guess' about the practical situation will be correct, and in proposing a formal description of what is meant by prior ignorance. On this latter point, see Edwards[3].

Bayes' ideas act as the springboard for the modern approach to inference known as *Bayesian inference*. This current expression of the earlier work was a long time in appearing (nearly 200 years) and we have considered possible reasons for this elsewhere (Chapter 1). Also, many might claim that Bayes would have difficulty in recognizing his own tentative offerings in the wealth of detail and interpretative sophistication of modern *Bayesian inference*. Nonetheless, the seeds of this approach were certainly sown by Bayes 200 years ago, and the dispute over the meaning of his ideas has now been transferred to the analogous concepts in their offspring. There is little point in going further into the details of the 'Essay'. We proceed instead to describe some of the principles and techniques of *Bayesian inference*, its position within the different approaches to statistical inference and decision-making, and a little of the external and internal controversy concerning basic concepts and criteria.

Much of the published material on the Bayesian idiom is concerned with its extension to decision-making problems through the vehicle of *decision theory* which we consider in Chapter 7. An elementary presentation of Bayesian inference (*per se*) is given by Lindley (1965) and the development of Bayesian techniques for (predominantly) regression and analysis of variance problems is described by Box and Tiao (1973). Zellner (1971) is concerned with Bayesian methods in econometrics. Lindley (1971a) provides an interesting, well motivated, introduction to Bayesian decision-making; Winkler (1972) gives an elementary treatment of 'Bayesian inference and decision' with some discussion of the problems of implementation and intercomparison of Bayesian and classical ideas. The review by Lindley (1971b) provides a wider, more advanced, discussion of the topics of the present chapter in the context of decision-making. Survey papers by Lindley[4, 5] are elementary and somewhat evangelical in style; Lindley[6] reviews recent developments in Bayesian statistics at a more advanced level, as does Bernardo, De Groot, Lindley and Smith (1980). Books on *decision theory* which discuss aspects of the Bayesian approach at an intermediate or advanced level include De Groot (1970), Ferguson (1967) and Raiffa and Schlaifer (1961). de Finetti (1974, 1975), Good (1950) and Jeffreys (1961) and relevant to the underlying philosophical considerations; see de Finetti [7].

6.2   THE BAYESIAN METHOD

We start with what is known as **Bayes' theorem**. At one level this may be regarded as a result in deductive probability theory.

Bayes' theorem. *In an indeterminate practical situation a set of events* A_1, $A_2, \ldots, A_k$ *may occur. No two such events may occur simultaneously, but* **at least** *one of them must occur (i.e.* $A_1, \ldots, A_k$ *are mutually exclusive and exhaustive). Some other event, A, is of particular interest. The probabilities,* $P(A_i)$ $(i = 1, \ldots, k)$, *of each of the* A_i *are known, as are the conditional probabilities,* $P(A|A_i)$ $(i = 1, \ldots, k)$, *of A given that* A_i *has occurred. Then the conditional ('inverse') probability of any* A_i $(i = 1, \ldots, k)$, *given that A has occurred, is given by*

$$P(A_i|A) = \frac{P(A|A_i)P(A_i)}{\sum\limits_{j=1}^{k} P(A|A_j)P(A_j)} \qquad (i = 1, \ldots, k). \qquad (6.2.1)$$

As expressed, *Bayes' theorem* finds wide application, and arouses no controversy. No difficulty arises in the philosophical interpretation of the probabilities involved in it. As the starting point for the study of *Bayesian inference*, however, it is necessary to extend its meaning in one particular respect. Rather than considering *events* A_i, we must work in terms of a set of hypotheses $H_1, \ldots, H_k$ concerning what constitutes an appropriate model for the practical situation. One, and only one, of these must be true. The event A becomes re-interpreted as an observed outcome from the practical situation: it is the sample data. Prior to the observation, the probability, $P(H_i)$, that H_i is the appropriate model specification, is known for all $i = 1, \ldots, k$. These probabilities are the **prior probabilities** of the different hypotheses, and constitute a secondary source of information. The probabilities, $P(A|H_i)$ $(i = 1, \ldots, k)$, of observing A, when H_i is the correct specification, are known also—these are simply the *likelihoods* of the sample data. We can correspondingly re-interpret *Bayes' theorem* as providing a means of updating, through use of the sample data, our earlier state of knowledge expressed in terms of the prior probabilities, $P(H_i)$ $(i = 1, \ldots, k)$. The updated assessment is given by the **posterior probabilities**, $P(H_i|A)$ $(i = 1, \ldots, k)$, of the different hypotheses being true after we have utilized the further information provided by observing A to occur. These *posterior probabilities* (*inverse probabilities*) are given by

$$P(H_i|A) = \frac{P(A|H_i)P(H_i)}{\sum\limits_{j=1}^{k} P(A|H_j)P(H_j)} \qquad (i = 1, \ldots, k). \qquad (6.2.2)$$

This is the essence of Bayesian inference: that *the posterior probability of* H_i *given A is proportional to the product of the prior probability of* H_i *and the likelihood of A when* H_i *is true* [the denominator of the right-hand side of (6.2.2) is merely a normalizing constant independent of i]. Prior information about the practical situation is in this way augmented by the sample data to yield a current probabilistic description of that situation. In this respect the Bayesian approach is *inferential*. It asserts that our current knowledge is fully described by the set of posterior probabilities, $\{P(H_i|A)\}$ ($i = 1, \ldots, k$).

It is interesting to note an immediate consequence of (6.2.2), providing a restatement of the fundamental principle of Bayesian inference. Suppose we are interested in two particular hypotheses, H_i and H_j. The ratio of their posterior probabilities—the **posterior odds ratio**—is given by

$$\frac{P(H_i|A)}{P(H_j|A)} = \frac{P(A|H_i)}{P(A|H_j)} \cdot \frac{P(H_i)}{P(H_j)}, \tag{6.2.3}$$

that is, by the product of the **prior odds ratio** and the *likelihood ratio*.

Example 6.1. A box contains equal large numbers of two types of six-faced cubical die. The dice are perfectly symmetric with regard to their physical properties of shape, density, etc. Type I has faces numbered 1, 2, 3, 4, 5, 6; whilst type II has faces numbered 1, 1, 1, 2, 2, 3. A plausible probability model is one which prescribes equal probabilities of $\frac{1}{6}$ for the uppermost face when any die is thrown. One die is picked at random from the box, thrown twice, and the uppermost face shows a 1 and a 3 on the separate (independent) throws. Having observed this, we wish to comment on whether the die was type I or type II. We denote these alternative possibilities as hypotheses H_I and H_{II}, and there seem to be reasonable grounds for assigning equal prior probabilities to each of these. Bayes' theorem then trivially yields posterior probabilities of $\frac{1}{4}$ and $\frac{3}{4}$ for the hypotheses H_I and H_{II}, respectively. This conclusion is an inferential statement about the underlying practical situation. It describes the relative chances that the die used was of type I or type II, respectively, as 3:1 in favour of it being type II. In Bayes' original formulation we would declare that a 'guess' that the die is type II has probability $\frac{3}{4}$ of being correct.

The change of emphasis in (6.2.2) is well illustrated by this example. It is now a statement about the plausibility of alternative models for generating the observed data; no longer a deductive probability statement. This re-interpretation raises some problems, which centre on the nature of the probability concept involved in the prior, and posterior, probabilities. In Example 6.1 the situation is straightforward. Both the prior and the posterior probabilities can be interpreted in frequency terms, within the larger experiment of choosing a die at random from the box containing equal numbers of each type. Also, the numerical values of the prior probabilities are derived directly from this 'super-experiment'. Their accuracy, of course, remains dependent on our assumptions about the super-experiment; that choice of the die to be used really is random from equal numbers of each type. The evaluation of the likelihoods also rests on the assumed randomness and independence of consecutive outcomes of throwing the die that is chosen. Strictly speaking there also remains the question of validating the assumptions about the super-experiment, but *if we accept them* there is no difficulty in interpreting the results of the inverse probability statement derived from *Bayes' theorem*.

Example 6.1 is a simple, but typical, illustration of real-life situations, for example in genetics. In general, however, further complications can occur. Consider two other examples, both superficially similar to Example 6.1, and both representative of practical situations.

Example 6.2. A box contains large numbers of two types of six-faced cubical die. The dice are perfectly symmetric with regard to their physical properties of shape, density, etc. Type I has faces numbered 1, 2, 3, 4, 5, 6; whilst type II has faces numbered 1, 1, 1, 2, 2, 3. There are more type I dice in the box than type II. One die is picked at random from the box, thrown twice, and the uppermost face shows a 1 and a 3 on the separate (independent) throws. What can we say about whether this die was type I or type II?

Example 6.3. A friend has one of each of the types of die described in Examples 6.1 and 6.2. He chooses one of these without revealing its type, but comments that he would always use that type in preference to the other. He throws the die twice and reports that the uppermost face shows a 1 and a 3 on the separate (independent) throws. What can we say about what type of die was used?

In Example 6.2, there still exists some credible notion of a 'super-experiment'. The probabilities $P(H_i)$ and $P(H_i|A)$ (i = I, II) have corresponding frequency interpretations. But our information about the super-experiment ('more type I than type II dice') is now insufficient to assign precise numerical values to the prior probabilities, $P(H_I)$ and $P(H_{II})$. We know only that $P(H_I) > P(H_{II})$, i.e. $P(H_I) > \frac{1}{2}$. To make inferences by Bayesian methods we are compelled to substitute numerical values for $P(H_I)$ and $P(H_{II})$. How is this to be done? We may have even less information; not knowing whether type I or type II dice are in the majority. The problem remains! What do we use for the values of $P(H_I)$ and $P(H_{II})$? We might describe this latter state as one of *prior ignorance*, and to use Bayesian methods we must quantify this condition. The *principle of insufficient reason* suggests that we assume that $P(H_I) = P(H_{II}) = \frac{1}{2}$, in that there is no evidence to favour type I or type II. But $P(H_I)$ is well defined in frequency terms; it has a specific value, albeit unknown. In declaring that $P(H_I) = \frac{1}{2}$ we are not really claiming that there are equal numbers of each type of die in the box. We are making a statement about our own attitude of mind; we cannot find grounds for believing, *a priori*, in there being a majority of one type of die rather than the other. The probability concept involved in the statement $P(H_I) = \frac{1}{2}$ has become a degree-of-belief one (either *subjective*, or *logical*).

In Example 6.3 even the super-experiment structure seems to have disappeared! Again on the principle of insufficient reason we may make the conventional assignment $P(H_I) = \frac{1}{2}$, possibly with a *logical* interpretation of the probability concept. Alternatively, personal opinions may enter the problem. We may feel that the friend is somewhat eccentric, and more likely to choose the 'odd' type II die. Is this feeling relevant to the need to assign a value to $P(H_I)$? Some would claim it is, and use a value of $P(H_I) < \frac{1}{2}$. Again arbitrariness enters in deciding what *precise* value ($< \frac{1}{2}$) to give to $P(H_I)$. Someone else faced with the same problem may regard the friend as a rather 'conservative' person, more inclined to choose the type I die. He would choose a value for $P(H_I) > \frac{1}{2}$. The prior probabilities may now need to be interpreted in *personal* terms, rather than logical, in the sense of the distinction drawn in Chapter 3.

We have pinpointed two difficulties:

(a) the interpretation of the probability idea implicit in a particular Bayesian analysis,

(b) the numerical specification of prior probabilities to be used in the analysis.

The first difficulty, (a), is at the heart of the criticisms that are made of the Bayesian approach in general, and of the internal divisions of attitude which exist. We take this up again in §6.8. We must recognize, however, that many statisticians regard problems such as those described by Examples 6.2 and 6.3 as squarely within the realm of legitimate enquiry (using the ideas of Bayesian inference). One thing is clear then; we shall not proceed far in studying the application of Bayesian methods unless we admit that a wider view of probability may be necessary than the frequency one. Lindley (1965) has chosen to present the subject entirely in degree-on-belief terms 'for it to be easily understood'. (See §1.6.) It is not necessary to be so specific for much of the following discussion. To demonstrate principles at an elementary level the term 'probability' may often be used in an intuitive way, leaving its interpretation dependent on the nature of any problem being studied and on the attitude of the investigator. But when trying to *interpet* ideas in Bayesian inference, this intuitive approach often will not do. It does seem (in agreement with Lindley) to be simplest to adopt a degree-of-belief attitude in such cases, and this is what has been done below. This is not to suggest, however, that there may not be a perfectly valid and direct frequency interpretation in certain situations (but see §6.8.1): it merely facilitates a general presentation of the subject.

The second difficulty, (b), cannot be postponed. Its resolution forms an essential part of the practical detail of Bayesian inference. Ideas on how to process prior information to yield numerical values for prior probabilities are not fully developed; much remains to be done in this area. Even the apparently simple case where the prior information consists of sample data from earlier observations of the same practical situation is not without its difficulties. (See §6.7 on **empirical Bayes' methods**.) We consider briefly what results are available on this matter in §§6.4, 6.5 and 6.6 which distinguish between the cases of **prior ignorance, vague**

prior knowledge, and **substantial prior knowledge**, respectively. For the moment, however, we take the attitude that numerical valued prior probabilities are available, and enquire in what way they are used in Bayesian inference.

6.3   PARTICULAR TECHNIQUES

The *rationale* of Bayesian inference may be summarized as follows.

Inferences are to be made by combining the information provided by prior probabilities with that given by the sample data; this combination is achieved by 'the repeated use of Bayes' theorem' (*Lindley*, 1965*b*, p. xi), *and the final inferences are expressed solely by the posterior probabilities.*

To see how this works out we revert to the parametric model used earlier, but some extra care is needed in its description.

We suppose that sample data, x, arise as an observation of a random variable, X. The distribution of X, specified by the probability model, is assumed to belong to some family, $\mathscr{P}$, indexed by a parameter θ. It is assumed that the probability (density) function of the random variable has a known form, $p_\theta(x)$, depending on θ; but that θ is unknown, except that it lies in a parameter space, Ω. For fixed x, $p_\theta(x)$ is again the likelihood function of θ. Knowledge of the 'true' value of θ, i.e. the value relevant to the current practical situation, would be all that is needed to describe that situation completely. [This assumes, of course, the adequacy of the family $\mathscr{P} = \{p_\theta(x); \theta \in \Omega\}$ as a general model for the practical situation. This may need to be examined separately.] As earlier, both θ and x may be either univariate or multivariate, discrete or continuous.

The aim of any inferential procedure must be to 'reflect' the unknown 'true' value of θ. This cannot be done with any certainty. All we can expect is some probabilistic statement involving θ, based on the information at our disposal. In the *classical approach* of Chapter 5 we saw how this was achieved by processing the *sample data*, as the only available information, to produce point or interval estimates of θ, with associated assessments of their accuracy. In Bayesian inference it is assumed that we have further information available *a priori*, i.e. before observing the sample data. To incorporate this, a wider view is taken of the nature of the parameter θ. It is assumed that any knowledge we have of the 'true' value of θ, at any stage, can be expressed by a probability distribution, or some 'weight function', over Ω. *The parameter θ is now essentially regarded as a random variable* in the sense that different values are possible with different probabilities, degrees-of-belief or weights! (We consider the implications of this in more detail in §6.8.1.) The prior knowledge of θ is expressed as a *prior probability distribution*, having probability (density) function, $\pi(\theta)$. Sampling increases this knowledge, and the combined information about θ is described by its *posterior distribution*. If the posterior probability (density) function of θ is denoted by $\pi(\theta|x)$ we have, from

Bayes' theorem,

$$\pi(\theta|x) = p_\theta(x)\pi(\theta)/\int_\Omega p_\theta(x)\pi(\theta).$$ (6.3.1)

The posterior distribution, $\pi(\theta|x)$, is regarded as a complete description of what is known about θ from the prior information and the sample data—it describes how we now assess the chances of the true value of θ lying in different parts of the parameter space Ω. Leaving aside the question of the interpretation of the probability concept involved in $\pi(\theta)$ and $\pi(\theta|x)$, we now appear to have a more direct form of inference than in the classical approach. We can answer directly such questions as 'what is the probability that $0.49 < \lambda < 0.51$?' in the radioactivity example of §1.3. This facility is not available in the classical approach, without reference to some larger framework than the current situation and its associated information. It was there necessary to consider sampling distributions defined in terms of a sequence of independent repetitions of the current situation under what were assumed to be identical circumstances. The effect of this, in providing only aggregate measures of accuracy, has been considered in some detail in Chapter 5. In contrast, inferences in the Bayesian approach contain their own internal measure of accuracy and relate only to the current situation. This distinction is crucial. It is the difference between *initial precision* and *final precision* (see §5.7.1). Classical statistics assesses initial precision; Bayesian inference, final precision. We accept this distinction for the moment and will consider some examples of it. Later (§6.8.1) we will need to re-examine it in the light of a closer study of the *'random'* nature of θ.

Whilst the posterior distribution, $\pi(\theta|x)$, constitutes the complete inferential statement about θ, there are situations where such a full description is not needed. Certain summary measures of $\pi(\theta|x)$ may suffice. For example, it may be enough to know what value of θ is most likely, or in what region θ is highly likely to fall, or even whether some specific statement about θ is credible. These needs lead to concepts in Bayesian inference parallel with the ideas of point estimates, confidence regions and hypothesis tests in the classical approach. It is convenient to describe these by the same names, but it must be remembered that their interpretation, and the numerical answers they provide, are likely to be quite different to those of the analogous quantities in classical inference.

For illustrative convenience, θ is assumed to be univariate unless otherwise stated.

Bayesian Point Estimates

There may be situations where it is convenient to choose a single value as an estimate of θ. It seems sensible to choose that value with highest posterior probability (density). Consequently we define as a *Bayesian point estimate* the quantity $\tilde{\theta}(x)$ which maximizes $\pi(\theta|x)$. See Figure 6.1.

In itself $\tilde{\theta}(x)$ has limited practical value; it provides a crude summary of the complete posterior probability distribution. It has, however, a direct interpretation as the most likely value for θ *in the current situation*, and the posterior distribution measures directly our remaining uncertainty about θ in this situation. There would seem to be no other useful criterion for choosing a single value to estimate θ than to use the most likely value (the *mode* of the posterior distribution), unless we incorporate further information on the consequences of incorrect choice of θ. (See §7.3.)

In classical inference the situation was quite different. A variety of alternative estimators may exist for θ. These will have the general form $\hat{\theta}(X)$, stressing that their interpretation is in terms of different sets of data which may potentially arise for the *fixed* current value of θ, rather than in terms of differing degrees-of-belief about θ for the present observed data x. Choice between alternative estimators, and assessments of their individual properties (bias, efficiency, etc.), were all derived from the behaviour of this sampling distribution of $\hat{\theta}(X)$; that is, in relation to repeated realizations of the current situation.

Although the criterion for the choice of the Bayesian estimate $\tilde{\theta}(x)$ seems incontrovertible, one difficulty does arise. The estimate will not be invariant with respect to transformations of the parameter space. Working in terms of $\phi(\theta)$, rather than θ, it does not necessarily follow that $\tilde{\phi} = \phi(\tilde{\theta})$ for any particular x. Thus different inferences may be drawn from the same data and prior information in alternative parameterizations of the model. This problem is sometimes resolved by claiming that there will usually be a 'natural' parameterization, and that inferences must therefore relate to this 'natural' parameter. This does not seem completely satisfactory, and all summary measures of the posterior distribution remain somewhat arbitrary in this respect. We will meet this concept of a 'natural' parameter again in relation to ways of describing prior ignorance (§6.4). Note that this use of the word 'natural' is *not* the same as its use in defining the *exponential family* of distributions (§5.3.2).

Bayesian Confidence Regions (Credible Regions)

A more informative summary of $\pi(\theta|x)$ for practical purposes is obtained by saying that θ lies in some region of Ω with a prescribed probability. The region $S_\alpha(x)$ is a $100(1-\alpha)$ per cent *Bayesian confidence region* (*credible region*) for θ if

$$\int_{S_\alpha(x)} \pi(\theta|x) = 1 - \alpha. \tag{6.3.2}$$

If $S_\alpha(x)$ can be chosen to satisfy (6.3.2) we call $(1-\alpha)$ the *Bayesian confidence* (*credibility*) *coefficient*. As in the classical case we may not be able to achieve this precisely, but can only ensure that $P[\theta \in S_\alpha(x)] \geqslant 1 - \alpha$. Then again we call $(1-\alpha)$ the *Bayesian confidence* (*credibility*) *level*. This occurs in cases where θ has a discrete component—in the classical approach it happened if X had a discrete component.

To be of practical value α will need to be chosen small, and we will again typically consider 90, 95, 99 per cent etc., confidence (credible) regions, although the actual choice is arbitrary. The region $S_\alpha(x)$ will not be unique, again in parallel with the classical confidence region. There may be several regions which contain a proportion $(1 - \alpha)$ of the posterior distribution, often an infinite number, and it is necessary to choose between them. In the classical approach we saw (§5.5) that the so-called *central confidence intervals* had certain optimality properties in special cases, and the idea of central intervals (cutting off equal tail-area probabilities) was extended to form a practical criterion for common use. This concept seems to have little relevance to the Bayesian situation, however. Any region of Ω omitted on this criterion may have small total probability, but can still contain values of θ with high probability (density) relative to some values of θ contained in $S_\alpha(x)$; we are perhaps excluding some θ which are more likely to be true than other θ included in $S_\alpha(x)$. Consequently a different criterion is usually adopted: $S_\alpha(x)$ *shall not exclude any value of θ more probable than any value of θ which is included.* It seems inevitable, on the Bayesian idiom, that this criterion should always be applied if the Bayesian confidence region is to be properly interpretable.

In many situations $\pi(\theta|x)$ will be unimodal, and this principle then yields a finite interval for $S_\alpha(x)$: a *Bayesian confidence interval* or *credible interval*. See Figure 6.1. Here $S_\alpha(x)$ takes the form $(\underline{\theta}_\alpha(x), \bar{\theta}_\alpha(x))$, where

$$\int_{\underline{\theta}_\alpha(x)}^{\bar{\theta}_\alpha(x)} \pi(\theta|x) = 1 - \alpha \tag{6.3.3}$$

and

$$\pi(\theta|x) \geqslant \pi(\theta'|x), \qquad \text{for any } \theta \in S_\alpha(x), \theta' \notin S_\alpha(x).$$

Under fairly general conditions $(\underline{\theta}_\alpha(x), \bar{\theta}_\alpha(x))$ is both unique—except for cases where several values of θ have equal probability (density)—and shortest amongst all Bayesian confidence regions of prescribed Bayesian confidence coefficient. But again variations may occur in alternative parameterizations!

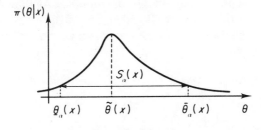

Figure 6.1

The difference of interpretation of the classical, and Bayesian, confidence intervals is obvious and striking. Within its framework the Bayesian confidence region has a direct probability interpretation,

$$P[\theta \in S_\alpha(x)] = 1 - \alpha, \tag{6.3.4}$$

and is determined solely from the current data x, and prior information. The classical confidence region is also expressed merely in terms of the current data, but its probability interpretation is as a *random* region containing the *fixed* value of θ. The assessment of its probability of actually enclosing θ is in terms of repetitions of the experimental situation. As we saw earlier (§5.5) there is no way of judging whether a *particular* classical confidence region does, or does not, include θ.

Example 6.4. Suppose a random sample of n independent observations is available from a normal distribution, $N(\mu, \sigma^2)$, with unknown mean μ, and known variance σ^2. (This latter assumption is introduced for convenience, hardly in pretence of reality.) The sample mean is $\bar{x}$. Anticipating ideas in the next section we suppose that nothing is known a priori about μ and that this is expressed by an ('improper') assignment of equal prior probability density over $(-\infty, \infty)$. It is easy to show that (6.3.1) yields for the posterior distribution of μ the normal distribution, $N(\bar{x}, \sigma^2/n)$. Thus we obtain the Bayesian point estimate, $\bar{x}$, for μ. Furthermore the $100(1 - \alpha)$ per cent Bayesian confidence interval (credible interval) for μ has the form $(\bar{x} - z_\alpha \sigma/\sqrt{n}, \bar{x} + z_\alpha \sigma/\sqrt{n})$, where z_α is the double-tailed α-point of $N(0, 1)$.

We know, however, that on the classical approach $\bar{x}$ and $(\bar{x} - z_\alpha \sigma/\sqrt{n}, \bar{x} + z_\alpha \sigma/\sqrt{n})$, are also the optimum point estimate and $100(1 - \alpha)$ per cent confidence interval, respectively, for μ.

This correspondence in the explicit expressions should not be allowed to conceal the quite different meanings of the results in the two cases. Neither should it be taken as indicating any general area of unanimity. We shall obtain identical expressions only with appropriate choice of the prior distribution. It is intriguing to question whether any such agreement, for cases of prior ignorance, lends respectability to the classical approach in that it produces the same answers as the Bayesian approach, or vice versa, or whether instead it serves to justify the particular expression used to represent prior ignorance. All three cases have been argued separately in the literature. Whether any of them has any justification depends on the personal attitudes of their proponents, and can hardly be assessed objectively. (See §6.4.)

Example 6.5. A random variable X has a Cauchy distribution centred at an unknown point, θ. Its probability density function is

$$f(x) = \frac{1}{\pi} \frac{1}{[1 + (x - \theta)^2]} \qquad (-\infty < x < \infty).$$

Two independent observations x_1, x_2 are drawn from this distribution and constitute the sample data, x; they are to be used in drawing inferences about θ. The likelihood of the sample is

$$p_\theta(x) = \frac{1}{\pi^2} \frac{1}{[1+(x_1-\theta)^2]} \cdot \frac{1}{[1+(x_2-\theta)^2]}.$$

On both the classical and Bayesian approaches some rather strange results arise when we try to draw inferences about θ. The sample mean $\bar{x}$ has intuitive appeal, but it has a sampling distribution with infinite mean and variance, and cannot lead to classical confidence intervals for θ. Maximum likelihood is no better! If x_1 and x_2 differ by more than 4, as is highly probable, $p_\theta(x)$ is symmetric about θ but double peaked. Typically it looks like this:

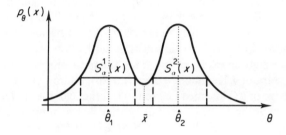

The sample mean is now at a relative minimum *of $p_\theta(x)$. The two points $\hat{\theta}_1$ and $\hat{\theta}_2$ where $p_\theta(x)$ has relative maxima, are*

$$\hat{\theta}_2, \hat{\theta}_1 = \bar{x} \pm \tfrac{1}{2}\sqrt{[(x_1-x_2)^2 - 4]},$$

and $p_\theta(x)$ is equal at these points, so no unique maximum likelihood estimator exists. In terms of likelihood there seems no good reason to distinguish between $\hat{\theta}_1$ and $\hat{\theta}_2$. $\bar{x}$ (or the sample median, which is the same thing here) has some intuitive appeal, but no statistical justification in view of its sampling properties.

On the Bayesian approach, again using an improper uniform distribution to express prior ignorance, the posterior distribution $\pi(\theta|x)$ is proportional to $p_\theta(x)$ so it has the same form as that shown above. No unique Bayesian point estimate exists, since $\pi(\hat{\theta}_1|x) = \pi(\hat{\theta}_2|x)$. One possibility might be to estimate θ by a value for which there is equal posterior probability of the true value being in excess of, or less than, this value. This is just the median of the posterior distribution, here $\bar{x}$. But it is surely not desirable on the Bayesian approach to estimate θ by what is one of the lowest *posterior probability (density) points! Of course, we can construct a Bayesian confidence region (credible region) for θ, but even this is somewhat strange in that it is quite likely to have 'a hole in the middle', e.g. $S_\alpha(x) = S_\alpha^1(x) \cup S_\alpha^2(x)$ in the figure above. This illustrates a general point: that Bayesian confidence regions even for a single parameter may consist of a set of intervals, rather than a single interval.*

This is an artificial example, of course. No one would expect to say much about the location parameter of a Cauchy distribution from two observations. But the anomalous behaviour it demonstrates is not entirely artificial. Similar difficulties can arise, in varying degrees, in quite realistic situations.

Bayesian Hypothesis Tests

Yet another method of summarizing the information in the posterior distribution leads to a further procedure paralleled in the classical approach: that of an hypothesis test. In view of the direct probability interpretation provided by the posterior distribution, the *Bayesian hypothesis test* has a simple form.

In a practical situation we may need to assess whether some statement about θ lying in a particular region of Ω is reasonable. For example, a biscuit manufacturer producing packets of biscuits is required to state on the packet a weight which is at least $\frac{1}{2}$ oz less than the mean weight, θ, of packets of biscuits produced by his process. He aims to produce 8-oz packets so quotes a weight of $7\frac{1}{2}$ oz on the packet, but needs to examine whether he is meeting the requirement of not stating a weight in excess of $\theta - \frac{1}{2}$. On the basis of some sample data and prior knowledge about θ, he obtains a posterior distribution $\pi(\theta|x)$ expressing his current knowledge of θ. This must be used to test the hypothesis $H:\theta < 8$ against the alternative hypothesis $\overline{H}:\theta \geqslant 8$. But on the Bayesian approach we have a direct evaluation of the probabilities of H and $\overline{H}$, in the form

$$P(H|x) = \int_H \pi(\theta|x) = 1 - P(\overline{H}|x). \tag{6.3.5}$$

If $P(H|x)$ is sufficiently small we will want to reject H, and this can again be expressed in terms of '*significance levels*' if required. Thus a 5 per cent Bayesian *test of significance* rejects H if $P(H|x) \leqslant 0.05$; similarly for other significance levels. Alternatively, the outcome of the test may again be expressed in terms of the significance level actually attained, $P(H|x)$; this is just the '*critical level*' of the test.

Note that the direct expression of the result of the test in the form $P(H|x)$ eliminates the asymmetric nature of the test observed in the classical approach. In particular there is no need to distinguish formally between the *null* (*working*) and the *alternative* hypotheses.

The test just described is a one-sided test in our earlier terminology. The idea of a two-sided test of a simple hypothesis also exists in Bayesian inference, but opinions differ on its appropriate form, and difficulties of interpretation arise.

Much of the literature of the 1960s stressed the presence, within the Bayesian approach, of a facility which paralleled that of the classical hypothesis test. Whilst the above form of one-sided test seems unexceptionable, attempts to formulate an equivalent two-sided test do not seem to be so successful (and the very notion of a Bayesian hypothesis test figures less prominently in recent descriptions of

Bayesian inference). However, it is of interest to consider some of the proposals in this matter.

Lindley (1965*b*, pp. 58–62) describes one form of such a test in terms very similar to those used to construct the classical hypothesis test. To test $H:\theta = \theta_0$ against $\overline{H}:\theta \neq \theta_0$ at a level α he suggests that we obtain the $100(1 - \alpha)$ per cent Bayesian confidence interval for θ and accept H if this interval contains θ_0, otherwise reject H. This procedure is limited to cases where the prior information on θ is vague (see §6.5): in particular where there is no prior discrete concentration of probability at $\theta = \theta_0$. As such, we no longer have any simple probability interpretation of the result of the test *per se*: we cannot now talk about $P(H)$. There is only the inferred interpretation arising from the confidence interval being an interval within which we have $100(1 - \alpha)$ per cent confidence that the true θ lies—and hence only 100α per cent confidence in values of θ outside this interval. This is somewhat analogous to the tail-area concept of probability on the classical approach. The situation becomes more confused when one-sided tests are constructed in a similar way from *one-sided confidence intervals*, which seems to have little meaning in the Bayesian approach!

Another method for testing a point hypothesis, $H:\theta = \theta_0$, is described by Jeffreys (1961, Chapter 5). The idea here is that the particular value θ_0 has a different order of importance to the other values of θ in Ω. It achieves this through having been singled out for particular study, usually due to extraneous practical considerations. As a result our prior information about θ should be in two parts; a prior discrete probability for θ_0, $P(\theta_0)$, together with some prior distribution $\pi(\theta)$ over $\theta \neq \theta_0$. The sample data, x, refine the probability that $\theta = \theta_0$, and the distribution of probability over $\theta \neq \theta_0$, to produce $P(\theta_0|x)$ and $\pi(\theta|x)$ for $\theta \neq \theta_0$. The decision on whether to accept, or reject, H is now taken on the basis of the posterior 'odds in favour of H', i.e.

$$P(\theta_0|x) \bigg/ \int_{\Omega - \theta_0} \pi(\theta|x).$$

Jeffreys proposes that in the absence of any prior knowledge about θ we should divide the prior probability equally between H and $\overline{H}$, by taking $P(\theta_0) = \frac{1}{2}$ and a uniform distribution of the remaining probability mass over the values of $\theta \neq \theta_0$. He states, 'In practice there is always a limit to the possible range of these values'.

An interesting illustration of this procedure is given by Savage (1962, pp. 29–33) in the context of the legend of King Heiro's crown. Pearson[8] offers a detailed critical re-examination of this example, with particular emphasis on its subjective elements.

Pratt[9] declares that the posterior probability that $\theta \leqslant \theta_0$ is a reasonable measure of the 'plausibility' of a null (working) hypothesis $H:\theta \leqslant \theta_0$, tested against the alternative hypothesis $\overline{H}:\theta > \theta_0$. He argues, however, that there is no such 'natural interpretation' in the case of a simple null hypothesis $H:\theta = \theta_0$, tested against the two-sided alternative, $\overline{H}:\theta \neq \theta_0$.

Bernardo[10] examines the basis of Bayesian tests, which he claims to be 'clear in principle': namely, that to test if data x are compatible with a working hypothesis H_0 it is appropriate to examine whether or not the posterior probability $P(H_0|x)$ is 'very small'. He considers the case of a 'non-informative' prior distribution (expressing prior ignorance about the parameter) and concludes that the posterior probability $P(H_0|x)$ gives a meaningful measure of the appropriateness of H_0 in the current situation only if H_0 and the alternative hypothesis H, are *both simple* or *both composite* (and of the same dimensionality). Otherwise, in the context of the Jeffreys type of test of a simple null versus a composite alternative, he states that an interpretable unambiguous conclusion necessitates letting the prior probability $P(\theta_0)$ depend on the assumed *form* of $\pi(\theta)$ over $\theta \neq \theta_0$—possibly a rather severe constraint in terms of the practical usefullness of the procedure.

Prediction in Bayesian Inference

Consider an industrial problem in which a large batch of components contains an unknown proportion θ which are defective. The components are packaged in boxes of 50, being selected at random for this purpose from the batch. It is of interest to be able to say something about how many defective components might be encountered in a box of 50 components, and this of course depends on the value of the proportion defective, θ, in the batch. Inferences about θ may be drawn in the various ways already described. We could draw a random sample of size n from the batch, observe the number of defectives, r say, and construct a classical confidence interval for θ or (more relevant to the concern of the present chapter) determine a posterior distribution for θ based on some appropriate choice of prior distribution.

If we knew the value of θ precisely we could proceed to the more important aspect of the current problem: that is, to describe the probability distribution of the number of defectives which might be present in a box of 50 components. This distribution is just binomial: $\mathbf{B}(50, \theta)$. But θ will not be known, and yet we still have the same interest in the probabilistic behaviour of the contents of the box. We encounter a new situation here: that of *predicting* the probability distribution of potential future sample data on the basis of *inferred* knowledge of θ obtained from earlier sample data (and perhaps a prior distribution for θ).

There has been growing interest in recent years in this type of problem and the Bayesian solution takes, in principle, a particularly simple form.

In the context of the general parametric model described earlier, suppose that x represents a set of data presently available and y is an independent set of potential future data. The problem of prediction amounts to obtaining an expression for $p(y|x)$: *the probability distribution of y conditional on the present data x and their implications in respect of the value of the parameter θ.* This distribution is known as the **predictive distribution** of y.

Specifically, we have

$$p(y|x) = \int_\Omega p_\theta(y)\pi(\theta|x), \qquad (6.3.6)$$

(where $\pi(\theta|x)$ is the posterior distribution of θ, given the data x) as the **predictive probability (density) function** of y.

Clearly $\pi(\theta|x)$ has been determined on the basis of some assumed prior distribution, $\pi(\theta)$, for θ, so that (6.3.6) can be written in the form

$$p(y|x) = \{\int_\Omega p_\theta(y)p_\theta(x)\pi(\theta)\}/\{\int_\Omega p_\theta(x)\pi(\theta)\}. \qquad (6.3.7)$$

$p(y|x)$ provides the complete measure of inferential import about the future data y, based on earlier data x and the prior distribution $\pi(\theta)$. However, as in other aspects of inference, it may be that some summary form of this information is adequate for practical purposes. Thus, for example, the mode of the predictive distribution might be thought of as the 'most likely' outcome for the future data set, y. Or we could obtain an interval or region with prescribed predictive probability content: a so-called **Bayesian prediction interval** (or **region**).

Example 6.6. Suppose that, in the industrial problem described at the outset of this discussion of Bayesian prediction, we adopt a beta prior distribution for θ of the form $\mathscr{B}_1(l, m)$ (see §2.3.1 and the discussion of conjugate prior distributions in §6.6 below). Then the posterior distribution of θ, having observed r defectives in a sample of n components, is $\mathscr{B}_1(l + r, m + n - r)$. If y is the number of defectives in a box of $N = 50$ components, then from (6.3.6) we can immediately determine the form of its predictive distribution. We find that y has predictive probability function

$$p(y|r) = \binom{N}{y}\frac{\mathscr{B}_1(G + y, H + N - y)}{\mathscr{B}_1(G, H)},$$

where $G = l + r, H = m + n - r$. This distribution is known as the beta-binomial distribution with parameters (N, G, H).

We could go further and determine the most likely number of defectives in a box $[(G - 1)(N + 1)/(G + H - 2)$ if $G \geqslant 1; 0$ otherwise] or obtain a Bayesian prediction interval of some prescribed probability content (but note that only certain probability levels can be achieved in view of the discreteness of y).

We have considered only the simplest form of the prediction problem. Refinements which introduce a loss structure in relation to incorrect prediction can be incorporated. There is also much interest in the corresponding formulation of the problem in classical (sampling theory) terms. (See Cox and Hinkley, 1974, Chapter 7; Guttman, 1970; Cox[11], Mathiasen[12]).

Aitchison and Dunsmore (1975) present a detailed treatment of prediction analysis from a (predominantly) Bayesian standpoint, with illustrative applications and an extensive bibliography.

6.4   PRIOR IGNORANCE

We must now consider in more detail the problem of the numerical specification of prior probabilities. At one extreme we need to recognize that practical situations may arise where we have no *tangible* (objective or subjective) prior information. To make use of Bayesian methods of inference we are, nonetheless, compelled to express our prior knowledge in quantitative terms; we need a numerical specification of the state of **prior ignorance**. A common approach is to invoke the Bayes–Laplace *principle of insufficient reason* expressed by Jeffreys (1961) in the following way:

> If there is no reason to believe one hypothesis rather than another, the probabilities are equal. . . . *to say that the probabilities are equal is a precise way of saying that we have no ground for choosing between the alternatives.* . . . The rule that we should take them equal is not a statement of any belief about the actual composition of the world, nor is it an inference from previous experience; it is merely the formal way of expressing ignorance. (pp. 33–34.)

Jeffreys discusses at length (1961, Chapter 3; and elsewhere) the extension of this principle from the situation of prior ignorance concerning a discrete set of hypotheses to the case of prior ignorance about a continuously varying parameter, θ, in a parameter space, Ω. The obvious extension for a one-dimensional parameter is to assign equal prior probability density to all $\theta \in \Omega$. Thus for a location parameter, μ, where the parameter space is the whole real line $(-\infty, \infty)$ we would choose $\pi(\mu)$ to be constant. See Example 6.4. This assignment of probability is *improper* in that we cannot ensure that $P(a < \mu < b) < 1$ for *all* intervals (a, b); but this presents no basic interpretative difficulty if we are prepared to adopt (as Jeffreys demands) a degree-of-belief view of the concept of probability. We will have no interest in making prior probability statements about μ, and $\pi(\mu)$ acts merely as a *weight function* operating on the likelihood $p_\mu(x)$ to produce, after normalization, the posterior distribution $\pi(\mu|x)$. This posterior distribution is, of course, proportional to the likelihood (cf. discussion of the generalized *likelihood inference* approach in §8.2).

Whilst recommending this approach for the derivation of the posterior distribution of μ, we have already seen (§6.3; in his concept of a Bayesian hypothesis test) that Jeffreys is not wedded to such a direct extension of the principle of insufficient reason for general application. Indeed, on the topic of hypothesis tests he remarks (1961):

> The fatal objection to the universal application of the uniform distribution is that it would make any significance test impossible. (p. 118.)

(By this he means that no odds, or probability, could be assigned to a point hypothesis.) Jeffreys proposes further limitations on the use of the uniform distribution to desribe prior ignorance about θ in inferential problems, suggesting that the appropriate distribution depends upon the nature of Ω. Using as criteria of choice the need to avoid anomalies from the improper nature of the distribution, and a desire to maintain certain invariance properties, he proposes the following principles.

(i) When $\Omega = (-\infty, \infty)$, we should use the prior uniform distribution for θ. He supports this choice by noting that if we are, on this basis, in a state of prior ignorance about θ, the same will be true for any linear function of θ.

(ii) When $\Omega = (0, \infty)$ we should choose a prior distribution proportional to $1/\theta$ since this implies that we are similarly in ignorance about any power, $\theta^{\alpha}(\alpha \neq 0)$. In this case $\phi = \log \theta$ has parameter space $(-\infty, \infty)$ and a prior uniform distribution, in accord with (i).

For bounded parameter spaces, for example where θ is the parameter of a binomial distribution, so that $\Omega = (0, 1)$, the original proposal of Bayes (and Laplace) was to use the uniform assignment of prior probability density. Jeffreys expresses dissatisfaction with this on intuitive grounds other than for 'a pure estimation problem' and suggests that in many problems it is more natural to assign discrete probabilities to specific values of θ and uniform probability density elsewhere (an idea originally proposed by Haldane[13]). Haldane has introduced an alternative specification of prior ignorance for the case where $\Omega = (0, 1)$, namely to take

$$\pi(\theta) \propto \theta^{-1}(1-\theta)^{-1}. \tag{6.4.1}$$

In spite of some attraction in terms of invariance properties, Jeffreys (1961) rejects (6.4.1) in that it gives 'too much weight to the extremes of Ω'. An opposite dissatisfaction with the Bayes—Laplace uniform distribution prompts the tentative proposal that we might compromise by taking

$$\pi(\theta) \propto \theta^{-1/2}(1-\theta)^{-1/2} \tag{6.4.2}$$

to express prior ignorance when $\Omega = (0, 1)$.

We have already noticed in the quality control example of Chapter 2, other reasons why (6.4.1) might constitute an appropriate specification when $\Omega = (0, 1)$, and we shall return to this in more detail when considering conjugate prior distribution in §6.6.

Jeffreys' proposals, (i) and (ii), for the doubly and singly infinite cases seem to meet with approval and are widely adopted. Any fundamental demonstration of their validity seems impossible, however, since we are once more in the area of personal judgements about the propriety or importance of any criteria (such as invariance) which are used as justification, and of the relevance of intuitive ideas of what is meant by ignorance. Inevitably criticisms are made of the Jeffreys proposals. Why should we use a concept of linear invariance when $\Omega = (-\infty, \infty)$, and power-law invariance when $\Omega = (0, \infty)$? To reply that $\phi = \log \theta$ is the 'natural' parameter in the latter case seems to beg the question! Kendall and Stuart (1979) remark:

> Sophisticated arguments concerning the distinction somehow fail to impress us as touching the root of the problem. pp. (165–166).

Again, Jeffreys insists on a degree-of-belief interpretation of prior distributions but it is easy to understand why some of his proposals might cause concern in frequency interpretable situations. In Example 6.2 we adopted the Bayes–Laplace idiom of assuming $P(H_1) = P(H_2) = \frac{1}{2}$. This implicitly declares that prior ignorance is equivalent to the assumption that the box contains equal numbers of each type of die. Might it not be more reasonable to describe ignorance in terms of a diffuse *meta-prior distribution* for the proportion, θ, of type I dice in the box; that is, to say that the proportion of type I dice is itself a random quantity in $(0, 1)$ about which we know nothing. This could be expressed by a prior uniform distribution for θ, or by (6.4.1) or (6.4.2), which is quite different to our earlier assumption of equality of $P(H_1)$ and $P(H_2)$ (amounting to the belief that $\theta = \frac{1}{2}$, *a priori*, with probability 1).

It is perhaps fortunate that from the practical point of view the specific method we adopt for describing prior ignorance *will seldom make any material difference to the inference we draw*. The reasons for this are outlined in the next section.

Example 6.7. A random sample of n observations $x_1, x_2, \ldots, x_n$, is drawn from a normal distribution with known mean μ, and unknown variance, θ. If nothing is known, a priori, about θ, the posterior distribution of nv/θ is χ^2 with n degrees of freedom; where $v = \sum_1^n (x_i - \mu)^2/n$.

Using the Jeffreys expression of prior ignorance for θ, we have $\pi(\theta)$ proportional to θ^{-1}. So

$$\pi(\theta|x) \propto \theta^{-(n/2)-1} \exp\left\{ -\tfrac{1}{2}\frac{nv}{\theta} \right\},$$

which implies that $Y = nv/\theta$ has p.d.f.

$$f(y) = \frac{e^{-y/2}\, y^{(n/2)-1}}{\Gamma(n/2)2^{n/2}}.$$

That is, Y has a χ^2 distribution with n degrees of freedom.

It is interesting to note that if both μ and θ are unknown and we express prior ignorance about them by *taking μ and $\log \theta$ to be independent and to both have a uniform distribution on $(-\infty, \infty)$*, then $(n-1)s^2/\theta$ is χ^2_{n-1} and $n^{1/2}(\mu - \bar{x})/s$ has a t-distribution with $(n-1)$ degrees of freedom $(\bar{x} = (1/n)\sum_1^n x_i, s^2 = \sum_1^n (x_i - \bar{x})^2/(n-1))$.

These results are similar in form, but not in interpretation, to those on the classical approach where we found that $(n-1)s^2/\theta$ is χ^2_{n-1}, and $n^{1/2}(\mu - \bar{x})/s$ is t_{n-1}. But here it is s^2 and $\bar{x}$, not θ and μ, which are the random variables. We should resist the temptation to see this similarity as adding 'respectability' to one approach from the viewpoint of the other, or indeed as justifying the particular choice of prior distributions which have been used!

In pursuing invariance considerations in relation to the expression of prior ignorance, Jeffreys (1961) produces an alternative specification for the prior distribution in the form

$$\pi(\theta) \propto \{I_s(\theta)\}^{1/2},$$

where $I_s(\theta)$ is Fisher's *information* function. (See §5.3.2.) This has been extended by others to multi-parameter problems. But the use of the information function in this way is much criticized, since its use of *sample space averaging* is anathema to many Bayesians who claim that the only legitimate expression of the data is through the likelihood of the actual realized value x.

Other principles for constructing prior distributions to represent ignorance have been advanced by Jaynes (again using *information*), Novick (see Lindley, 1971b, §12.6) and Villegas[14].

Neyman[15] is critical of the very idea of trying to represent prior ignorance in a formal way, referring to this aim as the 'easy way out' aspect of the Bayesian approach. On the use of *improper* prior distributions in general he rejects the suggestion of Cornfield[16] that we regard these as 'initial functions' rather than prior distributions. Indeed there is increasing dissatisfaction with the use of improper prior distributions. Later Cornfield[17] himself voices this. (See also the discussion in Lindley, 1971b, §8.) Dawid, Stone and Zidek[18] discuss at length the paradoxes that can arise in multi-parameter problems by using improper prior distributions and Stone[19] extends this discussion. For such multi-parameter problems, there is a growing tendency to represent prior ignorance in a variety of situations merely by declaring that our prior knowledge about the parameters may be expressed by de Finetti's idea of *exchangeability*. (§3.5.) An interesting illustration of this is given in the Bayesian analysis of the linear model presented by Lindley and Smith[20].

Quite clearly there is much dissatisfaction even *within* the Bayesian approach about how we should proceed if we know nothing a priori about θ. But this need not be a matter of crucial concern; it sometimes happens that the import of the data 'swamps' our prior information (however spare, or precise, this is) and the formal expression of the prior information becomes largely irrelevant, in a sense we now consider.

6.5 VAGUE PRIOR KNOWLEDGE

The aim of Bayesian inference is to express, through the posterior distribution of θ, the combined information provided by the prior distribution and the sample data. Inevitably, in view of the form (6.3.1) of the posterior distribution, we cannot assess what constitutes useful prior information other than in relation to the information provided by the data. Broadly speaking, the prior information increases in value the more it causes the posterior distribution to depart from the (normalized) likelihood. On this basis we would expect prior ignorance to lead to a posterior distribution directly proportional to the likelihood. When the prior uniform distribution is used this is certainly true. However, situations arise where it would be most unreasonable to claim prior ignorance about θ but, nonetheless, the information in the sample data 'swamps' this prior information in the sense that the posterior distribution is again essentially the (normalized) likelihood. In cases of this type we talk of having *vague prior knowledge* of θ.

This effect, of the prior information being outweighed by the sample data, is described by Savage (1962) as the **principle of precise measurement**.

> This is the kind of measurement we have when the data are so incisive as to overwhelm the initial opinion, thus bringing a great variety of realistic initial opinions to practically the same conclusion. (p. 20.)

Savage's remarks point to a major practical advantage of this principle: that conflicting opinions of the extent, and manner of expressing, prior information for a particular problem will often have little effect on the resulting conclusions.

The principle of precise measurement may be expressed in the following way (Lindley, 1965b, p. 21):

> . . . if the prior [ditribution of θ] . . . is sensibly constant over that range of θ for which the likelihood function is appreciable, and not too large over that range of θ for which the likelihood function is small, then the posterior [distribution] . . . is approximately equal to the [normalized] likelihood function . . .

Elsewhere (1965b, pp. 13–14) he gives a more formal mathematical statement of this principle for the case of sampling from a normal distribution.

We should note that whatever the extent of the prior knowledge this can in general be outweighed by the sample data for a sufficiently large size of sample. In this sense the principle of precise measurement leads to limiting results analogous to the classical limit laws, e.g. the *Central Limit Theorem*, although special care is needed in interpretation. The principle also enables us to interpret the (normalized) likelihood function as representing the information about θ available from the sample, *irrespective of prior knowledge* about θ. (See §6.8.)

Elsewhere Savage describes this principle as one of **stable estimation**. The basic idea seems to have been recognized for a long time. Neyman[15] attributes it to a Russian mathematician, Bernstein, and (independently) to von Mises, both during the period 1915–20.

This principle is easily illustrated by the example in Chapter 1 on the rate, λ, of radioactive decay of a substance (§1.2). We consider two different situations:

(a) 50 α-particles are observed in a period of 100 seconds;

(b) 5000 α-particles are observed in a period of 10 000 seconds.

Suppose that, on grounds of the chemical affinity of this substance to others with known properties, we have prior reason to believe that λ is somewhere in the range $0.45 < \lambda < 0.55$. Figure 6.2(a) and (b) shows, on appropriate scales, a typical prior probability density function for λ, together with the respective likelihood functions.

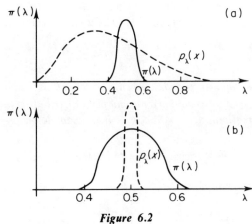

Figure 6.2

It is obvious that (a) and (b) represent distinctly different relationships between the prior density function and the likelihood function. The posterior distribution of λ is proportional to the product of these two functions. The case (b) provides a typical demonstration of the principle of precise measurement, where the prior distribution has negligible effect on the posterior distribution, which in turn is essentially proportional to the likelihood function. At the opposite extreme, in case (a), the major contribution to the posterior distribution comes from the prior distribution which is little modified by the more diffuse likelihood function. Thus we see that what appears at the outset to be tangible prior information about λ, varies in importance from one extreme to the other depending on the extent of the sample data.

6.6 SUBSTANTIAL PRIOR KNOWLEDGE

We must now consider the third possibility: that the prior information is **substantial** in the sense that it causes the posterior distribution to depart noticeably in form from the likelihood function. As with the concept of vague prior knowledge, this is an expression of the relationship between the amount of information in the data and the extent of the prior knowledge; rather than merely a reflection of the absolute import of the prior knowledge. When the prior information is quantified through a prior distribution there is no formal obstacle to the use of Bayesian methods; inferences are based on the posterior distribution (6.3.1) or its equivalent.

Example 6.8. In a biological experiment leaves infested with a certain type of insect are collected. A random sample, of size n, of these leaves shows n_i leaves with i insects on each of them (i = 1,2, . . .). The method of drawing the sample rules out the collection of any leaves with no insects on them. It is common to assume a Poisson distribution (with mean, m, say) for the unconditional distribution of the number of insects on a leaf, so that the probability that a leaf in the sample has i insects is

$$p_i = \frac{e^{-m}m^i}{i!} \bigg/ (1 - e^{-m}).$$

Prior information suggests, say, that m is in the vicinity of a value m_0, and that this is adequately expressed by assuming a normal prior distribution for m with mean, m_0, and standard deviation, $\sigma_0 \ll m_0$. So, if $\bar{x}$ is the sample mean, the likelihood is

$$p_m(x) = e^{-nm}m^{n\bar{x}} \bigg/ \left(\prod_i (i!)^{n_i}\right)(1 - e^{-m})^n$$

and the prior density of m is

$$\pi(m) = \frac{1}{\sigma_0\sqrt{(2\pi)}}\exp\left(\frac{-(m - m_0)^2}{2\sigma_0^2}\right).$$

Using (6.3.1) the posterior density of m is

$$\pi(m|x) \propto \lambda^n m^{n\bar{x}}\exp\left(\frac{-(m - m_0)^2}{2\sigma_0^2}\right),$$

where $\lambda = e^{-m}/(1 - e^{-m})$. For specific values of m_0, σ_0, m and $\bar{x}$ the constant of proportionality can be determined and $\pi(m|x)$ used to draw inferences about m.

Whilst this example is straightforward in principle, it nonetheless highlights three problems in the use of Bayesian methods when the prior information is substantial.

In the first place it is apparent that in such situations the final inferences will depend heavily on the assumed form of the prior distribution. This re-emphasizes the need to ensure that the prior distribution is an accurate and complete expression of the available prior information. Questions of the practical determination and validity of the prior distribution are at the heart of any criticism of the Bayesian approach. We have touched on this matter before, and must return to it later (§6.8).

The other two matters brought out by this example are more technical: they concern the mathematical manipulation of the posterior distribution, and its interpretation vis-à-vis the prior distribution. Even in such a simple example the derivation of the explicit form of $\pi(m|x)$ (i.e. determining the constant of proportionality) involves tedious, if not difficult, calculations. The same is true of the derivation of useful summary measures from $\pi(m|x)$ such as Bayesian confidence intervals or point estimates. Then again, in this type of example we lack any immediate 'feel' for the relative contributions of the prior information and the sample data to the combined information provided by $\pi(m|x)$. It would have been illuminating to have some simple means of comparing the contributions made by the two components.

It is in an attempt to resolve these two difficulties that the concept of *conjugate families of prior distributions* has been developed. This concept is of central importance in the *machinery* of Bayesian inference. Where applicable it can facilitate the mathematical calculations and provide a tangible comparison of the importance of the sample data and the prior information through the idea of an '*equivalent prior sample*'.

6.6.1 Conjugate Prior Distributions

As before, we are concerned with drawing inferences about some parameter θ which indexes the family of distributions $\mathscr{P} = \{p_\theta(x); \theta \in \Omega\}$, assumed as the model for the practical situation under study. Suppose the prior distribution of θ is a member of some parametric family of distributions, $\mathbf{P}$, with the property, in relation to $\mathscr{P}$, that the posterior distribution of θ is also a member of $\mathbf{P}$. If this is so we say that $\mathbf{P}$ is *closed with respect to sampling* from $\mathscr{P}$, or that $\mathbf{P}$ *is a family of* **conjugate prior distributions** *relative to* $\mathscr{P}$. More specifically if $\mathbf{P} = \{\pi_\alpha(\theta); \alpha \in A\}$, where α is the (possibly vector) parameter of the family of prior distributions, $\mathbf{P}$, with parameter space A, then the Bayesian inference process is represented simply as a mapping of A into itself. Thus if α_0 represents the prior information about θ, the sample data transform this to a new value α_1 representing the posterior information about θ. Symbolically we have

$$\alpha_0 \overset{\mathscr{P}}{\to} \alpha_1 \qquad (6.6.1)$$

and information at the *a priori* and *a posteriori* stages are measured by values of α in a common parameter space, A. An initial 'amount of information', α_0, has been enhanced through sampling to a final 'amount of information', α_1.

The potential advantages of this concept are self-evident. If only we can interpret the parameter α in terms of some properties of the sample data we have the dual advantages of being able to define the mapping (6.6.1) in simple terms, as well as measure the relative amounts of information in the prior distribution and in the sample data. For, writing α_1 as $\alpha_0 + (\alpha_1 - \alpha_0)$, we have $\alpha_1 - \alpha_0$ (expressed in terms of properties of the sample data) as a measure of the information in the sample data and can regard the prior information as that provided by an 'equivalent sample' yielding α_0. Furthermore, explicit expressions for posterior distributions involve little calculation. The posterior distribution is in the same family ⚈ as the prior distribution and all we eed to do is use our knowledge of how α relates to the sample to advance the parameter from α_0 to α_1. Such a sample-oriented interpretation of α can often be achieved, as is illustrated in the following simple example.

Example 6.9. An electronic component has a lifetime, X, with an exponential distribution with parameter θ. That is, X has probability density function

$$f_\theta(x) = \theta e^{-\theta x}.$$

A random sample of n components have lifetimes $x_1, x_2, \ldots, x_n$. The likelihood of the sample is thus

$$p_\theta(x) = \theta^n e^{-n\theta\bar{x}}.$$

Suppose the prior distribution of θ has the form

$$\pi(\theta) = \frac{\lambda(\lambda\theta)^{r-1} e^{-\lambda\theta}}{\Gamma(r)}. \tag{6.6.2}$$

then it is easy to show that the posterior distribution has precisely the same form, and is given by

$$\pi(\theta|x) = \frac{(\lambda+n\bar{x})[(\lambda+n\bar{x})\theta]^{n+r-1} e^{-\theta(\lambda+n\bar{x})}}{\Gamma(n+r)}. \tag{6.6.3}$$

We can immediately express the results of this example in the general terms above. The statistic $\bar{x}$ is sufficient for θ and essentially we are sampling at random from a gamma distribution with parameters n and $n\bar{x}$; we denote this distribution $\Gamma(n, n\bar{x})$. In this notation the prior distribution (6.6.2) is $\Gamma(r, \lambda)$ and the posterior distribution is $\Gamma(r+n, \lambda+n\bar{x})$. Thus the family of gamma prior distributions is closed with respect to sampling from a gamma distribution, and constitutes a family of conjugate prior distributions in this situation. The parameter α is the ordered pair (r, λ) and the rule of transformation (6.6.1) from prior to posterior distribution is

$$(r, \lambda) \rightarrow (r+n, \lambda+n\bar{x}), \tag{6.6.4}$$

which provides immediate access to the explicit form of the posterior distribution as well as an intuitive interpretation of the relative contributions of the prior distribution and the sample, to our posterior knowledge of θ. The transformation (6.6.4) suggests that we may regard the prior information as 'equivalent to' a 'prior sample of r observations from the basic exponential distribution yielding a sample total λ'. This concept of an *equivalent prior sample* is supported from another viewpoint. The Jeffrey's nil-prior distribution for θ (that is, his expression of prior ignorance about θ) can be represented as $\Gamma(0, 0)$. A sample of 'size' r with $n\bar{x} = \lambda$ will then, from (6.6.4), produce $\Gamma(r, \lambda)$ as the posterior distribution for θ. So to use $\Gamma(r, \lambda)$ as a *prior* distribution for θ it is as if we have started from prior ignorance about θ and taken a preliminary sample (r, λ) before obtaining the real sample $(n, n\bar{x})$. In this further respect the prior distribution (r, λ) is 'equivalent to a sample (r, λ)'. It might be tempting to feel that this argument also offers further support to Jeffreys' form of nil-prior distribution in this situation!

We must be careful not to read too much into Example 6.9 from the point of view of the *construction* of conjugate prior distributions. In the example it turned out that a singly sufficient statistic existed for θ, that the prior distribution was in the same family as the sampling distribution of the sufficient statistic and that α could be expressed in terms of the value of the sufficient statistic together with the sample size. It is not true that we will always encounter such a simple structure. No singly sufficient statistic may exist for θ, and the parameter space and sample space (even reduced by sufficiency) may be quite distinct, one from another. Even when a singly sufficient statistic exists it may not be enough merely to augment this with the sample size to construct an appropriate parameter, α, for the conjugate prior family, if we are to produce a plausible 'equivalent prior sample' concept. The role of *sufficiency* in Bayesian inference generally is somewhat less fundamental than it is in classical inference. It is obvious that the existence of sufficient statistics will be an advantage in 'boiling down' the data and in simplifying the derivation of the posterior distribution. This aspect of the importance of sufficiency in Bayesian inference (that it acts as an 'aid to computation', rather than as an obvious prerequisite for the existence of optimal, or desirable, inferential procedures) is presumably what prompts Kendall and Stuart (1979) to remark,

> . . . the only advantage of a small set of sufficient statistics is that it summarizes all the relevant information in the Likelihood Function in fewer statistics than the n sample values. As we have remarked previously, these sample values themselves *always* constitute a set of sufficient statistics, though in practice this may be only a comforting tautology. (p. 166.)

In the construction of families of conjugate prior distributions, however, sufficiency has a more important role to play. The existence of a sufficient statistic of fixed dimension independent of sample size ensures that a family of conjugate prior distributions can be found in that situation. In particular, conjugate prior distributions can be derived for sampling from any distribution in the *exponential family*, but they will exist for other distributions as well [for example, for the uniform distribution on $(0, \theta)$]. Raiffa and Schlaifer (1961, Chapter 2) develop a concept of 'Bayesian sufficiency' and discuss at length (in Chapter 3) its application to the construction of conjugate prior distributions. They develop in detail all the common families of conjugate prior distributions likely to be of practical value (in Chapter 3 in synoptic form, but an extended treatment is given in Chapters 7–13.)

Before leaving the topic of conjugate prior distributions it is useful to consider some further examples.

Example 6.10. If x represents a random sample of size n from a normal distribution with unknown mean, θ, and known variance σ^2 [denoted by $N(\theta, \sigma^2)$], and the prior distribution of θ is $N(\mu_0, \sigma_0^2)$ then the posterior distribution of θ is $N(\mu_1, \sigma_1^2)$ where

$$\mu_1 = \frac{n\bar{x}/\sigma^2 + \mu_0/\sigma_0^2}{n/\sigma^2 + 1/\sigma_0^2}, \qquad \sigma_1^2 = (n/\sigma^2 + 1/\sigma_0^2)^{-1}.$$

So for sampling from a normal distribution with known variance the family of conjugate prior distributions is the normal family. But some care is needed in developing a concept of an 'equivalent prior sample' here. We must redefine the parameters by putting σ_0^2 equal to σ^2/n_0. Then (6.6.1) has the form

$$(n_0, \mu_0) \to \left(n_0 + n, \frac{n_0 \mu_0 + n\bar{x}}{n_0 + n} \right) \qquad (6.6.5)$$

and we can now think of the prior information as equivalent to a sample of 'size' $n_0 \, (= \sigma^2/\sigma_0^2)$ from $N(\theta, \sigma^2)$ yielding a sample of mean, μ_0. Combining this equivalent sample with the actual sample produces a composite sample of size $n_0 + n$ with sample mean $(n_0\mu_0 + n\bar{x})/(n_0 + n)$, in accord with (6.6.5). Starting from prior ignorance [that is (0, 0), which is the improper uniform distribution] the 'equivalent sample' produces a posterior distribution $N(\mu_0, \sigma_0^2)$; the composite sample produces a posterior distribution

$$N\left[\frac{n_0 \mu_0 + n\bar{x}}{n_0 + n}, \sigma^2/(n_0 + n) \right],$$

which agrees with the results in Example 6.10. Note how the domain of the parameter n has had to be extended from the integers, to the half-line $(0, \infty)$. On other occasions it may be necessary to introduce *extra* parameters to facilitate the interpretation of the conjugate prior distribution. This is true of sampling from a normal distribution when *both the mean and variance are unknown*. (See Raiffa and Schlaifer, 1961, pp. 51–52.)

Another point is brought out by the following example.

Example 6.11. Binomial Sampling. Suppose r successes are observed in n independent trials, where the probability of success is θ.

If the prior distribution of θ is a beta distribution with parameters (α, β), i.e.

$$\pi(\theta) \propto \theta^{\alpha-1} (1-\theta)^{\beta-1} \qquad (\alpha > 0, \beta > 0),$$

then the posterior distribution of θ is also a beta distribution, having parameters $(\alpha + r, \beta + n - r)$.

So for binomial sampling, the family of conjugate prior distributions is the beta family. We have already observed this informally in Chapter 2. Interpreting the prior information as equivalent to a sample of 'size' $\alpha + \beta$, yielding α 'successes', seems to support Haldane's proposal (6.4.1) for an expression of prior ignorance in this situation. But the case of beta prior distributions raises certain anomalies for the concept of 'equivalent prior samples', concerned with the effects of transformations of the parameter space. (See Raiffa and Schlaifer, 1961, pp. 63–66.)

The dual advantages of mathematical tractability and ease of interpretation in the use of conjugate prior distributions are self-evident and make this concept one of major technical importance in Bayesian inference. But these potential advantages are of negligible value without an essential prerequisite. The function of the prior distribution is to express in accurate terms the actual prior information which is available. No prior distribution, however tractable or interpretable, is of any value if it misrepresents the true situation. We must rely here on the *richness* of the family of conjugate prior distributions, that is on the wide range of different expressions of prior belief which they are able to represent. When prior information consists of sparse factual measures augmented by subjective impressions, as is so often the case, a variety of different specific prior distributions may have the appropriate summary characteristics to encompass the limited information that is available. In such cases it should be quite straight forward to choose an appropriate prior distribution from the family of conjugate prior distributions. We have seen an example of this in the quality control problem discussed in Chapter 2. An interesting review of methods of quantifying subjective prior information is provided by the references numbered 3–11, inclusive, at the end of Chapter 2.

On rare occasions, however, the objective prior information may be so extensive as to essentially yield a detailed prior *frequency distribution* for θ. If in the current situation it is reasonable to assume that θ has arisen at random from the limiting form of this frequency distribution, then it is this frequency distribution which constitutes our best practical choice of prior distribution. Whether or not we may now take advantage of the desirable properties of conjugate prior distributions depends entirely on whether a member of this family *happens* to echo characteristics of the prior frequency distribution.

6.7 EMPIRICAL BAYES' METHODS; META-PRIOR DISTRIBUTIONS

Between the two extremes of prior information described at the end of the previous section there are other possibilities. Two such possibilities are the following. The prior information may consist of *limited sample data* from situations similar to the current one but of insufficient extent to build up a frequency distribution of accurate estimates of previous θ values. Alternatively, we may sometimes have available a sample of the previous *true θ values* for some similar situations. We shall consider these possibilities briefly in this section to illustrate the use of **empirical Bayes' methods**, and **meta-prior distributions**, respectively.

6.7.1 Empirical Bayes' Methods

Prominent amongst workers in this area are Robbins [21, 22], who pioneered the idea, and Maritz, who in his book on *empirical Bayes methods* (Maritz, 1970) gathers together a wide range of results, many of which derive from his own research efforts. In the Preface, Maritz broadly defines the *empirical Bayes' approach* as follows.

> [It] may be regarded as part of the development towards more effective utilization of all relevant data in statistical analysis. Its field of application is expected to lie where there is no conceptual difficulty in postulating the existence of a prior distribution that is capable of a frequency interpretation, and where *data suitable for estimation of the prior distribution* may be accumulated. (p. vii, italics inserted.)

Maritz describes the approach as a 'hybrid' one, in that whilst concerned with Bayesian inference it often employs classical methods of estimation for finding estimates of, for example, the prior distribution, based on the 'prior' sample data. We shall not be able to consider the approach in detail, or to discuss its application to estimating the prior distribution *per se*, or to testing hypotheses about θ. Furthermore, any question of the efficiency or optimality of the methods is outside our scope at present: these often involve *decision theory* ideas.

As an illustration of an empirical Bayes' procedure we shall consider a simple example concerning estimating the mean of a Poisson distribution. For Bayesian point estimation of a parameter θ we have so far suggested only the mode of the posterior distribution of θ, given the sample data, x. Choice of the mode rests on it being the value of θ having greatest posterior probability (density). But other summary measures of the posterior distribution might also constitute sensible point estimates of θ. In the next chapter we shall see that from the decision theory viewpoint, with a *quadratic loss structure*, the optimal point estimator of θ is the *mean* of the posterior distribution. That is, we would estimate θ by

$$\tilde{\theta}_\pi(x) = \int_\Omega \theta p_\theta(x)\pi(\theta) \bigg/ \int_\Omega p_\theta(x)\pi(\theta). \tag{6.7.1}$$

Suppose we apply this where θ is the mean of a Poisson distribution. Then for a single observation, x, $p_\theta(x) = e^{-\theta}\theta^x/x!$ and

$$\tilde{\theta}_\pi(x) = (x+1)\phi_\pi(x+1)/\phi_\pi(x), \tag{6.7.2}$$

where

$$\phi_\pi(x) = \int_\Omega p_\theta(x)\pi(\theta) = \frac{1}{x!}\int_\Omega \theta^x e^{-\theta}\pi(\theta), \tag{6.7.3}$$

i.e. the likelihood function smoothed by the prior distribution of θ.

So if the prior distribution were known we would have in (6.7.2) a reasonable estimator of θ. For example if θ has a prior gamma distribution, $\Gamma(r, \lambda)$, we find that $\tilde{\theta}_\pi(x)$ is $(r + x)/(1 + \lambda)$. But π is unlikely to be known. In the typical empirical Bayes situation we assume that we have, in addition to the current observation x when the parameter value is θ, a set of 'previous' observations $x_1, x_2, \ldots, x_n$ obtained when the parameter values were $\theta_1, \theta_2, \ldots, \theta_n$, say (these θ values being unknown). It is assumed that the $\theta_i(i = 1, 2, \ldots, n)$ arise as a random sample from the prior distribution, $\pi(\theta)$, and that the x_i $(i = 1, 2, \ldots, n)$ are independent sample observations arising under these values of θ. The previous observations 'reflect' the prior distribution, $\pi(\theta)$, and in the general empirical Bayes approach are used to estimate $\pi(\theta)$ for use in the Bayesian analysis. In some cases direct estimation of $\pi(\theta)$ is unnecessary and may be by-passed. This is so in the present example of estimating the mean, θ, of a Poisson distribution. Suppose that amongst our previous data the observation i occurs $f_n(i)$ times $(i = 0, 1, \ldots)$. The x_i may be regarded as a random sample from the smoothed likelihood function (6.7.3) since the θ_i are assumed to arise at random from $\pi(\theta)$. Thus a simple classical estimate of $\phi_\pi(i)$ is given by $f_n(i)/(n + 1)$ for $i \neq x$, or $[1 + f_n(x)]/(n + 1)$ for $i = x$ (including the current observation, x). The Bayes' point estimate (6.7.2) is then estimated by

$$\tilde{\theta}_\pi(n, x) = (x + 1)f_n(x + 1)/[1 + f_n(x)]. \tag{6.7.4}$$

This approach is due to Robbins[21]. For multiple observations in the previous and current situations obvious modifications will be necessary.

Any questions of the efficiency or optimality of such an empirical Bayes' procedure are complicated. They must take account of the possible variations in the parameter value itself which has arisen in the current situation, as well as sampling fluctuations in $\tilde{\theta}_\pi(n, x)$ arising from the different sets of previous data which might be encountered. Appropriate concepts of efficiency or optimality need to be defined and quantified, as do means of comparing empirical Bayes' estimators with alternative classical ones. Some of the progress and thinking on these matters is described by Maritz (1970), who also sounds a rather solemn note concerning practical application and assessment of empirical Bayes' methods:

> Regarding actual application of the techniques, many of them require rather laborious computations (p. vii),

and later:

> Unfortunately, assessment of the performance of [empirical Bayes'] methods for small amounts of prior data usually requires rather tedious and complicated

calculations, and no general results have been found. Consequently a number of important special cases have been examined in detail. (p. vii.)

Maritz remarks also on the paucity of published accounts of the application of empirical Bayes' methods, but concludes that the approach constitutes

> . . . a useful practical tool, because case studies indicate that its performance can be much better than that of conventional methods, and its application requires straight forward, albeit at times lengthy, computations. (p. viii.)

There has been a steady flow of published material on the empirical Bayes' method in the 10 years or so since the publication of Maritz's book. Applications include hypothesis testing, interval estimation, estimating a distribution function and the parameters in the binomial distribution, in the finite Poisson process and in multilinear regression.

It seems likely that we shall hear much more of this approach, though assessments of its value and interpretations of its basic nature vary from one commentator to another. Neyman[15] regards it as a major 'breakthrough' in statistical principle; Lindley (1971b, §12.1) declares that its procedures are seldom *Bayesian* in principle and represent no new point of philosophy.

6.7.2 Meta-prior Distributions

Another type of situation that Maritz regards as being within the sphere of empirical Bayes' methods is that where previous true values of θ are available relating to situations similar to the current one. He remarks that such problems have not received much attention. It is worth considering an example of such a situation to demonstrate a further extension of Bayesian methods.

Suppose some manufactured product is made in batches, for example on different machines or with different sources of a component. The quality of a product is measured by the value, x, of some performance characteristic: the corresponding quality of the batch by the mean value, θ, of this characteristic for the products in the batch. The parameter θ varies from batch to batch according to a distribution $\pi(\theta)$ which may be regarded as the prior distribution of the value relating to the batch being currently produced. We want to draw inferences about this current θ on the basis of a random sample of n products in the current batch having performance characteristics $x_1, x_2, \ldots, x_n$. There are circumstances in which our prior information may consist of exact values of θ, for previous batches, regarded as arising at random from $\pi(\theta)$. This would be so, for instance, if part of the final inspection of the products before distribution involved measuring x for each one, *and hence θ for each batch*. Suppose these previous values were $\theta_1, \theta_2, \ldots, \theta_s$. How are we to use this information in a Bayesian analysis?

To be more specific suppose we are prepared to accept that, within a batch, the x_i arise at random from a normal distribution, $N(\theta, \sigma^2)$, *where σ^2 is known*, and that $\pi(\theta)$ is also normal, $N(\theta_0, v)$, but where θ_0 and v are unknown.

If we knew θ_0 and v, then the posterior distribution of θ for the current batch, given $x_1, x_2, \ldots, x_n$, would be

$$N[(n\bar{x}/\sigma^2 + \theta_0/v)/(n/\sigma^2 + 1/v), (n/\sigma^2 + 1/v)^{-1}], \quad \text{where } \bar{x} = \sum_1^n x_i/n:$$

see Example 6.10. But θ_0 and v are not known; we have merely the random sample $\theta_1, \theta_2, \ldots, \theta_s$ from $\pi(\theta)$ from which to 'estimate' them. One intuitively appealing possibility would be to use $\bar{\theta}$ and s_θ^2, the sample mean and variance of the previous θ values, to estimate $\bar{\theta}_0$ and v. This yields

$$N[(n\bar{x}/\sigma^2 - \bar{\theta}/s_\theta^2)/(n/\sigma^2 + 1/s_\theta^2), (n/\sigma^2 + 1/s_\theta^2)^{-1}] \tag{6.7.5}$$

as an 'estimate' of the posterior distribution of θ.

But this cannot be entirely satisfactory! Sampling fluctuations will cause $\bar{\theta}$ and s_θ^2 to depart from θ_0 and v, and such departures will be more serious the smaller the size, s, of the θ-sample. The estimate (6.7.5) takes no account of this; it is irrelevant whether s is 2 or 20 000!

A more satisfactory approach might be to introduce a further preliminary stage into the inferential procedure. We can do this by declaring that $\theta_1, \theta_2, \ldots, \theta_s$ arise at random from a normal distribution, $N(\theta_0, v)$, where θ_0, v are *meta-prior parameters* having some *meta-prior distribution*, $\pi(\theta_0, v)$. We can then form the posterior distribution of θ_0 and v, given $\theta_1, \theta_2, \ldots, \theta_s$, denoted $\pi(\theta_0, v | \boldsymbol{\theta})$ and use this as '*post-prior*', distribution for the current situation, updating it by the sample data $x_1, x_2, \ldots, x_n$ to form the posterior distribution of θ: $\pi(\theta | \boldsymbol{\theta}, x)$.

Let us illustrate this for the present problem. Since our only prior information about θ is the normality assumption and the set of previous θ-values we can say nothing tangible about the meta-prior parameters θ_0 and v. The customary expression of this ignorance is to take

$$\pi(\theta_0, v) \propto 1/v. \tag{6.7.6}$$

(See §6.4.)

Using (6.7.6) we find that θ has a post-prior distribution which is $N(\theta_0, v)$ where (θ_0, v) have joint probability density function proportional to

$$v^{-(k/2)-1} \exp\{-[(s-1)s_\theta^2 + s(\bar{\theta} - \theta_0)^2]/2v\}. \tag{6.7.7}$$

Modifying this prior distribution of θ by the sample data $x_1, x_2, \ldots, x_n$, which involves averaging over (6.7.7) for θ_0 and v, finally yields the posterior density of θ as

$$\pi(\theta | \theta, x) \propto \phi(\bar{x}, \sigma^2/n) \left\{ 1 + \frac{s(\theta - \bar{\theta})^2}{(s^2 - 1)s_\theta^2} \right\}^{-k/2}, \tag{6.7.8}$$

where $\phi(\bar{x}, \sigma^2/n)$ is the density function of $N(\bar{x}, \sigma^2/n)$.

This shows just how the sampling fluctuations of θ and s_θ^2 affect the situation. As $s \to \infty$ it is easy to show that (6.7.8) tends to the density function corresponding to (6.7.5). For finite s, (6.7.8) is quite different in form to the distribution (6.7.5), although just how important this difference is from a practical point of view needs further study for special cases. (See Barnett[23].)

The Bayesian analysis of the linear model by Lindley and Smith[20] utilizes a somewhat similar meta-parametric structure, in a more fundamental manner. Prior ignorance in the multi-parameter situation is represented by assuming the parameters *exchangeable*, with prior distributions constructed hierarchically in terms of further 'hyperparameters', which in turn have their own prior distributions, and so on.

6.8   COMMENT AND CONTROVERSY

In concluding this brief survey of basic methods of Bayesian inference there are one or two further matters which need elaboration. We shall consider briefly the questions of the interpretation of the prior and posterior distributions, the roles of sufficiency and the likelihood function and the nature of the criticisms made of the Bayesian approach.

6.8.1   Interpretation of Prior and Posterior Distributions

In introducing the Bayesian approach in §6.3 for parametric models we suggested that the parameter value θ in the current situation may be thought of as *a value* (chosen perhaps 'by nature') *of random variable* with probability (density) function $\pi(\theta)$, the so-called prior distribution of θ. This prior expression of our knowledge of θ is augmented by sample data, x, from the current situation, through the application of Bayes' theorem, to yield the posterior distribution of θ, $\pi(\theta|x)$, as the complete expression of our total knowledge of θ from both sources of information.

Such a simple expression of the principle of Bayesian inference was sufficient for the development of the specific techniques and concepts discussed throughout this chapter so far. However, we have deliberately 'glossed over' the interpretation of the probability concept inherent in such an approach to inference and must now return to this matter. It is convenient to consider the prior and posterior distributions separately.

Prior Distribution. We have seen from examples how it is quite possible to encounter situations where the prior distribution both admits, and is naturally described by, a *frequency-based* probability concept. The quality control example of Chapter 2 is typical; the machinery of empirical Bayes' procedures presupposes such an interpretation. We are able to think of $\pi(\theta)$ as representing the relative frequency of occurence of the value θ in the 'super-experiment' of which the current situation is but a realization. But it is equally apparent that such a frequency interpretation will not always suffice. If θ is a measure on products made by some prototype machine, or the total rainfall during the coming month, it is difficult conceptually to define the 'super-experiment' (or 'collective' in von Mises' terms), enclosing the current situation, in order to obtain a frequency interpretation of $\pi(\theta)$. For example, suppose θ is a measure of the physical stature of Shakespeare and we wish to draw inferences about θ from 'data' derived from his allusions to men's stature in his writings. In any practical sense it cannot be right to attribute a frequency interpretation to $\pi(\theta)$. Conceptually also, the thought of an 'infinite sequence of Shakespeares, a proportion of whom had the value θ' is untenable, without a deal of 'mental juggling' (as Lindley puts it, see §1.6). To apply Bayesian methods in such a situation we are essentially forced to adopt a degree-of-belief interpretation of $\pi(\theta)$; to regard $\pi(\theta)$ as measuring the extent to which we support, from our prior experience, different values of θ as being the true one.

But it is not only in such personalistic situations that the frequency approach is untenable. We meet difficulties of interpretation even in such an apparently 'objective' problem as that of estimating the proportion, θ, of faulty items in a current batch of some product. It may be that this batch *can* be regarded as typical of a sequence of similar batches, so that there is meaning in the concept of relative frequencies of occurrence of different values of θ. Nonetheless we cannot use this frequency distribution as a prior distribution of θ *if we do not know its form*. In an extreme case we may have no tangible prior information on which to base our prior distribution, $\pi(\theta)$, and consequently choose to use a conventional expression of this prior ignorance. (See §6.4.) We are not pretending that this $\pi(\theta)$ corresponds to the relative frequency distribution of θ from batch to batch; we are merely saying that it expresses *our prior beliefs* about the *actual* value of θ in the *current* situation. The position is essentially the same if we quantify some limited objective and subjective information to form $\pi(\theta)$, perhaps by choice of a member of the appropriate family of conjugate prior distributions. The interpretation of $\pi(\theta)$ is again more naturally a degree-of-belief, rather than a frequency, one. For again we are *not* claiming that $\pi(\theta)$ *coincides* with the frequency distribution of θ from batch to batch.

Posterior Distribution. Here again it is difficult to accommodate a frequency interpretation of the probability concept, let alone insist on it. Even when the prior distribution has an immediate frequency interpretation, it is not self-evident that the posterior distribution, $\pi(\theta|x)$, can also be described in frequency terms.

Consider again the quality control problem just described, and suppose that the limiting form of the relative frequencies of occurrence of different values of θ is known precisely. We can take this as the *exact* form of the prior distribution of θ. Diagrammatically the situation is as shown in Figure 6.3. We have in mind an infinite sequence of similar batches of the product each with its corresponding

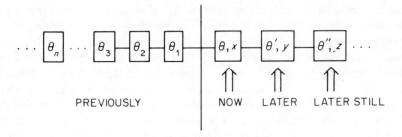

Figure 6.3

proportion, θ, of faulty items. A current batch (observed **NOW**) has unknown θ but we have extracted sample data x to reflect on θ. In the past (**PREVIOUSLY**) batches have been produced, each with its own proportion of faulty items, $\theta_1, \theta_2, \ldots, \theta_n, \ldots$, and this sequence extends indefinitely into the past determining $\pi(\theta)$, the prior distribution of θ values. The data, x, and prior distribution, $\pi(\theta)$, combine to produce the posterior distribution, $\pi(\theta|x)$, and the question that arises is how we are to interpret the probability concept inherent in $\pi(\theta|x)$. Since θ in the current situation is a unique, if unknown, value describing this situation, it seems natural (inevitable) that we should invoke a degree-of-belief interpretation of $\pi(\theta|x)$.

We might ask, however, if there is *any* sense in which we can give $\pi(\theta|x)$ a frequency interpretation. To answer this it helps to re-iterate a distinction drawn earlier (§2.3.1) on how the proper processing of *extra data* depends on whether the extra data arose in the *current* situation or in a *new* (but similar) situation.

Consider a further batch (**LATER**) with its corresponding parameter value θ'. *Suppose it yields sample data, y.* To draw inferences about θ', what should we use as the prior distribution of θ? Should it be $\pi(\theta)$ or $\pi(\theta|x)$ (since x constitutes prior information at this **LATER** stage) or perhaps something else entirely? We cannot use $\pi(\theta|x)$ for this *new* batch—this represents our views about the parameter value *in the batch which has given rise to x* (not to y). In fact we must again use $\pi(\theta)$ since this is assumed to be the *exact* distribution of θ from batch to batch, and knowing this there is nothing more to know about the random mechanism which has produced θ' for this new batch.

But the use of $\pi(\theta)$ for the new batch needs to be qualified in two respects. If y had arisen not **LATER** from a new batch with a new parameter value θ', but as an independent sample **NOW** extending the earlier data x, then it would obviously have been necessary to use $\pi(\theta|x)$ *not* $\pi(\theta)$ as the prior distribution to apply to y. This is typical of a sequential inference situation and we shall say more about this in the discussion of Bayesian methods in *decision theory* (Chapter 7). Secondly, had $\pi(\theta)$ not been a complete and precise expression of variation in θ from batch to batch then the data x *would* have provided more information on this variation. We should still *not* use $\pi(\theta|x)$ as the prior distribution in the **LATER** situation, but would want to augment the 'incomplete' $\pi(\theta)$ *appropriately* by the extra information, x, before applying it to the new data, y. What is meant by appropriately is far from clear, however, except perhaps in an empirical Bayes' problem. In this case $\pi(\theta)$ has been estimated from previous data. We can go back to this previous data, extend it with x, and derive a corresponding better estimate of $\pi(\theta)$ to apply to y.

These observations give the clue to a possible *frequency* interpretation of $\pi(\theta|x)$. Consider yet further batches (LATER STILL . . .) with their own parameter values, θ'', . . ., and sample data, z, ... Within this indefinite sequence, there will be a subsequence in which the data was the same as in the NOW situation, that is x. We can regard this subsequence as a 'collective' and develop a frequency interpretation of $\pi(\theta|x)$, in which $\pi(\theta|x)$ is the limiting relative frequency of batches *with data x* for which the parameter value is θ. But such an interpretation has to be viewed on the same level as that of the *classical* confidence interval. In both cases inferences relate to a single determined quantity, yet are interpreted in the wider framework of 'situations which may have happened'. The criticisms of the confidence interval concept (§5.7.1) must apply equally here, and it is likely that most Bayesians would, for similar reasons, reject such a frequency interpretation in favour of a degree-of-belief view of $\pi(\theta|x)$.

When no 'collective' can be defined in the practical manner illustrated above, *any* frequency interpretation of $\pi(\theta|x)$ becomes very contrived, and a degree-of-belief view inevitable.

6.8.2 Sufficiency, Likelihood and Unbiasedness

In Chapter 5 we discussed at length the central nature of these concepts in the classical approach to inference. For comparison we should consider their role in Bayesian inference.

We have already discussed *sufficiency* (§6.6) and seen it serving essentially utilitarian needs. The existence of a small set of sufficient statistics reduces the computational effort in applying Bayesian methods generally. More specifically it is a prerequisite for the existence of interpretable families of conjugate prior distributions. This latter service reflects a real importance of sufficiency in this approach to inference. There is little question here, as in the classical approach, of it being a precondition for the existence of optimal inferential procedures (but see Chapter 7 on *decision theory*).

On the other hand the *likelihood function* acts as the cornerstone of the Bayesian approach, whereas in the classical approach it acts more as a tool for the construction of *particular* methods of estimation and hypothesis testing. The real importance of the likelihood in Bayesian inference is in its function as the *sole* expression of the information in the sample data. This function is expressed through the **likelihood principle**, which has already been discussed in §5.6.

As we remarked earlier (§5.7.3), this principle is a direct consequence of Bayes' theorem. One implication of the likelihood principle is that inferences about θ will depend only on *relative variations* in the likelihood function from one value of θ to another. This leads (in the *strong* form of the likelihood principle) to the effect described as '*the irrelevance of the sampling rule*', and which constitutes one of the major philosophical ·distinctions between the Bayesian and classical approaches. Let us reconsider this by means of a specific example.

Consider a sequence of independent Bernoulli trials in which there is a constant probability of success, θ, on each trial. The observation of, say, 4 successes in 10 trials could arise in two ways; either by taking 10 trials yielding 4 successes, or by sampling until 4 successes occur which happens to require 10 trials. On the Bayesian approach this distinction is irrelevant; the likelihood is proportional to $\theta^4(1 - \theta)^6$ in each case and inferences about θ will be the same provided the prior distribution is the same in both cases. This is not true on the classical approach. The direct sampling procedure can produce quite different results to the inverse sampling one. For example, a 95 per cent upper confidence bound for θ is 0·697 in the first case, and 0·755 in the second case. For comparison, using the Haldane form (6.4.1) to express prior ignorance about θ we obtain a 95 per cent upper Bayesian confidence bound of 0·749.

Reaction to this basic distinction between the classical and Bayesian approaches will again be a matter of personal attitude. The Bayesian will have no sympathy with any prescription which takes account of the method of collecting the data, in view of the centrality of the likelihood principle in the Bayesian approach. In contrast the classicist is likely to see this as a fundamental weakness of Bayesian methods: that they cannot take account of the sampling technique. See §5.7.3 for a more fundamental discussion of this point and of the central role played by the concept of *coherence* in this debate.

The Bayesian view that the likelihood function conveys the total import of the data x rules out any formal consideration of the sample space $\mathscr{X}$ [except as the domain over which $p_\theta(x)$ is defined]. Inferences are conditional on the realized value x; other values which *may* have occurred are regarded as irrelevant. In particular, no consideration of the *sampling distribution* of a statistic is entertained; sample space averaging is ruled out. Thus there can be no consideration of the *bias* of an estimation procedure and this concept is totally disregarded. A Bayesian estimator $\tilde{\theta}(x)$ relates in probability terms to the posterior distribution of θ given the particular data x; it cannot be regarded as a typical value of $\tilde{\theta}(X)$ having a probability distribution over $\mathscr{X}$. (But again see the decision theory applications in Chapter 7.)

6.8.3   Controversy

It is necessary to recognize that there are many statisticians who feel unable to sympathize with, or adopt, the Bayesian approach to inference. This attitude is often justified by claims of 'lack of objectivity' and 'impracticality', but we need to cut through the emotive nature of such criticism to appreciate the substance of the dissatisfaction. Some objections are raised to the use of the concept of 'inverse probability' as a legitimate tool for statistical inference. Fisher was particularly vehement in his rejection of this concept as we have observed earlier (§1.6). But most current criticism concerns the basic nature of the prior distribution, and its quantification. Dissatisfaction is expressed with the use of prior distributions where the essential form of the problem precludes a frequency interpretation. We have seen (§1.6) that von Mises, whilst committed to the Bayesian approach, conceived of its application only in frequency-interpretable situations; Hogben (1957, Chapters 5 and 6) likewise. Others would claim that the *whole approach* is untenable since it sometimes requires a subjective or degree-of-belief concept of probability. In this sense it is not 'objective' and thus not appropriate for a formal scientific theory of inference!

Then again, even ignoring such philosophical complaints, the Bayesian approach meets opposition and rejection on the grounds that the prior information itself is often subjective—its quantification correspondingly 'arbitrary'. How can we adopt an approach where the conclusions may depend critically on ill-formulated or imprecise information—which may vary from individual to individual, or time to time, in its formal expression?

Criticism of this type is well illustrated by Pearson[8], relatively free from the emotive undertones often associated with such expressions of opinion:

Let me illustrate some of my difficulties [with subjective Bayesian methods] very briefly.

a) We are told that "if one is being consistent, there is a prior distribution". "A subjectivist feels that the prior distribution means something about the state of his mind and that he can discover it by introspection". But does this mean that if introspection fails to produce for me a stable and meaningful prior distribution which can be expressed in terms of numbers, I must give up the use of statistical method?

b) Again, it is an attractive hypothesis that Bayesian probabilities "only differ between individuals because individuals are differently informed; but with common knowledge we have common Bayesian probabilities". Of course it is possible to define conceptual Bayesian probabilities and the "rational man" in this way, but how to establish that all this bears a close relation to reality?

It seems to me that in many situations, if I received no more relevant knowledge in the interval and could forget the figures I had produced before, I might quote at intervals widely different Bayesian probabilities for the same set of states, simply because I should be attempting what would be for me impossible and resorting to guesswork. It is difficult to see how the matter could be put to experimental test. Of course the range of problems is very great. At one end we have the case where a prior distribution can be closely related to past observation; at the other, it has to be determined almost entirely by introspection or (because we do not trust our introspection) by the introduction of some formal mathematical function, in Jeffreys' manner, to get the model started. In the same way utility and loss functions have sometimes a clear objective foundation, but must sometimes be formulated on a purely subjectivist basis.

To have a unified mathematical model of the mind's way of working in all these varied situations is certainly intellectually attractive. But is it always meaningful? I think that there is always this question at the back of my mind: can it really lead to my own clear thinking to put at the very foundation of the mathematical structure used in acquiring knowledge, functions about whose form I have often such imprecise ideas?

Pearson's remarks reflect the nature of most criticism of the Bayesian approach. It is not part of our purpose to take sides on such issues—any approach to inference involves personal judgements on the relevance and propriety of its criteria. We have seen equally forceful criticism of the classical approach. In essence it revolved around the same issues of how well the approach meets up to its practical aims; how 'arbitrary' are its criteria, concepts and methods. In broad terms, one can find criticisms of the two approaches which are basically the same criticism. Again see §5.7.3.

The principle justification of the Bayesian approach advanced by its advocates is its '*inevitability*': that if we accept the ideas of *coherence* and *consistency* then prior probabilities (and utilities) *must* exist and be employed. We shall take this up in more detail in the next chapter.

In conclusion we might remind ourselves of one or two matters.

(i) We encountered in §1.3 a powerful plea from Savage for the incorporation of *subjective* information in statistics.

(ii) The 'principle of precise measurement' (§6.5) is crucial. It implies that in a large number of situations detailed quantitative expression of the prior information is unnecessary. The Bayesian approach in such situations is *robust*; the data 'swamps' the prior information and different observers must arrive at essentially the same conclusions.

(iii) In situations where the prior information is quantitative and relates to a prior distribution admitting a frequency interpretation, one hears little objection expressed to the principle, or practice, of Bayesian inference.

References

Textbooks on Bayesian inference have been discussed in §6.1. The professional journals over the last 30 years have contained a vast amount of material on Bayesian inference, often related to decision theory. One particular area of activity should be singled out for special mention; that is, the attempt to reconcile, or contrast, results in Bayesian inference with analogous results in other approaches. In this context we have, amongst other work, that by Pearson[8], Pratt[9], Bernardo[10], Lindley[24] (Bayesian and fiducial interval estimation, see

§8.1), Peers[25] (confidence intervals) and Bartholomew[26], as well as the references numbered 10, 16, 17, 19, 20 and 22 at the end of Chapter 1. Interesting applications of Bayesian methods to different practical problems have been considered by Mosteller and Wallace (1964) (on the authorship of published material), by Freeman[27] (on the existence of a fundamental unit of measurement in megalithic times) and by Efron and Thisted[28] ('How many words did Shakespeare know?').

1. Bayes, T. (1963). 'An essay towards solving a problem in the doctrine of chances', *Phil. Trans. R. Soc.*, **53**, 370–418.

† 2. Barnard, G. A. (1958). 'Thomas Bayes—a biographical note' (and reproduction of reference 1), *Biometrika*, **45**, 293–315.

† 3. Edwards, A. W. F. (1978). 'Commentary on the arguments of Thomas Bayes', *Scand. J. Statist.*, **5**, 116–118.

4. Lindley, D. V. (1972). 'Bayesian statistics', *Bull. Inst. Math. Applic.*, **8**, 183–187.

† 5. Lindley, D. V. (1975). 'The future of statistics—a Bayesian 21st century', *Adv. appl. Probab.*, Suppl. **7**, 106–115.

† 6. Lindley, D. V. (1978). 'The Bayesian approach', *Scand. J. Statist.*, **5**, 1–26.

7. de Finetti, B. (1974). 'Bayesianism: its unifying role for both the foundations and applications of statistics', *Int. Statist. Rev.*, **42**, 117–130.

† 8. Pearson, E. S. (1962). 'Some thoughts on statistical inference', *Ann. Math. Statist.*, **33**, 394–403.

† 9. Pratt, J. W. (1976). 'A discussion of the question: For what use are tests of hypotheses and tests of significance?', *Commun. Statist. –Theor. Meth.*, **A5,** 779–787.

10. Bernardo, J. M. (1980). 'A Bayesian analysis of classical hypothesis testing', in Bernardo, De Groot, Lindley and Smith (1980).

† 11. Cox, D. R. (1975). 'Prediction intervals and empirical Bayes confidence intervals', in Gani, J. (ed.) (1975) *Perspectives in Probability and Statistics*. London: Academic Press (for the Applied Probability Trust).

12. Mathiasen, P. E. (1979). 'Prediction functions', *Scand. J. Statist.*, **6**, 1–21.

13. Haldane, J. B. S. (1932). 'A note on inverse probability', *Proc. Camb. Phil. Soc.*, **28**, 58–63.

14. Villegas, C. (1975). 'On the representation of ignorance', *J. Amer. Statist. Assn.*, **72**, 651–654.

† 15. Neyman, J. (1962). 'Two breakthroughs in the theory of statistical decision making', *Rev. Int. Statist. Inst.*, **30**, 11–17.

16. Cornfield, J. (1961). See Neyman[15].

† 17. Cornfield, J. (1969). 'The Bayesian outlook and its application' (with Discussion), *Biometrics*, **25**, 617–657.

18. Dawid, A. P., Stone, M. and Zidek, J. V. (1973). 'Marginalization paradoxes in Bayesian and structural inference' (with Discussion), *J. Roy. Statist. Soc. B*, **35**, 189–233.

† 19. Stone, M. (1976). 'Strong inconsistency from uniform priors', *J. Amer. Statist. Assn*, **71**, 114–116.

20. Lindley, D. V. and Smith, A. F. M. (1972). 'Bayes estimates for the linear model' (with Discussion), *J. Roy. Statist. Soc. B*, **34**, 1–41.

† 21. Robbins, H. (1955). 'An empirical Bayes approach to statistics', in *Proceedings of the Third Berkeley Symposium on Mathematical Statistics and Probability*, Vol. 1. Berkeley: University of California Press, pp. 157–164.

† 22. Robbins, H. (1964). 'The empirical Bayes approach to statistical decision problems', *Ann. Math. Statist.*, **35**, 1–20.

23. Barnett, V. D. (1958). 'Bayesian, and decision theoretic, methods applied to industrial problems', *The Statistician*, **22**, 199–226.

24. Lindley, D. V. (1958). 'Fiducial distributions and Bayes' theorem', *J. Roy. Statist. Soc. B*, **20**, 102–107.

† 25. Peers H. W. (1968). 'Confidence properties of Bayesian interval estimates', *J. Roy. Statist. Soc. B*, **30**, 535–544.

† 26. Bartholomew, D. J. (1971). 'A comparison of frequentist and Bayesian approaches to inferences with prior knowledge', in Godambe and Sprott (1971), pp. 417–434.

† 27. Freeman, P. R. (1976). 'A Bayesian analysis of the megalithic yard' (with Discussion), *J. Roy. Statist. Soc. A*, **139**, 20–55.

† 28. Efron, B. and Thisted, R. (1976). 'Estimating the number of unknown species: How many words did Shakespeare know?', *Biometrika*, **63**, 435–447.

CHAPTER 7

Decision Theory

The third major component in our survey of distinct attitudes to statistical inference and decision-making is **decision theory**. This differs from *classical statistics* or *Bayesian inference* in a variety of fundamental respects.

The first, and prime distinction, is that its function is solely a decision-making one. In so far as it is applied to such apparently inferential problems as parameter estimation, these must be re-expressed in a 'decision-making' context. Its basic premise is that some action must be taken, chosen from a set of well defined alternatives, and that we seek in some sense the best action for the problem in hand. Certain features of the situation in which action must be taken are unknown. Depending on the prevailing circumstances the different actions have differing merit. It is assumed that some numerical value can be assigned to the combination of circumstance and action to represent its merit. Such values, perhaps termed *utilities* in the spirit of Chapter 4 or *losses* if we prefer to work in terms of demerit (or disadvantage), provide the basis for assessing how reasonable (or unreasonable) a particular action is under given circumstances.

But the actual circumstances that the action encounters will not be known in general. Whilst unknown, it may be possible, however, to obtain *sample data* which provide some information about the prevailing circumstances; we may even have some prior knowledge concerning the propensity for different circumstances to arise. We shall see that decision theory provides a means of exploiting any such information to determine a reasonable, or even a best, course of action.

It is on the assessment of the relative advantages of different prescriptions for action that we encounter the second major distinction with alternative approaches to statistics. Although the methods proposed for choice of action do take into account probability considerations, in the sense of any prior probabilities we can assign to the different possible circumstances or of any probabilistic mechanism governing the generation of the sample data, the success of the methods is not expressed in direct probability terms. This is in distinct contrast to Bayesian inferences which are entirely probabilistic: the final inference is a probability distribution. In decision theory any prescribed course of action is seen to be better than some other to the extent that its *average* utility is higher (or *average* loss lower). Such an average is taken with respect to both the prior information and the potential sample data, if these are appropriate information components for the problem in hand. The actual difference between these average values tells us how much better (in recognizable, quantified, terms) one course of action is than some other, *on average*. There is no recognition in the final result that the relative advantages of the one or other course of action may vary probabilistically, that sometimes one will be best, sometimes the other.

Note that in averaging over the different probability structures all the familiar 'bones of contention' must arise: concerning the quantification of prior information, the interpretation of the probability concept and the use of aggregate or long-term considerations in relation to sample data. At the same time there is an obvious appeal in a basic system designed *specifically* as a prescription for action in the face of uncertainty, rather than merely as a means of delimiting the uncertainty without indicating how this knowledge might guide any necessary action. We shall need to return to such questions as whether this extended function is a proper responsibility of the statistician, and, more important, whether it is a viable practical possibility or an idealistic abstraction.

It is important to recognize that even within the limited time-span of the development of statistical theory and practice, decision theory is very much an infant progeny.

Although both Laplace and Gauss regarded errors of observation as 'losses' and justified the *method of least squares* on the basis of minimizing such losses (and Gauss considered the idea of a 'least disadvantageous game') no substantial development of principles for statistical decision-making arose until the work of Abraham Wald in the 1940s.

This individual approach represented a unique departure from earlier statistical ideas, and immediately awakened an excited response that has hardly abated in its enthusiasm to the present time. As a result an impressive range of applications and developments have appeared in the literature: both as fundamental research contributions and as textbooks at various levels of sophistication.

Wald's pioneering book (1950) presented the foundations of the subject, with extensions and generalizations, as 'an outgrowth of several previous publications of the author on this subject' in the professional journals. He explains how decision theory was motivated by dissatisfaction with two serious restrictions in current statistical theories: the assumption of fixed sample sizes (the sequential approach to statistics was very much in its infancy) and the limited scope of decision-making procedures as represented by classical hypothesis testing and estimation.

> The general theory, as given in this book, is freed from both of these restrictions. It allows for multi-stage experimentation and includes the general multi-decision problem. (p. v.)

Wald makes it quite clear how decision theory as a general principle for guiding the making of decisions *in the face of uncertainty* flows naturally out of the (then) recently developed *deterministic* 'theory of games'—essentially from the two-person zero-sum games strategies developed in the work of von Neumann and Morgenstern (1953; first published 1944).

By any standards the lone contribution made by Wald is most impressive. Subsequent work has built on his sound and broad basis; in extending the particular areas of application; generalizing, interpreting and analysing basic features; refining the Bayesian contribution; following through the tortuous details of the sequential aspects of the theory; reassessing classical ideas in the new light and arguing about philosophical implications and practical application. (It is only to be regretted, however, how little real-life *case study* material seems to have been published.) We shall review in the following sections some of the results of the feverish activity of the last 30 years or so.

Among many excellent modern books on decision theory, De Groot (1970) figures as a detailed and comprehensive treatment at an intermediate level. See also Berger (1980). Elementary texts of merit include Aitchison (1970), Chernoff and Moses (1959), Lindgren (1971), Lindley (1971a), Raiffa (1968), Schlaifer (1959) and Winkler (1972). The more advanced work by Raiffa and Schlaifer (1961) contains a wealth of detailed material on decision theory and Bayesian inference still not readily accessible elsewhere.

7.1 AN ILLUSTRATIVE EXAMPLE

In anticipation of the more formal development of the later sections of this chapter, it is convenient to demonstrate the principles and techniques of decision theory by considering a simple practical example in some detail. This is of the 'finite state-space, finite action-space' variety commonly encountered in elementary texts. Whilst such examples inevitably greatly oversimplify the practical decision-making problems they claim to represent, they do serve as a useful introduction. Let us consider (somewhat uncritically) the following situation.

Cheap wrist watches are sold through a chain of department stores and supermarkets. No guarantee is given on these watches but as a mild form of compensation the manufacturer agrees to service any watch once only at a modest charge if it does not operate satisfactorily. Customers return faulty watches with a remittance of the service charge and a simple indication of the nature of the fault. No detailed repairs are carried out—the watch is either cleaned, or the works replaced, or *both* (if cleaning proves inadequate). The servicing facility is primarily a public relations exercise and is not intended to serve a commercial purpose in its own right. It is obviously desirable to keep the cost of the service as low as possible, subject to the manufacturer's assessment of what he is prepared to pay for the promotion advantages of running the scheme.

For any watch a decision must be made on whether to replace the works, or merely to clean the watch in the hope of remedying the trouble by this cheaper procedure. The manufacturer needs to determine a policy for deciding between the two actions of initially cleaning or immediately replacing the works. He also needs to determine an appropriate level for the service charge.

We can set this situation in a decision-theory framework. Consider what happens when a watch is received for service. Two *actions* are possible, denoted a_1 and a_2.

a_1: initially clean the watch,
a_2: immediately replace the works.

The set of possible actions is called the **action-space**; here it is very simple, $\mathscr{A} = \{a_1, a_2\}$.

Let us suppose that there are just two possible faults, either that dust has got into the works, or that the watch has been mechanically deranged (a cog buckled, or a spring broken). These are the unknown circumstances under which action must be taken. They may be thought of as the possible '*states of nature*'.

θ_1: dust in works,
θ_2: mechanical damage.

Thus we have a **state-space**, $\Omega = \{\theta_1, \theta_2\}$, which may equivalently be thought of as a parameter space: the *parameter* θ takes values θ_1 or θ_2.

It is apparent that the different actions will have different implications depending on the state of nature. (Cleaning the watch is clearly insufficient if it is mechanically damaged.) Suppose it costs 2 units on some monetary scale to clean the watch, 5 units to replace the works. We can construct a **table of losses** to represent this. The entries $L(a_i, \theta_j)$ are the **losses** (what it costs) if action a_i is taken when the state of nature happens to be θ_j. The losses $L(a_i, \theta_j)$ illustrate the idea of *utilities* defined in Chapter 4 where numerical values were assigned to *consequences* (i.e. the conjunction of an action and the circumstances it encounters). We shall consider in more detail later the relationship between losses and utilities. All that Table 7.1 is really saying is that it costs 2 units to clean a watch, 5 to replace the works.

Table 7.1

	θ_j	
a_i	θ_1	θ_2
a_1	2	7
a_2	5	5

Now if we knew θ for any watch there would be no decision problem. For a damaged watch we would immediately replace the works; for one with dust in the works we would merely clean it. From the table of losses these are the most economical actions to take in each case. But the state of nature will not be known and let us suppose that any detailed investigation to determine it is uneconomical. What now should we do when a watch comes in for service?

Suppose that although θ is not known for any particular watch, experience has shown that only about 30 per cent of watches that are received are suffering from damaged mechanisms. Thus we have some *prior probability* for θ_2, and hence for θ_1. This provides us with useful information, since we can now determine that the *average loss* from taking action a_1 would be $2 \times 0.7 + 7 \times 0.3 = 3.5$, whilst that from taking action a_2 would be $5 \times 0.7 + 5 \times 0.3 = 5$. So in the long run it is going to be cheaper (by 1·5 units per watch) merely *to clean each watch in the first instance* rather than to immediately replace the works.

Note how the best policy changes with the prevailing prior probabilities of θ_1 and θ_2 (proportions of watches in the two categories). If it happened that as many as 70 per cent of the watches received had in fact been mechanically damaged then the average losses become:

$$\text{for } a_1, 5.5; \quad \text{for } a_2, 5.0.$$

Here it would make sense always to replace the works rather than initially to clean the watch, although the advantage is far less dramatic than in the earlier situation.

The information that is available about θ in the form of the prior probabilities, $\pi(\theta_1)$ and $\pi(\theta_2)$, enables a choice of action to be determined once and for all with the assurance that *on average* it is cheaper, and with a measure of the extent of its advantage. (Of course it would be unfortunate if, when operating a_2 in this spirit, an enormous batch of dusty watches arrived. But that's life, we play for the average in decision theory!) With the loss structure of Table 7.1 prescribed for this problem it is apparent that there is some neutral prior probability $\pi(\theta_1)$, namely $\pi(\theta_1) = 0.4$, for which the average losses for actions a_1 and a_2 are *both* 5 units. Thus the choice of action is irrelevant if it should happen that 40 per cent of watches received for service were only suffering from exposure to dust.

Apart from any knowledge of the propensity for the two states of nature to arise, there is an additional source of information in the remarks made by the customer concerning his dissatisfaction. Suppose these fall into three categories: the watch has ceased functioning completely, is erratic in its accuracy or will not run for very long without being rewound. Imagine that this information exists for *each* watch; it plays the role of *sample data*. For any watch we have an observation x taking one (only) of the forms

x_1: watch has stopped operating,
x_2: watch is erratic in its timekeeping,
x_3: watch only runs for a limited period.

Such information must give some indication of the state of nature, through the *likelihood* function. This may well be known quite accurately from past records, which yield estimates of the probability distributions $\{p_\theta(x); x = x_1, x_2, x_3\}$ for $\theta = \theta_1, \theta_2$. Suppose these are as follows.

Table 7.2 Likelihoods $p_\theta(x)$

	x_1	x_2	x_3
θ_1	0·1	0·4	0·5
θ_2	0·7	0·2	0·1

We might now decide to let the observation x guide what action should be taken, by considering different **decision rules** (or **strategies**) for action. *A decision rule $\delta(x)$ tells us what action to take if we observe x.* In this simple example there are just eight possibilities.

Table 7.3. Decision rules (or strategies)

	δ_1	δ_2	δ_3	δ_4	δ_5	δ_6	δ_7	δ_8
x_1	a_1	a_1	a_1	a_1	a_2	a_2	a_2	a_2
x_2	a_1	a_1	a_2	a_2	a_1	a_1	a_2	a_2
x_3	a_1	a_2	a_1	a_2	a_1	a_2	a_1	a_2

But how are we to choose between these on the basis of the losses, and the likelihood functions? *In terms of likelihood alone* (see Table 7.2) x_1 supports θ_2, whilst x_2 and x_3 support θ_1. So from the loss table (Table 7.1) $\delta_5 = (a_2, a_1, a_1)$ appears intuitively reasonable; in contrast δ_4 seems perverse whilst δ_1 and δ_8 essentially ignore the data. But although x_1 has higher probability under θ_2 than under θ_1, this does not imply that θ_2 *must* prevail. To take account of this feature (particularly important if we observe x_2) it is usual to represent the different decision rules in terms of their *average losses over different possible values of x in each state of nature*. Thus the decision rule $\delta(x)$ is represented by the pair of values

$$R(\delta, \theta) = \sum_x L[\delta(x), \theta] p_\theta(x) \qquad (\theta = \theta_1, \theta_2).$$

$R(\delta, \theta)$ is called the **risk function** (over θ) for the decision rule $\delta(x)$. For example, for δ_5 we have $(R_{5,1}, R_{5,2})$ where

$$R_{5,1} \equiv R(\delta_5, \theta_1) = L(a_2, \theta_1) p_{\theta_1}(x_1) + L(a_1, \theta_1) p_{\theta_1}(x_2) + L(a_1, \theta_1) p_{\theta_1}(x_3)$$

$$= 5 \times 0·1 + 2 \times 0·9$$

$$= 2·3,$$

and, similarly,

$$R_{5,2} = 5 \times 0 \cdot 7 + 7 \times 0 \cdot 3$$
$$= 5 \cdot 6.$$

Table 7.4 shows the values of the two components in the risk functions for all eight different decision rules.

Table 7.4. Risk functions $(R_{i,j})$

θ_j	δ_1	δ_2	δ_3	δ_4	δ_5	δ_6	δ_7	δ_8
θ_1	2·0	3·5	3·2	4·7	2·3	3·8	3·5	5·0
θ_2	7·0	6·8	6·6	6·4	5·6	5·4	5·2	5·0

So we can assess the decision rule δ_i by considering the risks $R_{i,1}$ and $R_{i,2}$ under the two states of nature. But unfortunately (and typically) none of the decision rules is *uniformly* best in having both component risks simultaneously as small as possible. Figure 7.1 shows the risk functions in graphical form.

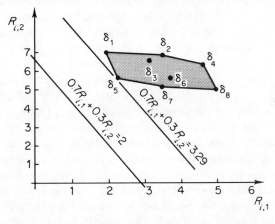

Figure 7.1 Risks for the different decision rules

Apparently there is no clear-cut choice of a best decision rule on the basis of their risk functions, although some (δ_2, δ_3, δ_4 and δ_6) may be immediately ruled out since others have *both* risk components which are smaller. Such rules are said to be **inadmissible**. One principle sometimes advanced (arising from *games theory* considerations) for choice among those remaining is to choose that decision rule for which *the maximum risk is as small as possible*. This **minimax principle** yields the decision rule δ_8 for our problem, which declares that *we should always replace the works, irrespective of the customer's indication of the nature of the malfunction*. On this principle the same data are irrelevant, although the probability distributions of Table 7.2 are not. But the minimax principle is a poor compromise; it implies a thoroughly pessimistic outlook. Why should we assume that the *worst* possible eventuality will arise and act on this basis? Admittedly it provides the basis for a unique choice of decision rule for action, but it can be a dearly bought reconciliation.

Suppose, for example, that the vast majority of watches were merely in need of a clean; then why incur an inevitable cost of 5 units for each watch, when in most cases we need only expend 2 units on cleaning (and very occasionally 7 units, if the watch happens to need a new mechanism). This brings us back to the question of the possible additional source of information that might be provided by prior probabilities $\pi(\theta_1)$ and $\pi(\theta_2)$, if they happen to be known. If we *combine the prior information on θ with the sample data* we can avoid both of the unsatisfactory alternatives of indecision or pessimism. All we need do is to further average our losses with respect to the prior information, and now represent each decision rule δ by a single measure

$$r(\delta, \pi) = R(\delta, \theta_1)\pi(\theta_1) + R(\delta, \theta_2)\pi(\theta_2).$$

We shall see later that it makes sense to call $r(\delta, \pi)$ the **posterior expected loss** for the decision rule δ with resect to the prior distribution $\{\pi(\theta_1), \pi(\theta_2)\}$. It seems reasonable now to choose as the best decision rule, the so-called **Bayes' decision rule** (or **Bayes' solution**), that one which minimizes $r(\delta, \pi)$. The corresponding minimum value of $r(\delta, \pi)$ is called the **Bayes' risk**.

Thus if we incorporate the earlier values $\pi(\theta_1) = 0.7$, $\pi(\theta_2) = 0.3$ we find values of $r(\delta, \pi)$ for $\delta_1, \ldots, \delta_8$ as 3·50, 4·49, 4·22, 5·41, **3·29**, 4·28, 4·01, 5·00, respectively, so that the optimum decision rule is δ_5 which declares that *we should take note of the customer's indications and initially clean the watch, unless it has stopped operating at all*. The Bayes risk for this policy is 0·21 units less costly than the most attractive option (ignoring the data) of always initially cleaning the watch. Note how the minimax decision rule is now seen as almost the worst possible one!

By virtue of the fact that the Bayes risk is 0·21 units less than the average loss for the optimum decision rule based on the prior information alone, we obtain a measure of the *value of the sample data*. Merely to take note of the customer's comments and communicate them from department to department may involve administrative costs in excess of 0·21 units. If so, the customer's comments are best ignored and the 'no-data' solution of *always cleaning the watch in the first instance* should be adopted, at an average cost of 3·5 units. This also provides the basis for an economical costing of the service of re-instating customers' watches to an adequate running condition.

One simple way of representing the above analysis, which clearly demonstrates

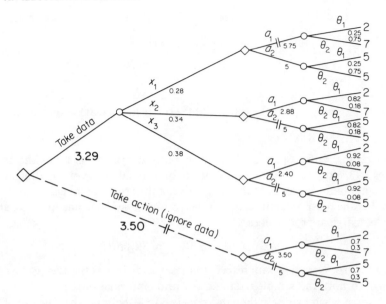

Figure 7.2 Decision tree

its operation, is to construct what is called a **decision tree**. This is merely a diagrammatic representation of the sequence of alternative possible actions and the circumstances which they encounter. Figure 7.2 shows the decision tree for the current problem where, working from left to right, the diamond nodes ($\diamondsuit$) generate the actions and the circles ($\circ$), the random circumstances. Thus the first action is to take note of the data, or ignore them and take immediate action to renovate the watch. If we ignore the data we are faced with a choice of action, either a_1 or a_2. In each case we encounter one of the states of nature: either θ_1 or θ_2. If we take account of the data, we encounter x_1, x_2 or x_3 and in each case we have a choice of action a_1 or a_2. Either choice encounters one of the states of nature, θ_1 or θ_2.

The qualitative function of the decision tree is self-evident. It enables us to trace out in a logical way the full sequence of actions and circumstances. But the decision tree will obviously have limited utility. Unless the numbers of alternative actions, states of nature and possible data outcomes are small it will just not be feasible to construct a decision tree. We are not far off this limit in the current simple problem.

The decision tree once constructed may be used to determine the merit of different possible courses of action, or decision rules. The **decision analysis** proceeds in reverse order (right to left) by assigning losses (or average losses) to the different actions, or probabilities to the different circumstances. The probabilities attaching to the states of nature θ_1 and θ_2 are, of course, the posterior probabilities $\pi(\theta_i|x_j)$ if we have taken note of the data. Thus, for example,

$$\pi(\theta_1|x_1) \propto p_{\theta_1}(x_1)\pi(\theta_1),$$

$$\pi(\theta_2|x_1) \propto p_{\theta_2}(x_1)\pi(\theta_2),$$

so that

$$\pi(\theta_1|x_1) = (0{\cdot}1 \times 0{\cdot}7)/[(0{\cdot}1 \times 0{\cdot}7) + (0{\cdot}7 \times 0{\cdot}3)] = 0{\cdot}25$$

$$\pi(\theta_2|x_1) = \hspace{6cm} 0{\cdot}75.$$

Averaging with respect to these probability distributions yields the average losses 5·75, 5 etc., for the different actions, conditional on the observations x_1, etc. (or on not taking account of the data). We can thus rule out (indicated ‖) for each x_i all but the most economical action. The probabilities attaching to x_1, x_2 and x_3 are the marginal probabilities

$$p(x_i) = p_{\theta_1}(x_i)\pi(\theta_1) + p_{\theta_2}(x_i)\pi(\theta_2),$$

and a further averaging with respect to these yields 3·29 as the average loss (with respect to both the sample data and the prior information).

Thus we reach precisely the same conclusion as before—that the use of both sources of information yields an optimum decision rule (a_2, a_1, a_1) with average loss 3·29, whilst ignoring the data we should take action a_1 for an average loss of 3·50. So if the handling of the data costs ξ units, all data-based actions must have their average losses increased by ξ and the use of the data can only be justified if $\xi < 0{\cdot}21$.

One general principle is illustrated in this form of decision analysis: that in order to determine the Bayes' solution we do not need to consider all possible decision rules. It suffices to determine for each observation x_i the *best action* $a(x_i)$ in the sense of that which has minimum expected loss with respect to the posterior probability distributions $\{\pi(\theta_1|x_i), \pi(\theta_2|x_i)\}$. This is indeed the principle that is adopted in practice. Some interesting worked examples using decision trees are given by Raiffa (1968), Lindley (1971a) and Aitchison (1970). All three of these books, in their distinct styles, are interesting modern elementary treatments on decision-making with few mathematical demands on their readers. Lindley, in particular, addresses his remarks directly to a lay audience of 'business executives, soldiers, politicians, as well as scientists; to anyone who is interested in decision-making'. In the same spirit of popular enlightenment are two articles by Moore and Thomas[1] which appeared in the *Financial Times* of London, demonstrating in simple terms the construction and use of decision trees. The *Sunday Times* of London, on 12 February 1978, reported on a rather disturbing example of elementary decision theory on the matter of the relative costs to a company of calling in a large number of cars to modify a design fault, or of not doing so.

Returning to our example, let us consider how the choice of the 'best' decision rule would change in this problem for different possible prior distributions

$\{\pi(\theta_1), \pi(\theta_2)\}$. Obviously the *minimax rule* is unaffected, since it takes no account of $\pi(\theta_1)$. But the *Bayes' solution* will change. We saw above that only δ_1, δ_5, δ_7 and δ_8 are worth considering, and indeed as $\pi(\theta_1)$ varies from 1 to 0 the Bayes' solution proceeds through these in order. The Bayes' solution is that δ for which $r(\delta, \pi)$ is a minimum. Any decision rule satisfying

$$R_{i,1}\pi(\theta_1) + R_{i,2}\pi(\theta_2) = c$$

has the same posterior expected loss: namely c. Thus to determine the Bayes' solution we need to find that δ_i which satisfies this equation for the smallest possible value of c. This minimum value of c is then the Bayes' risk. Geometrically, in relation to Figure 7.1, this amounts to choosing that δ_i from which a line with slope $-\pi(\theta_1)/\pi(\theta_2)$ has minimum intercept on the $R_{i,1}$ axis. Thus when $\pi(\theta_1) = 0.7$ we obtain δ_5 with $c_{min} = 3.29$ as required. (See Figure 7.1.) As we change the slope [i.e. $\pi(\theta_1)$] we obtain different Bayes' solutions. These turn out to be

$$\delta_1: \text{for } \pi(\theta_1) \geqslant 0.824,$$

$$\delta_5: \text{for } 0.250 \leqslant \pi(\theta_1) \leqslant 0.824,$$

$$\delta_7: \text{for } 0.118 \leqslant \pi(\theta_1) \leqslant 0.250,$$

$$\delta_8: \text{for } \pi(\theta_1) \leqslant 0.118.$$

This dependence of the optimum rules $\delta_1, \delta_5, \delta_7$ and δ_8 on the value of $\pi(\theta_1)$ can be illustrated by a graph of the posterior expected losses for each of the four decision rules. This is shown as Figure 7.3. Note that when $\pi(\theta_1)$ is particularly small the Bayes' solution is the same as the minimax rule. It is easy to confirm also

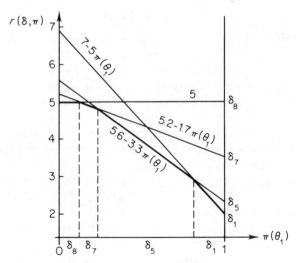

Figure 7.3 Posterior expected losses

that the prior distribution for which the Bayes' risk is as large as possible is given by $\pi(\theta_1) = 0$: that is, when the two 'optimum rules' coincide. This illustrates a general principal which we shall consider later.

The above analysis of a simple problem has illustrated many of the basic features of decision theory, and many of the properties of optimum decision rules. The following sections discuss these in a more formal manner.

7.2 BASIC CONCEPTS AND PRINCIPLES

We have seen in the discussion of utility theory in Chapter 4 a possible basis for the choice of action in the face of uncertainty. In §4.6 there was outlined a principle of choosing that action, a, which has maximum expected utility over the set of possible circumstances which might be encountered. In the current context the 'circumstances' correspond to the different states of nature, θ, which can arise. Let us briefly review this earlier discussion.

Each combination of a possible action, a, and state of nature, θ [what was described as a *consequence* (a, θ)], was assigned a numerical value $U(a, \theta)$, its *utility*. Any action, a, becomes interpretable as a *prospect* if we think of it in terms of a probability distribution, where values $U(a, \theta)$ arise with probabilities $\pi(\theta)$, which are the probabilities of the different states of nature. The probability for each θ remains the same whatever action is contemplated, it is the set of utilities which varies with a. Since θ is unknown we cannot just use the set of utilities $\{U(a, \theta); \theta \in \Omega\}$ for direct choice of a best action, a. However, we have a prior distribution, $\pi(\theta)$, and utility theory proposes that any action a should be assessed in terms of the *expected* value of $U(a, \theta)$ with respect to the prior distribution, i.e.

$$\int U(a, \theta)\pi(\theta), \qquad (7.2.1)$$

which is defined to be the *utility of the prospect*, or *action*, a. It is then appealing to *choose that action which has the largest (expected) utility* in this sense.

Note how utility theory, through Property I of §4.4, advances (7.2.1) as the *inevitable* and proper quantitative measure of the action, a. Outside the confines of utility theory, however, it becomes meaningful to ask why the single summary measure, the *mean value* of $U(a, \theta)$ over Ω, should be chosen. Why restrict our assessment of any possible action to this extent, or indeed to this particular single measure? The full assessment of the action, a, is in terms of the complete set of utilities and probabilities $\{U(a, \theta), \pi(\theta); \theta \in \Omega\}$. We want to choose that action, a, for which this set is most attractive. The use of (7.2.1) for this purpose must rest on the credibility of formal utility theory as an appropriate vehicle for expressing, and contrasting, our preferences.

Two further points arise. The first concerns the probabilities, $\pi(\theta)$. It is quite likely that we do not know these, or that at best we have only some vague (possibly subjective) knowledge. In adopting the *expected utility* standpoint in these circumstances we encounter all the problems discussed in Chapter 6 of expressing intangible prior information, or prior ignorance, in quantitative form. Also, no provision is made in basic utility theory for incorporating any extra information about θ that might be available in the form of sample data.

The second point concerns the relationship between the utilities $U(a, \theta)$, in utility theory, and the more familiar measures of *loss*, $L(a, \theta)$, in the Wald-style decision theory formulation. It is appealing to think of the losses as merely *negative* utilities, and to replace the principle of maximizing expected utility with one of minimizing expected loss. But this is not entirely satisfactory. Lindley's proposal (§4.8) that utilities should be so normed as to admit a direct probability interpretation will not necessarily be satisfied for the negative losses in any practical problem. However, the norming is purely conventional; ultilities are unique only up to linear transformations, and the optimum action in the utility theory sense is obviously invariant with respect to such transformations. So it is still possible that the losses are just negative utilities. But there is nothing in decision theory (*per se*) which *guarantees* that losses are so interpretable; that if we change their sign we obtain, for the problem in hand, that unique set of utilities whose existence is implied by the tenets of utility theory. Decision theory *assumes* that losses exist; utility thoery *implies* that utilities exist.

In less fundamental terms we should note that additional assumptions are sometimes made concerning the losses $L(a, \theta)$. One attitude commonly adopted is for any particular state of nature, θ, what is important is the loss *relative to the best action*. This implies that for each θ there is an action a with $L(a, \theta) = 0$. Such a scheme can be produced from any loss structure by redefining the losses as

$$L(a, \theta) - \inf_{\mathscr{A}} L(a, \theta),$$

for each θ. Chernoff and Moses (1959) refer to such relative losses as **regrets**. See also Lindgren (1971). In terms of utilities, the *regret* may be taken as simply

$$\sup_{\mathscr{A}} U(a, \theta) - U(a, \theta)$$

If we choose the best action in the sense of minimizing expected loss over Ω it is clearly unimportant whether we operate in terms of loss or relative loss. For in decision theory the basis for optimum choice amounts to choosing a either to minimize

$$\int L(a, \theta)\pi(\theta) \tag{7.2.2}$$

in absolute loss terms, or to minimize

$$\int \left[L(a, \theta) - \inf_{\mathscr{A}} L(a, \theta)\right]\pi(\theta) \tag{7.2.3}$$

in relative loss terms. Since $\inf_{\mathscr{A}} L(a, \theta)$ depends on θ alone, the resulting choice of optimum action must be the same in either situation. On the other hand, alternative criteria for choice of the best action will not necessarily yield the same results if we work with absolute loss rather than regret. The *minimax* procedure is one such example.

7.2.1 *The Decision Theory Model*

In decision theory we make the following assumptions. In relation to any problem we assume that there is a well defined set of possible actions, a, jointly constituting the **action space,** $\mathcal{A}$. In addition, we assume that different specified states of nature, θ, might prevail. The set of possible states of nature is assumed to be known and comprises the **state-space** (or *parameter space*), Ω. Furthermore, there is a function $L(a, \theta)$ defined on the product space, $\mathcal{A} \times \Omega$. This is the **loss function**; the individual values $L(a, \theta)$ measuring the **loss** which arises if we take action a when the prevailing state of nature is θ. (Whether this loss is measured in absolute or relative terms is immaterial as far as structure is concerned; it is a matter of what seems most appropriate to the problem in hand.) The aim is to choose a best action with respect to the loss function $L(a, \theta)$. What constitutes a 'best action' in terms of its form and properties will depend on the extent, and basis, of any information we might have about the prevailing state of nature. The extreme (if unrealistic) possibility is that θ is known. Here the choice of action is straightforward. When θ is not known, we may have available some prior distribution $\pi(\theta)$ to guide the choice of action. Alternatively some sample data may be available to throw light on the value of θ. We shall consider these possibilities separately, discussing first the no-data situation, then the situation where sample data exists.

7.2.2 *The No-data Situation*

Here we must distinguish between different possibilities, depending on the extent of our information about θ.

θ **known.** If θ were known the choice of the best action is simple. The sole measure of how different actions compare is provided by the loss function $L(a, \theta)$. *We should choose that action for which $L(a, \theta)$ is as small as possible.* The case of θ known is of little practical importance. We include it merely for completeness.

θ **unknown.** In this case we may have some information about θ in the form of a *prior distribution*, $\pi(\theta)$. This might be objectively based on past experience of similar situations, reflecting the relative incidence of different values of θ (as in the watch repair example in §7.1). Alternatively, it may represent degrees-of-belief in different values of θ based on some mixture of subjective judgements and objective evidence. Then again, $\pi(\theta)$ may have been chosen to reflect the relative importance of safeguarding our actions for different possible values of θ; it is merely a general weight function. Finally, it may seem that we have no tangible information about θ, and $\pi(\theta)$ has been chosen as a conventional expression of this state of prior ignorance.

Having specified $\pi(\theta)$, it now seems sensible to assess our possible actions in terms of the *prior expected losses*

$$\int L(a, \theta)\pi(\theta),\qquad\qquad(7.2.4)$$

and to *choose as the best action that one which has minimum expected loss* (7.2.4).

This is directly analogous to the utility theory principle of maximizing the expected utility.

Of course, all the familiar points of dispute arise concerning how we should construct $\pi(\theta)$, and of the dependence of the optimum choice on the assumed form of $\pi(\theta)$. But these have been effectively aired elsewhere in this book (notably in Chapter 6).

In this situation decision theory advances no criteria for choice of action other than that above: namely, of minimizing the single summary measure provided by the mean value of $L(a, \theta)$ with respect to $\pi(\theta)$. If we are not prepared to entertain a prior distribution over Ω, there seems no formal basis for choice of action in decision theory terms, unless some alternative form of information about θ is available (or we adopt the *minimax* principle).

7.2.3 Sample Data

In certain circumstances we may have some information, about θ, provided in the form of sample data obtained from conducting an experiment whose outcomes depend on the value of θ. Suppose the data are x, with likelihood function $p_\theta(x)$. This alternative source of information about θ might provide some assistance in our choice of action. Again we consider the two cases, θ known and θ unknown.

θ **known.** This is trivial! If we know θ, x can tell us nothing extra and we merely act as in the *non-data* situation.

θ **unknown.** Here there is much to be considered. We need to distinguish between the possibilities of having no prior information about θ [and no desire to express this in conventional terms through an appropriate form for $\pi(\theta)$], or of having a prior distribution for θ to augment the information provided by the sample data.

Let us consider first of all the way in which sample data *alone* are used to assist in the choice of action. This is somewhat similar to their use in estimation or hypothesis testing in *classical statistics*. There we considered mappings from the sample space, $\mathcal{X}$, to the parameter space, Ω. In decision theory we seek to identify regions of the *action space*, $\mathcal{A}$, that warrant consideration, rather than regions of the parameter space. It is natural, therefore, to consider mappings from $\mathcal{X}$ to $\mathcal{A}$. Suppose $\delta(x)$ identifies an action in $\mathcal{A}$ corresponding to the data x. Then $\delta(X)$ is such a mapping, and is known as a **decision rule**. It tells us which action to take when we encounter the sample data x. Of course, there are many possible decision rules, and we need to determine which of them are advantageous as a basis for choice of action. Indeed it would be most desirable if we could identify a *best* decision rule in some sense.

How are we to assess the value of any particular decision rule? It is suggested that we do so in terms of the long-run *expected loss*; that is, as the average loss with respect to different data which might arise. Thus for any decision rule $\delta(X)$ we consider

$$R[\delta(X), \theta] = \int_{\mathcal{X}} L[\delta(x), \theta] p_\theta(x). \qquad (7.2.5)$$

$R[\delta(X), \theta]$ is of course a function of θ, and is commonly known as the **risk function**. [A slight confusion of terminology arises here. Aitchison (1970) uses a quite individual terminology throughout his elementary treatment of decision theory. Chernoff and Moses (1959) reserve the term 'risk' for sample-average *regrets*, that is for relative losses, and talk of 'expected loss' for the case of absolute losses.]

Thus any decision rule, $\delta(X)$, has its corresponding risk function $R[\delta(X), \theta]$. We might hope to choose between the different decision rules in terms of these risk functions, preferring those with smallest risk! But we immediately encounter difficulties. Suppose θ varies continuously over $\mathbf{R}^1$. Then for four decision rules, $\delta_1(X), \delta_2(X), \delta_3(X)$ and $\delta_4(X)$, their risk functions may well appear as in Figure 7.4.

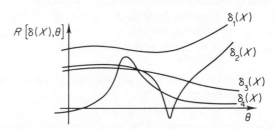

Figure 7.4 Typical risk functions

How are we to choose between these? None of them has smallest risk *simultaneously for all* θ. Certainly $\delta_1(X)$ can be ruled out since it is uniformly worse than $\delta_2(X)$, $\delta_3(X)$ and $\delta_4(X)$. But as far as $\delta_2(X)$, $\delta_3(X)$ and $\delta_4(X)$ are concerned, for some values of θ one is better than another whilst for other values the reverse is true. This is essentially the same difficulty that we encountered in discussing UMP tests in Chapter 5. Once again, in decision theory, we typically do *not* encounter decision rules with uniformly minimum risk (except in very special circumstances). This was true even in the very simple example of §7.1.

If we have no grounds for distingushing between different possible values of θ (in terms of prior probabilities, or of it being more important to minimize risks in certain regions of Ω) there is little more that can be done to effect a choice between the different decision rules. Certainly some [such as $\delta_1(X)$ above] can be immediately rejected, but the remainder are essentially indistinguishable. One purely intuitive criterion for a unique choice is afforded by the **minimax** procedure which seeks to ensure that *the worst that can happen is as good as possible*. The *minimax* choice from among $\delta_1(X)$, $\delta_2(X)$, $\delta_3(X)$ and $\delta_4(X)$ is the decision rule $\delta_3(X)$. But this implies a most conservative attitude; after all $\delta_4(X)$ seems much more attractive except in the local area of its maximum. In complete opposition to the *pessimistic* nature of the minimax criterion, Chernoff and Moses suggest that we might sometimes be prepared to act as if the best circumstances will prevail and that we should accordingly choose the **minimin** decision rule which minimizes the *minimum* risk. In our example this would be $\delta_2(X)$, but again the fruits of false optimism are most bitter! [Note the extreme contrast between the two decision rules, $\delta_4(X)$ and $\delta_2(X)$.]

But suppose in addition that we do have some information about θ, in the form of a prior distribution $\pi(\theta)$. We now have a basis for further distinction between the different decision rules, through weighting the risk function $R[\delta(X), \theta]$ by $\pi(\theta)$. Some summary measure of the probability distribution of $R[\delta(X), \theta]$ over Ω might now be singled out as a basis for choosing between the different $\delta(X)$. For example we could use the *median*, or *mean*, value of $R[\delta(X), \theta]$ for this purpose (if they exist). Conventionally it is the *mean* that is used, and *the best decision rule is defined as that one which has minimum mean risk with respect to variations in θ*. This is known as the **Bayes' decision rule**, and is that decision rule $\delta(X)$ which minimizes

$$r(\delta, \pi) = \int_\Omega R(\delta, \theta)\pi(\theta) = \int_\Omega \pi(\theta)\{\int_{\mathcal{X}} L[\delta(x), \theta]p_\theta(x)\}. \qquad (7.2.6)$$

Its resulting mean risk, $\min_\delta r(\delta, \pi)$, is called the **Bayes' risk**, and it provides an absolute basis against which to assess the value of other (non-optimum) decision rules.

The use of the *mean* value of $R[\delta(X), \theta]$ warrants some comment. Apart from wanting a decision rule which has *on average* a low value for $R[\delta(X), \theta]$, we might also be concerned that the range of values of $R[\delta(X), \theta]$ over Ω should not be excessive. So perhaps some constraint should be imposed on this variation in values of $R[\delta(X), \theta]$. One possible scheme of this type has been described by Lehmann (1959), who suggests restricting attention only to those $\delta(X)$ for which $R[\delta(X), \theta] < G$ over Ω, for some choice of

$$G > \min_\delta \max_\Omega R[\delta(X), \theta].$$

From this restricted class we again choose that rule which minimizes (7.2.6) to obtain the **restricted Bayes' decision rule**.

So far all the proposed criteria have been given merely an intuitive justification. However, the *Bayes' decision rule* has a more formal justification in certain circumstances; namely *when the (negative) losses* $\{-L(a, \theta)\}$ *may be interpreted as utilities* in the sense of utility theory. For if we accept the premises of utility theory, this renders the losses unique (up to linear transformations) and implies that preferences between decision rules *(as prospects) must* be assessed in terms of mean risks: the lower the mean risk the more attractive the decision rule. Thus the *Bayes' decision rule* is inevitably the optimum choice. There can be no such formal justification of the *restricted Bayes'* principle.

Furthermore, we cannot justify such principles as the *minimax* one on the grounds of having no prior information about θ. In the dual development of utility and subjective probability by Ramsey (see Chapter 4) there is implied *the existence* of a prior distribution $\pi(\theta)$, over Ω, so that we inevitably return (on this standpoint) to the Bayes' decision rule as the optimum choice for action under uncertainty.

Thus the Bayes' decision rule has a double claim to respectability.

In the first place, in utility theory terms it is 'inevitably' the best procedure. To accept utility theory as an appropriate model for describing the way people should react to uncertainty is to induce an essentially unique utility (or loss) structure and prior probability distribution and to compel choice of action to be made in relation to mean risk. This implies that the Bayes' decision rule is optimum. The actual determination of the appropriate losses and *subjective* prior probabilities in any particular situation is fraught with the difficulties already discussed in Chapters 3 and 4. Apart from having to accept the *subjective* basis of $\pi(\theta)$, its determination (and that of the loss structure) must be approached by such introspective aids as the individual's reactions to hypothetical betting situations. The structure and the choice of the best course of action remain entirely personal to the individual facing the problem. Some would claim that this is entirely appropriate—that the decision-maker must be responsible solely to himself in terms of what is judged the best course of action. Others see the utility theory model as invalid as a general prescription for action, and regard its personalistic basis as inappropriate to the study of practical problems.

Secondly, even without the formal justification of utility theory the Bayes' decision rule has some appeal in practical terms. If losses are well defined, in being closely tied to measurable economic factors, and the prior distribution is accessible (as, for example, where it relates to the relative incidence of different θ-values in earlier situations of a kind similar to the current one), then few would object to the decision theory model as a plausible one for choice for action. As such, the *mean* risk seems at least a *sensible* way of summarizing the value of a decision rule. (On this level, of course, it is not unique in this respect.) Correspondingly, the Bayes' decision rule seems a reasonable basis for choosing the best course of action. We shall see later, through the idea of *admissibility*, an added practical support for this.

Note how criticism of decision theory may operate at two distinct levels. It is either philosophical in nature, in rejecting utility theory and its subjective basis. Alternatively it may take a purely practical form, in denying the possibility (in all but the most trivial problems) of adequately specifying the action space or of eliciting the appropriate loss structure or prior probabilities. It was precisely this latter concern which directed Neyman, Pearson, Fisher and others away from considering prior probabilities and consequential costs to a sole concern for sample data. (See Chapter 5, Introduction.)

We have seen in this section what are the basic ingredients of decision theory, how they interrelate and in what terms a choice of action is to be made. The structure is complex. There are three basic spaces: of *actions, states of nature* and *sample data.* We aim to choose an action in the light of how the actions vary in their propriety under different states of nature, as represented by the *loss structure.* Not knowing the true state of nature we utilize information provided from prior knowledge and sample data. The resulting prescription for action takes the form of a mapping from the sample space to the action space, chosen to yield the best average return with respect to possible variations in the state of nature and sample data. This system is represented diagrammatically in Figure 7.5.

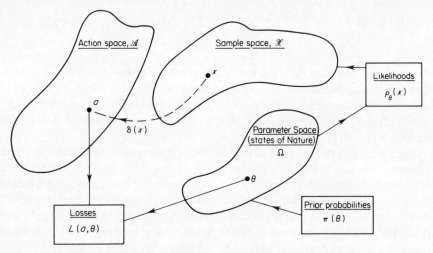

Figure 7.5   The superstructure of decision theory

7.3   ATTAINMENT AND IMPLEMENTATION

We have seen what the decision theory approach proposes as a general basis for the choice of action under uncertainty. We must now consider in more detail the essential properties of reasonable decision rules, and how such rules are determined.

7.3.1   *Admissibility and Unbiasedness*

The fundamental problem of decision theory is to choose the best decision rule $\delta(X)$ from the class of all possible decision rules. We have seen how, without introducing a prior distribution $\pi(\theta)$, it is unlikely that any decision rule will have *uniformly* minimum risk over Ω. This being so, it is interesting to ask if there are any additional criteria that might be invoked to reduce the range of different decision rules which should be considered. A similar attitude was taken in respect of hypothesis tests in Chapter 5, where the field was limited by considering specific types of probability distribution (for instance, those with monotone likelihood ratio) or by imposing extra practical requirements such as *unbiasedness* or *invariance*.

An initial reduction may be achieved without extra conditions, as we have already seen in § 7.2. Referring to the example illustrated by Figure 7.4 we see that the decision rule $\delta_1(X)$ may be discarded because it is *uniformly worse than some other decision rule*. There is no point in considering such decision rules.

This prompts the following notions. For any two decision rules $\delta(X)$ and $\delta'(X)$ we say that $\delta'(X)$ is **dominated** by $\delta(X)$ if

$$R[\delta'(X), \theta] \geqslant R[\delta(X), \theta] \qquad (\text{all } \theta \in \Omega),$$

and, for some $\theta \in \Omega$

$$R[\delta'(X), \theta] > R[\delta(X), \theta].$$

We need only consider those decision rules which are not so *dominated*. The decision rule $\delta(X)$ is **admissible** if there is no other decision rule which dominates it. A class of decision rules is **complete** if for any $\delta'(X)$ *not* in this class there is a $\delta(X)$ in the class which dominates it. The class is **minimal complete** if it does not contain a complete subclass.

The minimal complete class of decision rules in any problem contains all these decision rules which are worth considering, and we can obviously reduce the scale of our enquiries by restricting attention to this class. But how are we to determine the minimal complete class in any situation? If this involves considering *all* decision rules and rejecting the inadmissible ones we have gained nothing. In fact this is not necessary, since under certain conditions it is possible to characterize the *admissible* decision rules in a rather special way. We have the important result, due to Wald, that: *any admissible decision rule $\delta(X)$ is a Bayes' decision rule with respect to some prior distribution $\pi(\theta)$*. For this to be true we need to include the possibility that the required $\pi(\theta)$ might be *improper* (see § 6.4), but this is not unprecedented and we have discussed its practical implications elsewhere. In a special situation, namely when the state space and action space both contain a finite number of elements, we have an even stronger result. This says that *any* Bayes' solution with respect to a *strictly positive* prior distribution $\pi(\theta)$, is admissible.

This is an important link between *admissibility* and *Bayes' decision rules*. It provides a mechanism which assists in determining the minimal complete class. But it also implies a special practical importance for Bayes' decision rules, which transcends philosophical arguments about the existence of prior probabilities (or of their personalistic basis). It suggests as a pragmatic policy in decision making that *we should consider only the range of decision rules generated as Bayes' decision rules with respect to different prior distributions.* This is not because we wish to incorporate specific prior information about θ, but because in adopting this policy we will generate all possible *admissible* decision rules for consideration. A fuller discussion of this topic, at an intermediate level, is given by De Groot (1970, Chapter 8); or in greater detail by Wald (1950).

One point must be emphasized. Any decision rule that is inadmissible, is so *with respect to the particular loss function which is being used.* Thus to reject a decision rule on grounds of inadmissibility is to place crucial emphasis on a particular loss structure as an appropriate description of costs and consequences for the problem being studied. This conditional nature of admissibility is important!

The idea of admissibility is well illustrated by a simple example on drawing conclusions about the location parameter of a distribution. [A somewhat similar example is discussed by Chernoff and Moses (1959).]

Example 7.1. *A manufacturer markets a product on which he is required to state its weight. The manufacturing process is assumed to be in a controlled state; a random sample of n items is chosen and the individual items weighed. The information gained from this sample data, x, is to be used to specify a mean weight for the product. We shall assume that the weights of individual components have a normal distribution,* $\mathbf{N}(\theta, \sigma^2)$*. This estimation problem can be set in a decision theory mould by augmenting the sample data with a loss function representing consequential costs incurred by stating an incorrect value for the mean, θ. Indeed, this loss structure may be most important. Depending on circumstances, costs of understatement or overstatement may be more serious (compare packets of biscuits with weights of containers used for air freight). There may even be legal obligations to be met in respect of the quoted weight, making the need to safeguard against weight discrepancies in one direction, or the other, paramount.*

In decision theory terms the action space here coincides with the parameter space. The action we must take is to state the value of θ; a decision rule $\delta(X)$ is an estimator of θ. In such estimation problems it is common to take a **quadratic loss function**

$$L[\delta(x), \theta] = [\delta(x) - \theta]^2.$$

(Note how this is unbounded, in contrast to the common utility theory assumption.) Such a loss function is unlikely to be adequate in general. We must at least allow it to be scaled in some way. Indeed the remarks above militate against its implicit symmetry. Purely for illustration we shall consider a loss function

$$L[\delta(x), \theta] = h(\theta)[\delta(x) - \theta]^2.$$

which makes some slight concession to reality through the scaling function $h(\theta)$.

Suppose we consider just three possible decision rules, or estimators: the sample mean

$$\delta_1(X) = \overline{X},$$

the sample median

$$\delta_2(X) = M,$$

and the single value θ_0 (independent of the data),

$$\delta_3(X) = \theta_0.$$

The risk functions are easily determined. They are

$$R_1(\theta) = h(\theta)\,\sigma^2/n,$$

$$R_2(\theta) \doteq 1{\cdot}57\,h(\theta)\,\sigma^2/n,$$

$$R_3(\theta) = h(\theta)\,(\theta - \theta_0)^2,$$

and are shown in typical form in Figure 7.6.

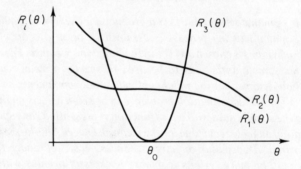

Figure 7.6   Risk functions for mean, median and θ_0

We see immediately that the median can be rejected. It is inadmissible, since it is dominated by the mean. But then the mean is not uniformly the best. The strange estimator θ_0 will be better if θ happens to be close to θ_0. So we are in a dilemma. The estimator θ_0 is intuitively unattractive. Surely the data are relevant! And yet the local behaviour of this estimator, when θ happens to be in the region of θ_0, makes it impossible to adopt the sample mean on any formal basis.

But suppose the reason we considered θ_0 was because of some prior information about θ: say, that θ has a prior normal distribution, $N(\theta_0, \sigma_0^2)$. To fix ideas consider the simple case of

$$h(\theta) = \alpha + \beta\theta.$$

Then the posterior expected losses are just

$$r_1 = (\alpha + \beta\theta_0)\sigma^2/n,$$

$$r_2 = 1\cdot57(\alpha + \beta\theta_0)\sigma^2/n,$$

$$r_3 = (\alpha + \beta\theta_0)\sigma_0^2.$$

We see that $\delta_3(X) = \theta_0$ might now have a real claim. Indeed $r_3 < r_1$ if it happens that

$$n < \sigma^2/\sigma_0^2.$$

In other words, unless our sample is of sufficient size we seem best advised to ignore the data and merely adopt the prior mean value of θ, namely θ_0, rather than using $\overline{X}$ or M. (This typically needs quite extensive prior knowledge of θ, with $\sigma_0^2 \ll \sigma^2$.)

Returning to our theme of enquiring how we might reduce the number of decision rules that need to be cosidered, we encounter another possibility other than the criterion of *admissibility*. As in classical statistics, we might consider imposing external constraints supported by practical considerations. In fact there has been proposed a general criterion of **unbiasedness** of decision rules. This is applied under special conditions.

Suppose that for each $\theta \in \Omega$ there is a *unique* correct action $a \in \mathscr{A}$, and that each action, a, is correct for some θ. (This is true, for example, in 'estimation' problems such as the one just described.) Suppose further that $L(a, \theta_1) = L(a, \theta_2)$ for all a whenever the same action is correct for both θ_1 and θ_2. Then $L(a', \theta)$ depends only on the contemplated action a', and the action $a(\theta)$ which is *correct* for θ, and can be regarded as a metric on $\mathscr{A}$ with values $L[a', a(\theta)]$ measuring 'how far apart' a' and $a(\theta)$ are. In such cases a decision rule $\delta(X)$ is **unbiased** if for all $\theta \in \Omega$, and $a' \in \mathscr{A}$,

$$R[\delta(X), a'] \geq R[\delta(X), a(\theta)].$$

That is, $\delta(X)$ is unbiased if on average it is at least as far from any incorrect action as it is from the correct one.

Example 7.2. Consider again an estimation problem. We want to estimate $\theta \in \Omega$, using a quadratic loss function

$$L(a, \theta) \propto (a - \theta)^2.$$

Here $\mathscr{A} \equiv \Omega$, and the conditions are satisfied for an unbiased decision rule to be constructed. Suppose $\delta(X)$ is unbiased. Then if θ is the true value, and θ' any false value,

$$\int_{\mathscr{X}} [\delta(x) - \theta']^2 p_\theta(x) \geq \int_{\mathscr{X}} [\delta(x) - \theta]^2 p_\theta(x) \quad (all\ \theta, \theta').$$

Thus

$$2(\theta' - \theta) \int_{\mathcal{X}} \delta(x) p_\theta(x) \leqslant (\theta')^2 - \theta^2.$$

Considering separately values θ' less than, or greater than, θ, but arbitrarily close to θ, leads to the conclusion that

$$\int_{\mathcal{X}} \delta(x) p_\theta(x) = \theta,$$

so that unbiasedness of $\delta(X)$ is equivalent to the usual classical concept of an unbiased point estimator.

The idea of unbiasedness seems to have some intuitive appeal, and certainly restricts the number of decision rules which would need to be considered. Yet it appears to play little part in the development of decision theory—in distinct contrast to the centrality of the corresponding idea in classical point estimation or hypothesis testing. One contributory factor in this neglect is the predominance of the subjective Bayesian attitude in much of the study of decision theory. The idea of 'repetition under essentially identical conditions' implied in averaging over the sample space, $\mathcal{X}$, is anathema to this viewpoint. We shall see later that the construction of Bayes' decision rules (also described in terms of such averaging) is not such an obvious inconsistency of attitude as it might appear at first sight!

7.3.2 Determination of Bayes' Decision Rules

There are three factors supporting the use of the *Bayes' decision rule* as a reasonable prescription for action. These are

   (i) its formal *optimality* in utility theory terms,

   (ii) its inevitable *admissibility*,

   (iii) its intuitive appeal when tangible prior information about θ is available

If we accept the utility theory model and (negative) losses are equivalent to utilities then the Bayes' decision rule is the best one to use—determined with reference to the 'inevitable' subjective prior distribution in the Ramsey sense.

Without the utility theory justification, admissibility compels the use of *some* Bayes' decision rule (under fairly generous conditions). If in addition we can meaningfully adopt a particular prior distribution, $\pi(\theta)$, to describe our prior information about θ, then the Bayes' decision rule with respect to that $\pi(\theta)$ becomes the natural choice. [Strictly speaking we would need to check that it is admissible. It will be so if the 'generous conditions' are satisfied: in particular if there are finite numbers of actions and states of nature, and $\pi(\theta)$ is strictly positive.]

This raises the question of how we should construct the Bayes' decision rule with respect to $\pi(\theta)$ (ignoring for the moment the problem of how we are to derive an appropriate action space, loss structure and prior distribution in any practical situation).

We seek that decision rule which minimizes $r(\delta, \pi)$, as defined by (7.2.6). This involves choosing a minimizing *function* in the right-hand side of (7.2.6) which is a notoriously difficult task in practice, except in certain simple situations such as the small-scale finite type of problem described in §7.1. Fortunately, it is often possible to adopt an alternative approach which is much more tractable. This amounts to interchanging the orders of integration, and advancing the minimization stage. It will certainly be possible always to obtain the Bayes' decision rule in this way when the loss function is bounded and $\pi(\theta)$ is proper, although the boundedness condition is not essential as we shall see in Example 7.3 later.

We have to choose the function $\delta(x)$ to yield

$$\min_{\delta} \int_{\Omega} \pi(\theta) \left\{ \int_{\mathcal{X}} L[\delta(x), \theta] p_{\theta}(x) \right\}.$$

But this can be rewritten as

$$\min_{\delta} \int_{\Omega} \int_{\mathcal{X}} L[\delta(x), \theta] p_{\theta}(x) \pi(\theta).$$

Note that $p_{\theta}(x) \pi(\theta)$ is just $\pi(\theta|x) p(x)$ where $\pi(\theta|x)$ is the *posterior distriution* of θ, given x, and $p(x)$ is the marginal distribution of X. So we want to choose $\delta(x)$ to obtain

$$\min_{\delta} \int_{\Omega} \int_{\mathcal{X}} L[\delta(x), \theta] \pi(\theta|x) p(x) = \int_{\mathcal{X}} p(x) \min_{a} \left[\int_{\Omega} L(a, \theta) \pi(\theta|x) \right] \tag{7.3.1}$$

The effect of interchanging the order of integration in (7.3.1) is to simplify the task greatly. We have now only to choose that *single action*, $a(x)$, for any set of data, x, which *minimizes the expected loss with respect to the posterior distribution* $\pi(\theta|x)$.

This modified approach is known as the **extensive form** of analysis, in distinction to the **normal form** which seeks a function to minimize $r(\delta, \pi)$. Detailed discussion of this distinction of approach and its implications is given by Raiffa and Schlaifer (1961); where the terms *extensive* and *normal* first appeared.

But apart from its computational convenience, the *extensive* form of decision theory analysis has a special philosophical appeal in Bayesian terms. We have already remarked on several occasions how the Bayesian approach has little sympathy with the von Mises' concept of the '*collective*': as the framework of essentially similar situations against which a particular problem is to be assessed. Probability manipulations based on use of a probability model over the sample space constructed in these terms are regarded as largely irrelevant other than in the sense that the likelihood $p_\theta(x)$ represents the import of the actual realized data, x. In using the extensive form for determining the Bayes' decision rule this attitude is upheld. We have a principle for determining the best action *in relation to the current data alone*. Observing x, and knowing $\pi(\theta)$, the loss structure implies that $a(x)$ is the best action to take! The sample space averaging of the normal form of analysis is unnecessary for determining the Bayes' decision rule (in cases where the *extensive* and *normal forms* yield the same result) and thus causes no embarrassment in fundamental Bayesian terms. This is not to say that the normal form is entirely ruled out in the Bayesian approach. A frequency interpretation of it is so rejected but it is seen to be reasonable if:

the decision maker considers the situation before the data is available.

(Lindley, 1971*b*, p. 20.)

Presumably, any averaging over different potential sets of data will need to be interpreted, however, in terms of degrees-of-belief attaching to the different x that might arise in the present situation. Some see this as an example of the Bayesian approach wanting to 'have its cake and eat it'—though it does not seem to be inconsistent with a purely subjective, utility based, view of decision theory.

Leaving philosophy aside for the moment, however, it must be recognized that the *extensive form* of analysis is a vital practical aid to determining the Bayes' decision rule. Indeed, without it this determination is generally much more laborious, sometimes prohibitively so.

Example 7.3. A radio-telescope receives signals of two distinct types, coded 0 and 1, independently with probabilities $1 - \theta$ and θ, respectively. The radio-telescope is operated until the signal 1 has occurred r times; this happens on the nth signal. We know nothing about θ initially, and choose to represent this by a uniform distribution on $(0, 1)$. We must make a decision about the value of θ. If we decide it is a we incur a loss,

$$L(a, \theta) = (a - \theta)^2 / [\theta^2 (1 - \theta)],$$

and we wish to determine the corresponding Bayes' decision rule. Our data x are the pair (r, n) and

$$p_\theta(x) = \binom{n-1}{r-1} \theta^r (1 - \theta)^{n-r}.$$

Thus

$$\pi(\theta | x) \propto \theta^r (1 - \theta)^{n-r},$$

and we need to choose a to minimize

$$\int_0^1 (a - \theta)^2 (1 - \theta)^{n-r-1} \theta^{r-2} \, d\theta.$$

Hence, given (r, n), the best value to take for θ is a^* where (on differentiating the posterior expected loss with respect to a, and equating to zero)

$$2n(n-1)a^* - 2n(r-1) = 0.$$

That is,

$$a^* = (r - 1)/(n - 1).$$

The Bayes' risk can be calculated by substituting a^* in the right-hand side of (7.3.1) and it turns out to be $(r - 1)^{-1}$.

Several interesting points arise from this example. It is easily confirmed that the Bayes' decision rule is *unbiased*. It is apparent also that the sampling rule is irrelevant. It makes no difference whether we needed to observe n signals to obtain 1 on r occasions or whether we chose to observe n signals and happened to obtain 1 on r occasions. (The *likelihood principle* of Bayesian inference applies to this decision theory determination.) Finally it is easy to confirm what an enormous economy of effort is provided by the extensive form of analysis. We have only to try the *normal* form for comparison!

To round off this discussion of Bayes' decision rules, there are two further matters that deserve some mention.

Firstly, it is an easily confirmed fact that the Bayes' decision rule is invariant with respect to an overall change of scale or origin of the loss function, $L(a, \theta)$. In particular, it is irrelevant from this standpoint whether we work in terms of absolute *loss*, or the *relative* concept, *regret*.

Secondly, decision theory is often criticized on the grounds that where little real prior information exists about θ the Bayes' decision rule (as the best guide to action) is nonetheless crucially dependent on the adopted form of $\pi(\theta)$. The prior distribution may appear to be an 'unjustified' formalization of vague subjective views about θ, or may even have been chosen by convention or mathematical expediency (as in representing prior ignorance, or utilizing the conjugate family of prior distributions). This is the familiar criticism of general methods of Bayesian inference, and it engenders the usual response. (See Chapter 6.) Within the Bayesian approach it is compulsory that some $\pi(\theta)$ is used, and the individual must be responsible for ensuring that it is the most appropriate form in terms of his own assessment of the situation. Without the philosophical commitment to the Bayesian method *per se*, however, the criticism has some substance in the case of limited data. But this must be tempered by the *robustness* that arises by virtue of the *principle of precise measurement*. If the data are extensive the prior distribution has little effect, and different $\pi(\theta)$ will lead to essentially the same Bayes' decision rule.

7.3.3 Minimax Decision Rules

We saw above how, in the absence of a specified prior distribution for θ, a principle is sometimes advanced, for singling out a decision rule, which consists of choosing that decision rule for which the maximum risk over Ω is as small as possible. This is the **minimax** decision rule. Let us consider briefly what is the formal status of this principle. Can we attribute any desirable properties to minimax rules?

It would seem that in Bayesian terms the minimax principle can have little appeal since the specification of a prior distribution is obligatory (even in the case of prior ignorance). But the general concept of decision theory proposed by Wald has an attraction in its own right, and a range of attitudes exist to the question of prior information about θ. Thus, some would claim that only if there is an objective basis for specifying $\pi(\theta)$ (von Mises insisted on a *frequency* interpretation) should we operate in terms of the Bayes' decision rule. Otherwise the risk function in its entirety provides the assessment of any decision rule. From this viewpoint we are compelled to effect a choice of a single decision rule in *some* manner. The minimax principle is one such possibility. It is far from ideal in its pessimistic preoccupation with the worst that might happen, but no other serious alternatives have been suggested.

Leaving aside any personal attitudes to the principle itself, there is one important property that it possesses. That is, that in the case of a finite state space and a finite action space (at least) it is a Bayes' decision rule with respect to some prior distribution, and is hence *admissible*. In fact it is the Bayes' decision rule with respect to the '**least favourable prior distribution**' for θ; that is, the one which has *highest* Bayes' risk. (We saw this illustrated for a simple case in §7.1) Note how the extreme pessimism of the minimax rule is again demonstrated in this relationship!

To express fully the relationship between the Bayes' and minimax rules in the 'finite' case, we really need to extend the notion of a decision rule to incorporate the idea of *randomization* of several decision rules. We shall consider this further in §7.4.

7.3.4 Estimation and Hypothesis Testing

We have already had examples of parametric estimation problems discussed in decision theory terms. These gave some insight into how the classical point estimation principle might be re-interpreted. We might also enquire whether the hypothesis test can be given any new interpretation through the ideas of decision theory.

There has been a fair amount of activity on this problem of re-interpretation with textbooks developing (or motivating) classical statistical methods through the decision theory model. See, for example, De Groot (1970), Ferguson (1967), Lindgren (1971), Lindley (1965b) or Mood, Graybill and Boes (1973), for varying levels of treatment, and differing balance between the decision-theoretic, or classical, principles. The research literature also abounds with attempts to reconcile, and to identify genuine points of distinction between, the two approaches, as reference to the bibliographies of, for example, De Groot (1970), Lindley (1971*b*), or Winkler (1972) clearly demonstrate.

We shall consider here just two minor aspects of the possible equivalence of classical and decision theory methods.

Point Estimation

Here we wish to determine some point estimator $\tilde{\theta}(X)$ (a mapping from $\mathscr{X}$ to Ω) for the purpose of estimating θ in the light of sample data. In classical statistics, criteria are proposed for assessing the virtues of the estimator; we have discussed these in Chapter 5.

In decision theory terms an action consists of assigning a value $\delta(x)$ to θ guided by any prior information we have about θ, by the information provided through the data in the form of the likelihood $p_\theta(x)$, and in consideration of the loss $L[\delta(x), \theta]$ incurred when the true value is θ. This is most satisfactorily achieved when $\delta(X)$ is the Bayes' decision rule.

In this situation the action space $\mathscr{A}$ is identical with the parameter space Ω, so that once again we are effectively considering mappings from X to Ω, at least in terms of the *normal form* of decision theory analysis. Any distinction between the best $\delta(X)$ and the optimum classical $\tilde{\theta}(X)$ will arise because of the different constraints and criteria that are employed in the two cases. It is interesting to consider whether the two will coincide in certain circumstances. This is indeed possible, as we see in the following example.

Example 7.4. Suppose we have a random sample of size n from a normal distribution $N(\mu, \sigma^2)$, *with* σ^2 *known. We wish to decide on an appropriate value for* μ. *Invoking unbiasedness, and with the need for simplicity, we might consider restricting attention to decision rules of the type*

$$\delta(X) = \alpha_1 X_1 + \ldots + \alpha_n X_n \qquad \left(\sum_1^n \alpha_i = 1\right),$$

in relation to a quadratic loss function

$$L[\delta(x), \mu] = h(\mu)[\delta(x) - \mu]^2.$$

The risk function is simply

$$R[\delta(X), \mu] = h^2(\mu)\text{Var}[\delta(X)] = \left(\sum_1^n \alpha_i^2\right)h^2(\mu)\sigma^2.$$

Thus, whatever the value of μ, *the risk is minimized by choosing*

$$\alpha_1 = \alpha_2 = \ldots = \alpha_n = 1/n.$$

Hence the sample mean $\overline{X}$ is the optimum decision rule within the class considered. It was also seen earlier to be the optimum classical point estimator. In both cases we are minimizing the variance of an unbiased linear estimator.

This example prompts some discussion of the fairly widespread use of loss functions (in continuous parameter spaces where $\mathscr{A} \equiv \Omega$) which are *quadratic* in form. What justification can there be for adopting a **quadratic loss function**? Practical situations may exist where this is quite genuinely a proper description, either in the simple form

$$L(a, \theta) = h(a - \theta)^2 \qquad (7.3.2)$$

or, admitting some θ-weighting, as

$$L(a, \theta) = h(\theta)(a - \theta)^2.$$

Such situations are rare, however, and the loss function (7.3.2) is by no means so restricted in its use. Even when quantification of losses is difficult it is commonly employed. The reasons for this are a mixture of convention, pragmatism and mathematical convenience.

(i) A case can be made for (7.3.2) being a reasonable approximation to the true loss structure in many situations. Often we require losses to increase (at least locally) in relation to the distance of a from θ, i.e. $|a - \theta|$. (7.3.2) is a simple expression of this. De Groot (1970) justifies this choice by showing that if we work in terms of *regret* (which is no constraint as far as the Bayes' decision rule is concerned) and seek a loss structure which depends only on $(a - \theta)$ then the leading term in the loss function is inevitably (7.3.2). If we can be reasonably sure of a being close to θ, then the higher order terms are unimportant and (7.3.2) alone suffices. He goes on to discuss the alternative possibility

$$L(a, \theta) = h|a - \theta|. \qquad (7.3.3)$$

(ii) Using (7.3.2) has the dual effect of simplifying mathematical calculations and providing an interpretation in terms of ideas of *classical* statistics. For example, in the single parameter situation the Bayes' decision rule with respect to (7.3.2) takes the form of an unbiased estimator of θ, with Bayes' risk a multiple of the variance of the estimator (if this variance exists). Thus a correspondence arises between the classical criterion of minimizing the variance of unbiased estimators and the decision theory criterion of minimizing posterior loss. [Similarly (7.3.3) has parallels through the idea of median unbiasedness.]

But we must bear in mind that the correspondence between the two approaches which emerges in (ii) is entirely due to the quadratic form of the loss structure. It would be wrong to regard this correspondence as a legitimate reason for adopting the loss function (7.3.2) in some particular problem. Quite often it is clear from practical considerations that (7.3.2) is inadequate; and any appeal to its interpretative or mathematical convenience will not overcome such inadequacy. Unfortunately this lesson is not always learnt, and some decision theory developments rest on the convenience of (7.3.2) with no regard for applicability.

This tenuous relationship between classical results and decision theory results in point estimation has a further twist to it. It appears to be exploited on occasions as grounds for criticism of certain *classical* estimation procedures. We find such procedures criticized because they prove to be *inadmissible* in decision theory terms. For example, in Lindley and Smith[3] we read

> ... many techniques of the sampling-theory [that is, classical] school are basically unsound . . . . In particular the least-squares estimates are typically unsatisfactory: or, in the language of that school, are inadmissible in dimensions greater than two.

Here is a strange juxtaposition! Classical principles are condemned for being *inadmissible*—a concept which is defined *relative to some assumed loss structure*. Classical statistics pays no formal regard to losses, so it entertains no concept of admissibility in the decision theory sense. Yet the criticism of classical least-squares is endorsed as being expressed 'in the language of that school'. Kempthorne[4] makes a related point in response to the remarks of Lindley and Smith:

> ... with a reasonable use of language, any estimate is 'typically unsatisfactory' because it will at best be admissible only for a particular loss function or a small class of such functions. . . . The present authors indicate rather definitely that 'inadmissible' implies 'unsatisfactory'. . . . I . . . question . . . [the] relevance [of admissibility] to problems of interpretation of a given set of data. I would also like to register the plaint that 'inadmissible with respect to a particular loss function' becomes through journal space exigencies merely 'inadmissible', and then it is quite an easy step to replace this word by 'unsatisfactory'.

One disadvantage of using a quadratic loss function (apart from its possible failure to represent the real-life situation) arises from its unboundedness, which may render corresponding decision rules inadmissible. Lindley and Smith[3] suggest a transformed version of the quadratic form which restores the boundedness, and (at least for proper prior distributions) the admissibility of the Bayes' decision rule. But again the practical relevance of such a loss structure needs to be confirmed in any particular situation.

Hypothesis Tests

Let us consider just one point in relation to this topic. Is there any sense in which the classical hypothesis test can be interpreted in decision theory terms?

As a decision theory problem it is very simple in one respect. Suppose the working hypothesis is $H{:}\theta \in \omega$, and the alternative hypothesis is $\overline{H}{:}\theta \in \Omega - \omega$. There are *two possible actions*: to reject H or to accept H. So the action space is of the simplest kind; it contains just two points (a_o, a_1) corresponding to these two possibilities. Any decision rule $\delta(X)$ obviously partitions the sample space $\mathscr{X}$ into two regions $\{S_0, S_1\}$, where

if $x \in S_0$ we take action a_0,
if $x \in S_1$ we take action a_1.

So far, this is precisely the structure of the *classical* test. But to determine the best decision rule (choice of S_0 and S_1) we need to specify a loss structure. The appealing prospect that $L(a, \theta)$ should be zero when a is the correct action, and that it increases as (in some appropriate sense) a becomes further from the correct action will patently not yield the same principles as those employed in the hypothesis test. Simple examples readily confirm this.

An alternative simpler (if less realistic) loss structure presents a greater hope of reconciliation. This assigns zero loss to correct actions, constant losses to incorrect actions. Thus

$$L(a_0, \theta) = \begin{cases} 0 & (\theta \in \Omega - \omega), \\ \xi & (\theta \in \omega), \end{cases}$$

$$L(a_1, \theta) = \begin{cases} \eta & (\theta \in \Omega - \omega), \\ 0 & (\theta \in \omega). \end{cases}$$

The risk function here is

$$R(\delta, \theta) = \begin{cases} \xi P(X \in S_0) & (\theta \in \omega), \\ \eta P(X \in S_1) & (\theta \in \Omega - \omega). \end{cases}$$

Seeking $\delta(X)$ to uniformly minimize $R(\delta, \theta)$ amounts to seeking a partition $\{S_0 . S_1\}$ which uniformly minimizes $P(X \in S_0)$ over ω, and $P(X \in S_1)$ over $\Omega - \omega$. So, as in the *classical* situation, the principle of optimization relates only to the probabilities of the two kinds of error. But there is still a difference in that the decision theory formulation seeks to uniformly minimize *both*. The level-α hypothesis test seeks to minimize $P(X \in S_1)$ over $\Omega - \omega$, subject to the constraint that $P(X \in S_0) \leqslant \alpha$ over ω.

To produce a direct parallel between the optimality principles in the two cases we could assume that the loss function is made up of two components $L_1(a, \theta)$ and $L_2(a, \theta)$, where

$$L_1(a_0, \theta) = \begin{cases} 0 & (\theta \in \Omega - \omega), \\ \xi & (\theta \in \omega), \end{cases}$$

$$L_1(a_1, \theta) = 0,$$

and

$$L_2(a_0, \theta) = 0$$

$$L_2(a_1, \theta) = \begin{cases} \eta & (\theta \in \Omega - \omega), \\ 0 & (\theta \in \omega). \end{cases}$$

If we were to regard the two types of loss as being of different levels of importance, and chose to try to uniformly minimize the risk associated with L_2, whilst restricting that associated with L_1 to a value no greater than $\alpha\xi$, we would be seeking precisely the UMP level-α test. But this principle seems to have limited appeal in decision theory terms!

Lehmann (1959) has discussed this type of relationship between the two-action decision problem and the test of significance.

A much more detailed study of the decision-theoretic interpretation of classical point estimators and tests of significance for a range of specific situations has been considered by Raiffa and Schlaifer (1961), Ferguson (1967) and De Groot (1970).

7.4 PROBLEMS WITH FINITE NUMBERS OF ACTIONS, AND STATES OF NATURE

If it should prove adequate for us to express some practical problem in terms of a decision theory model with *finite numbers* of possible actions, and states of nature, certain special features will arise. We saw in §7.1 how, in a very simple case, calculations were straightforward and useful diagrammatic aids were available (graphical representations of risks, geometric determination of the Bayes' decision rule, tree diagrams for the extensive form of analysis, immediate assessment of the value of sample data). The minimax decision rule can also be obtained geometrically.

The finite model also ensures a one-to-one relationship between admissible, and Bayes', decision rules (with respect to strictly positive prior distributions over Ω). Furthermore, the minimax decision rule becomes interpretable as the Bayes' decision rule with respect to the least favourable prior distribution (provided we extend the idea of a decision rule to random mixtures of the basic rules so far considered).

The ease of calculation, and the advantages of pictorial devices, progressively reduce as the numbers of possible actions, and states of nature, increase beyond 2 or 3 in each case; likewise with increase in the complexity of the sample space. However, the essential features remain and it is interesting to consider the nature of these. Mathematical considerations revolve around the idea of *convexity* in finite dimensional spaces.

7.4.1 *The No-data Problem*

Suppose there are just k possible actions, $a_1, a_2, \ldots, a_k$ and l states of nature $\theta_1, \theta_2, \ldots, \theta_l$. The loss function is $L(a, \theta)$. Then the different possible actions need to be compared in terms of the associated *vectors* of losses

$$\mathbf{L}(a) = [L(a, \theta_1), L(a, \theta_2), \ldots L(a, \theta_l)]' \qquad (a = a_1, a_2, \ldots, a_k).$$

Each of these can be represented as a point in Euclidean l-space: $\mathbf{R}^l$. Thus if $l = 2$, for example, we can obtain a two-dimensional graphical representation (Figure 7.7).

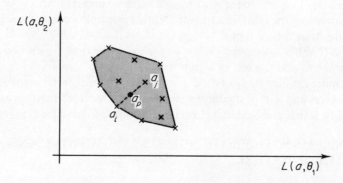

Figure 7.7   Losses with two states of nature

Consider now the idea of a **mixed action**, which consists of taking action a_i with probability p_i, or a_j with probability $p_j = 1 - p_i$. The loss function for this mixed action is simply

$$L(a_p, \theta) = p_i L(a_i, \theta) + p_j L(a_j, \theta).$$

Thus, the mixed action is represented by $p_i \mathbf{L}(a_i) + p_j \mathbf{L}(a_j)$: a point in $\mathbf{R}^l$ a proportion p_i of the distance along the line segment from $\mathbf{L}(a_i)$ to $\mathbf{L}(a_j)$. When $l = 2$, we see the effect of this in Figure 7.7.

In reverse, *any* point on the line segment joining the vector losses for any two actions may be viewed as the vector for an appropriate mixed action. But we could further mix the *mixed* actions! In this way it is obvious that any point within, or on the boundary of, the *convex hull* of the set of losses $\mathbf{L}(a)$ (for $a = a_1, a_2, \ldots, a_k$) corresponds to a mixed action. Thus we fill up the whole of this region, and conclude that the *convex set* so generated represents the losses corresponding to all possible actions (single or mixed) in our decision problem. The convex set for the illustrative example with $l = 2$ is shown shaded in Figure 7.7.

Note how the idea of a mixed action need not be restricted to a finite, or even countable, number of possible actions. Any probability distribution applied to the action space yields a mixed action, whose loss function is the expected value of the loss function with respect to that probability distribution. Properties of convexity are again encountered and influence the processing of the decision theory model. We shall not consider this general case in any detail.

7.4.2 The Use of Sample Data

Suppose we now have available sample data x with likelihood function $p_\theta(x)$. This is to be used to construct decision rules $\delta(X)$, taking the form of mappings from $\mathcal{X}$ to $\mathcal{A}$. The assessment of $\delta(X)$ is in terms of the *risk function*

$$R(\delta, \theta) = \int_{\mathcal{X}} L[\delta(x), \theta] p_\theta(x),$$

and again this can be regarded as a point in $\mathbf{R}^l$: that is as a vector $\mathbf{R}(\delta)$. The special case when only a finite number, m, of data sets can be encountered deserves special mention. There will now be only l^m simple decision rules. (In §7.1 for instance we had $l = 2$ and $m = 3$, and considered eight possible decision rules.) These are sometimes called **pure strategies**, or **pure decision rules.** Whether or not $\mathcal{X}$ is finite, all deterministic mappings from $\mathcal{X}$ to $\mathcal{A}$ can be thought of as *pure decision rules*, and the totality of risks $\mathbf{R}(\delta)$ over $\mathbf{R}^l$ represents their overall assessment.

As an example, when $l = 2$ and $\mathcal{X}$ is finite we obtain a set of points in $\mathbf{R}^2$ which will typically appear in a similar form to Figure 7.7 but with the axes measuring *risk* rather than *loss*, and the points corresponding to the finite number of *pure decision rules* rather than *actions*.

But again, we could mix decision rules, operating δ_i with probability p_i and δ_j with probability $p_j = 1 - p_i$. This mixture of pure decision rules is known as a **randomized decision rule**. Mixtures of several (or even an uncountable number of) pure or mixed decision rules can be similarly countenanced and we arrive at a conclusion akin to that in the no-data case: that *the totality of pure and mixed decision rules is represented as a dense convex set of risks in $\mathbf{R}^l$ enclosed by the convex hull of risks for the pure decision rules.*

Thus the choice of a particular decision rule for application amounts to singling out some member of a convex set in $\mathbf{R}^l$. Bases for this choice include admissibility, and the determination of the Bayes', or minimax, decision rule. It is interesting to see how these principles appear against the background of the convex set representation. To illustrate this we shall consider only the two-dimensional case ($l = 2$) with an assumed finite number of pure decision rules, and with all losses (hence all risks) assumed to be non-negative. Extensions to more than two states of nature, non-finite $\mathcal{X}$ and more general losses are easily obtained in qualitative terms.

Let the risk function for any (pure or randomized) decision rule δ consist of two components $R_1(\delta)$ and $R_2(\delta)$. Admissibility immediately severely limits the number of δ worth considering. Figure 7.8 shows a typical convex configuration $\mathcal{R}$ of risks, with the black dots corresponding to *pure* decision rules.

It is apparent that the only admissible decision rules are those *on that part of the boundary of $\mathcal{R}$ nearest to the origin contained between the two supporting lines to from the origin.* This set is shown by a heavy line in Figure 7.8.

The idea of a *supporting line* (or more generally a supporting hyperplane) also provides a means of identifying a Bayes' decision rule. The Bayes' decision rule

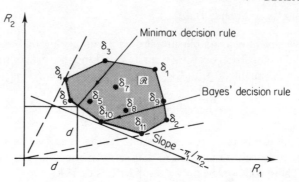

Figure 7.8   Risks with two states of nature

with respect to prior probabilities π_1 and π_2 is that decision rule for which $\pi_1 R_1 + \pi_2 R_2$ is a minimum. Clearly this is the decision rule on the *admissible boundary* which has a supporting line of slope $-\pi_1/\pi_2$. (We saw this illustrated in §7.1.) Note how this line either determines a unique *pure* decision rule as the Bayes' decision rule, or else lies along a line segment of the admissible boundary. In either case we see that for any prior distribution there exists a *pure* Bayes' decision rule; if there are two, then all intermediate randomized decision rules are also optimum. We observe also the correspondence between Bayes' and admissible decision rules. (Generalizations to more complex situations require some care!)

Finally, how can we identify the minimax decision rule? This is that decision rule for which $\max_{1,2}(R_i)$ is as small as possible. We can obtain this as follows. Construct a square of side d with the positive R_1 and R_2 axes as two sides. If we increase d until this square first meets the admissible boundary, the point of meeting must determine the minimax decision rule. This will typically not be a *pure* strategy, and may be an extremal point on the admissible boundary as in the example of §7.1. On the other hand, any randomized decision rule is equivalent in terms of posterior expected loss (for a suitably chosen prior distribution) to the two pure decision rules at the extremes of the line segment on which it lies. Thus the minimax decision rule is admissible, is a Bayes' decision rule, and is equivalent in terms of posterior expected loss to a *pure* Bayes' decision rule.

It is self-evident that changes of scale or origin in the loss function can affect the minimax decision rule, but will not influence the Bayes' decision rule for given (π_1, π_2).

The ideas outlined in this section are more fully illustrated by Chernoff and Moses (1959). De Groot (1970) provides a more formal treatment.

7.5   EXTENSIONS AND MODIFICATIONS

As in the treatments of classical statistics and Bayesian inference, the discussion of *decision theory* in this chapter has concentrated on broad aims and principles.

In the last 30 years or so there has been hectic activity in extending or modifying the basic ideas to deal with quite complex problems. Special situations have been studied in much detail, and modified principles have been advanced to study situations which do no fit immediately into the simple decision theory scheme described above. The various quoted texts, and their bibliographies, give clear indication of the range of this work. Cox and Hinkley (1974) review some of the important developments and potentialities (as well as providing a clear and concise description of the basic notions of decision theory). It is not appropriate to consider the details here, but the following short comments illustrate some of the applications, extensions and modifications which have been considered.

(a) *Multi-parameter problems, multivariate data.* It is clear that the implicit assumption of a single parameter θ is unrealistic for many situations. A large range of practical problems are traditionally represented by *linear* (multi-parameter) *models*, and the decision theory counterparts of the classical methods for studying *regression* and *analysis of variance* situations have been widely discussed. More complicated non-linear models have also been considered. The treatment of *nuisance parameters* requires special attention, largely in the specification of the loss structure.

Multi-dimensionality of the sample data, x, requires no formal extension of the basic ideas above, in view of the liberal definition of x and its sample space $\mathscr{X}$. However, in decision theory studies of particular 'multivariate analysis' problems, the fine detail must inevitably recognize that the data often takes the form of independent vectors from some prescribed multivariate probability distribution.

(b) *Specific study of particular distributions.* Much work has been done on decision theory analysis when $p_\theta(x)$ has particular standard forms (when θ is one- or multi-dimensional). Thus the special problems of sampling from normal, exponential, gamma, beta (etc.) distributions have been considered. As in the classical approach the wider specification of sampling from the *exponential family* has attracted attention in view of its relative simplicity.

(c) *Classical statistical methods.* Apart from the general topics of point estimation and hypothesis tests (briefly discussed in §7.3), much work has been done on re-interpreting, or providing an alternative *decision theory* formulation for, some of the particular methods of classical statistics. This includes the study of regression and analysis of variance as examples of the use of the linear model, methods of multivariate statistics, *finite sampling methods, design of experiments* and *sequential analysis*. The application of decision-theoretic ideas to interval estimation is considered by Winkler[5]. These decision theory counterparts have been applied to a range of practical problems in many fields of application including economics, education, industry and medicine. On the industrial front, for example, methods of analysis have been proposed for life-testing, quality control, inventory regulation and general prediction and control problems. It would appear that industry has again provided a stimulus for the development of decision theory methods, as it did for classical statistics.

(d) *Generalized action spaces and utility functions.* There are certain types of problem in which it becomes necessary to consider rather more complicated specifications of the action space and utility function. In particular it may be necessary to allow these to depend on, or be modified by, observational data.

One such example is in the decision theory approach to *the design of experiments*, viewed as a preliminary to a decision-making problem. The formal development is reasonably straightforward, but requires a generalized utility function. Lindley (1971b) describes this in the following way. We seek to utilize data x to effect a choice of action a from an action space $\mathscr{A}$. But initially we ask what is the best *experiment* to perform to produce our sample data. An experiment E is defined as a triplet $\{\mathscr{X}, \Omega, p_\theta(x)\}$, where E ranges over some space of experiments $\mathscr{E}$, with fixed Ω. There are two decision stages now; the choice of E from $\mathscr{E}$ and the subsequent choice of a decision rule in terms of the resulting sample data. An appropriate loss function must now include the cost of the data, and of the experiment which yields it, and will be of the general form $L(a, \theta, E, x)$. The corresponding Bayes' decision rule is easily defined, but little detailed application has been achieved. This remains an inevitable growth area for decision theory.

Another most important example is in the study of problems where the data may be obtained in stages, rather than once and for all through some prescribed experiment. The action space now needs to be extended to include *intermediate* actions of seeking more data if the cost considerations justify this, as well as the set of basic actions from which a *terminal* choice is to be made. Such intermediate actions may be dependent on the current data, and the loss function will also reflect the sequential way in which the data arise. Thus $\mathscr{A}$ becomes dynamic as the process proceeds; both $\mathscr{A}$ and the loss function must be allowed to vary with the data. Such a decision theory approach to *sequential analysis* is both valuable in concept, and of much potential importance. It seems eminently reasonable that we should not be compelled to make a decision on the basis of some prescribed experiment (which may be inadequate on the one hand, or extortionate on the other), but that we should be able to 'play things by ear'. We seek data in successive stages in total regard of what they cost and what they convey until we are in an optimum position to make a final choice of action. Optimum sequential procedures may greatly improve on the corresponding optimum fixed sample-size ones.

A vast amount of elegant theory has been derived on this topic. Interesting papers are by Wetherill[6] and Whittle[7,8]. A most illuminating elementary explanation of sequential decision-making is given by Aitchison (1970). More general formulations have been widely discussed, and Raiffa and Schlaifer (1961) and De Groot (1970) devote several chapters to these. The latter provides a lengthy list of references. See also Lindley (1971*b*). Part of this work appears under the heading of **dynamic programming** [on which the book by Bellman (1957) remains an important reference]. A somewhat different form of sequential decision analysis is represented in developments of *empirical Bayes, procedures*, a simple form of which was described in §6.7.1.

Some indication of the range of recent developments in decision theory is given by the conference proceedings edited by Gupta and Yackel (1971) and Gupta and Moore (1977).

7.6 CRITICAL COMMENT

We need not spend too much time discussing the nature of any criticisms of the decision theory approach. It inevitably encounters the range of controversy surrounding *both* classical methods and Bayesian inference, in view of its incorporation of both sample data and prior information. Such controversy has been considered at length throughout the book, particularly in Chapters 5 and 6. Its form and relevance in decision theory has also been discussed at different stages in the development of ideas in decision theory within the current chapter. It suffices to summarize briefly the apparent advantages and disadvantages.

The major appeal is utilitarian. It would seem most attractive to operate a system of analysis which clearly distinguishes the decision-making function, is designed solely for the purpose of decision-making, and exploits to this end *all three forms* of relevant information contemplated in our basic *rationale* of statistics discussed in the introductory chapter.

There seems to be little dispute about the great value of decision theory principles in practical situations which admit a 'straightforward' specification of losses and prior probabilities (usually expressed in financial, and frequency, terms, respectively). Controversy appears when prior probabilities are inaccessible, when the probability concept is of a subjective form, or when losses are introduced in what might appear to be purely conventional or expedient form. We recall for instance how the pioneers in classical statistics resisted the appeal of an approach based on losses and prior probabilities on the grounds that these could seldom be quantified, and could therefore not form part of an 'objective' or 'universal' theory of statistics.

Even so, opinion is divided and whilst many castigate such 'non-objectivity', others regard the *personal* formulation as entirely appropriate. Indeed, justification for *Bayesian* decision theory is advanced on philosophical grounds: in the inevitability of prior probabilities and utilities in utility theory terms. For instance, Lindley[9] remarks

In the context of a decision problem the principle of consistency or coherence clearly demonstrates that decision-making must be based on a Bayesian analysis, using a (prior) distribution and a utility function, and selecting the optimum decision by maximizing expected utility. There are numerous justifications for this;. . . . No substantial counter argument is known to me.

In replying to Lindley's paper, Bross[10], flatly repudiates the 'Principle of Coherence' and hence the whole framework of utility–theoretic decision-making. See also Cox[11], and the discussion of coherency in §5.7.3. In contrast, Kempthorne[12] mounts an enthusiastic attack on the *impracticality* of decision theory. (Indeed, he seems so to reject, in fine rhetoric style, almost all recognizable approaches to inference and decision-making!)

No purpose is served in reproducing fragments of what is often a vitriolic exchange of criticisms, rebuttals and counter-criticisms. The mood of this is well represented in the most interesting published proceedings of the Symposium on the Foundations of Statistical Inference (Godambe and Sprott, 1971) from which the last three references are taken.

Instead let us briefly review some of the main 'bones of contention' in the form of an imaginary debate.

The Action Space

Dispute: In practice it often makes no sense to try to specify at the outset of an enquiry the complete set of actions which are likely to be contemplated. In many scientific problems the spectrum is constantly changing as our knowledge accumulates.

Response: In any situation we must entertain action solely in relation to our current horizon. This is what we do in our personal behaviour, and it should carry over to scientific enquiries. *Sequential* decision theory allows for modification of the action space in the light of experience. The statement of a set of possible actions is no different in principle to a specification of a sample space.

Loss functions

Dispute: Except in rare circumstances where losses are readily assessible, the demand for a specified loss function may lead to an unjustifiable form being adopted with resulting arbitrary decisions being made. The choice of a loss function is often based on convenience rather than relevance.

Response: Information on losses always exists if only in a subjective form. This is *relevant* information and must be elicited and employed. In utility theory terms a unique specification exists. Even if our methods of determining this are far from perfect, the effort should be made.

Prior Probabilities

Dispute: As for loss functions.

Response: As for loss functions. Additionally, we have the safeguard of robustness properties arising from the *principle of precise measurement* in situations where the sample data are extensive.

Sample Data

Dispute: The prescription of a sample space, $\mathscr{X}$, and family of distributions $\{p_\theta(x); \ \theta \in \Omega\}$ is fraught with difficulties. Sample-space averaging is inappropriate to a *unique* situation.

Response: Such specification problems exist in any approach. We must always face up to the validation of the model. No such use of the sample space is required in the *extensive form* of decision analysis.

Subjectivity

Dispute: *Personal* judgement of loss or prior probability may lead to choice of action of no *external* importance or relevance. We are not interested in the idiosyncrasies of some individual, but in a valid prescription for action based on tangible and universally acceptable information ingredients.

Response: It is usually an individual who must consider a problem and be responsible for *his* or *her* actions. Such an assessment, whilst personal, will reflect any 'external, tangible' information. Indeed it is often predominantly based on such information, so that most individuals will draw essentially similar conclusions. Where they differ, it is entirely appropriate that they should. The principle of coherence totally justifies the personal nature of probability and utility, and demands that an individual acts to maximize *personal* expected utility.

Utility Theory

Dispute: I do not accept the utility-theory model as a valid framework for decision-making under uncertainty.

Response: This is to deny the principle of coherence to which we should all strive. Such an attitude is incoherent!

. . . But this takes us firmly into the depths of philosophical debate, and even into the sphere of disputes over semanticism in the formulation of statistical concepts and principles. See Bross[13] on the thesis that 'the foundations of statistical inference are a myth' for an initial exposure to this latter area of debate.

References

1. Moore, P. G. and Thomas, H. (1972). 'How to measure your risk', *Financial Times*, London, 18 Jan. 1972.

2. Moore, P. G. and Thomas, H. (1972). 'A tree to tease your mind', *Financial Times*, London, 19 Jan. 1972.

† 3. Lindley, D. V. and Smith, A. F. M. (1972). 'Bayes estimates for the linear model' (with Discussion), *J. Roy. Statist. Soc. B*, **34**, 1–41.

4. Kempthorne, O. (1972). Contribution to the discussion on Lindley and Smith[3].

† 5. Winkler, R. L. (1972). 'Decision-theoretic approach to interval estimation', *J. Amer. statist. Assn*, **67**, 187–191.

6. Wetherill, G. B. (1961). 'Bayesian sequential analysis', *Biometrika*, **48**, 281–292.

7. Whittle, P. (1964). 'Some general results in sequential analysis', *Biometrika*, **51**, 123–141.

8. Whittle, P. (1965). 'Some general results in sequential design' (with Discussion), *J. Roy. Statist. Soc. B*, **27**, 371–394.

† 9. Lindley, D. V. (1971). 'The estimation of many parameters', in Godambe and Sprott (1971).

10. Bross, I. D. J. (1971). Contribution to published comments on Lindley[9].

11. Cox, D. R. (1978). 'Foundations of statistical inference: the case for eclecticism', *Austral. J. Statist.*, **20**, 43–59.

† 12. Kempthorne, O. (1971). 'Probability, statistics and the knowledge business', in Godambe and Sprott (1971).

† 13. Bross, I. D. J. (1971). 'Critical levels, statistical language, and scientific inference', in Godambe and Sprott (1971).

CHAPTER 8

Some Other Approaches

We have adopted a primary division of the different approaches to inference and decision-making into three major categories: *classical statistics, Bayesian inference* and *decision theory*. In terms of function the first was seen as essentially inferential but with decision-making implications, the second was described in specifically *non*-decision-making terms, the third appeared solely concerned with decision-making. A major distinction between the approaches was in the interpretation and application of the probability concept—entirely *frequency* based in classical statistics, inevitably incorporating a *subjective* element in most expressions of Bayesian inference, and (depending on attitude and application) involving either frequency or subjective views in decision theory. The corresponding statistical principles, procedures and interpretations reflected, or were constrained by, these distinctions of function and probability basis.

This tripartite division greatly oversimplifies the vast range of conflicting attitudes. We have found it a convenient basis for illustrating essential distinctions. But even within any of the three approaches we must recognize a multitude of different emphases and interpretations. To some, decision theory is appealing only in situations where the specification of costs and prior probabilities is well supported in non-subjective terms—to others its *rationale* is securely based in utility theory and subjective elements are paramount. A similar distinction of attitude exists in Bayesian inference. Then again, the separation of Bayesian inference and decision theory will be to some people an artificial one. They see the loss structure analysis as just the natural extension of the Bayesian idiom. Such distinctions (both within and between the approaches) have been illustrated to some extent throughout the earlier chapters; the references to books and articles provide access to a much deeper study.

But what must also be recognized is that attitudes are encountered which do not fit at all into the three major categories described above. In striving to find an acceptable basis for drawing inferences and making decisions, an individual may well not be able to categorize his attitudes. A man's statistical *mores* are largely personal and unique. To categorize them is a convenience rather than an intrinsic reality. It aids communication, but who is to say that one man's Fisher is not another's Bayes! Nonetheless we *must* attempt to communicate, and the classical–Bayes–decision theory classification often serves for this purpose. Where it does not, it is still possible that some other range of principles and policies acts as a convenient summary framework within which to fit an individual's approach to statistics.

In this chapter we shall briefly consider a few of these alternative approaches. The first is largely historical in its interest, the others represent attitudes which are currently of more active concern to their proponents and antagonists. Many attempts have been made to justify, denigrate or interrelate these approaches; or to suggest that they are merely re-expressions of ideas in the three basic approaches. We shall consider some aspects of this activity, providing references for more detailed study.

8.1 FIDUCIAL INFERENCE

In the spate of developments of *classical* statistics in the 1920s and 1930s Neyman and Pearson (and others) proposed a method of interval, or region, estimation related to their ideas of significance tests. As we have seen, such estimates were termed *confidence intervals*, or *confidence regions*. At the same time Fisher, interested more in estimation than in hypothesis testing, was considering the construction of interval or region estimates from an alternative standpoint. Fisher proposed a method based on the idea of **fiducial probability**, leading to what are called **fiducial intervals** (or regions). The first reference to this topic is in a paper in 1930 (Fisher[1]) forcefully rejecting the idea of 'inverse probability' basic to Bayesian inference. Subsequent papers (Fisher[2,3]) develop the idea more fully through illustrations for particular situations.

The fiducial argument stands out as somewhat of an enigma in classical statistics. It contradicts a basic tenet in proposing a *probability distribution* as an inferential statement about a parameter θ. Fisher sometimes calls this the *posterior fiducial distribution* for θ, but vehemently denies (Fisher[1–3]) that this is the same in principle or in detail as a *Bayesian* posterior distribution. His attitude is summarized in his *Design of Experiments* (1966), first published in 1935, where he explains the 'different logical basis' as stemming from the fact that Bayesian methods

> . . . require for their truth the postulation of knowledge beyond that obtained by direct observation. (p. 198.)

Whilst admitting that such information is sometimes available (for example in some problems in genetics) and that the Bayesian argument is then the appropriate one to use, he otherwise rejects

> . . . its introduction by axiom . . . [as a] mathematical sleight-of-hand. (p. 198.)

in favour of using

> ... similar probability statements *a posteriori* ... inferred by the fiducial argument. (p. 198.)

In a much later (Fisher, 1956) reconsideration of fiducial probability, Bayesian and fiducial methods are admitted, in certain situations, as dual methods of inference: the former to be used when prior information exists, the latter when it does not. The probability concept is declared to be entirely identical with the classical probability of early writers, such as Bayes; the fiducial argument merely changes the 'logical status' of the parameter

> ... from one in which nothing is known of it, and no probability statement about it can be made, to the status of a random variable having a well-defined distribution. (p. 51).

We shall later consider more recent intercomparisons of the Bayesian and fiducial arguments.

Just how is a fiducial probability distribution defined? Unfortunately Fisher is nowhere specific on this matter, nor has any fully accepted formal definition been subsequently proposed by later advocates of the principle. Fisher contented himself with general statements of its applicability and advantages, with indicating limitations in its use, with descriptions of methods of obtaining fiducial distributions and with deriving these in a few special situations. He restricted its use to the processing of sample statistics with *continuous* distributions, an arbitrary restriction from the statistical viewpoint, but convenient mathematically for the particular methods proposed for constructing fiducial distributions. (See also Structural Inference, §8.4.) Lack of an earlier knowledge of the exact form of such sampling distributions, he claimed, 'stood in the way of the recognition' of a principle which

> ... leads in certain cases to rigorous probability statements about the unknown parameters of the population from which the observational data are a random sample, without the assumption of any knowledge respecting their probability distributions *a priori*.

(Fisher[3].)

Fisher further suggested that fiducial distributions may only be meaningfully derived from *sufficient* statistics, and that to do otherwise would be equivalent to rejecting part of the data arbitrarily.

In the original discussion (Fisher[1]) the argument proceeds in essentially the following terms. Suppose $\tilde{\theta}(X)$ is the *maximum likelihood estimator of* θ, *is continuous and has a distribution function* $F_\theta[\tilde{\theta}(x)]$. (This is customarily used to make probability statements about $\tilde{\theta}$ for the relevant θ.) According to Fisher, $F_\theta(\tilde{\theta})$ can equivalently be employed in certain circumstances *to make probability statements about* θ conditional on the actual value, $\theta(x)$, obtained from the data, x. Thus we have an inferred probability distribution (the **fiducial distribution**) over Ω.

This is illustrated by the following typical example of Fisher's application of the principle, taken from Fisher[3].

Example 8.1. Suppose we draw a random sample of size n from a normal distribution, $N(\mu, \sigma^2)$ *with* μ *and* σ^2 *unknown. Let* $\bar{x}$ *and* s^2 *be the sample mean, and unbiased variance estimate*

$$\left[\frac{1}{n-1} \sum_1^n (x_i - \bar{x})^2 \right],$$

respectively. Then Student's t-statistic

$$t = (\bar{x} - \mu)\sqrt{n}/s$$

has a known sampling distribution which is continuous. The inequality

$$t > t_1$$

holds with some known probability in this sampling distribution. But this can be rewritten

$$\mu < \bar{x} - st_1/\sqrt{n} \tag{8.1.1}$$

which 'must be satisfied with the same probability', to be called the fiducial *probability that* μ *is less than* $\bar{x} - st_1/\sqrt{n}$. *A similar argument (changing* t_1 *and keeping* $\bar{x}$, s^2 *fixed) yields the 'probability' that* μ *is less than any value, or (equivalently) in any interval, in the light of the sample data. Thus we obtain the fiducial distribution for* μ, *given* $\bar{x}$ *and* s^2 *as 'parameters'.*

Throughout Fisher's work on this topic the discussion remains intuitive and imprecise. As illustrated by Example 8.1 the crux of the argument rests on transferring the probability measure from X to μ in the informal inequality (8.1.1). This is the source of most criticism of the fiducial method. No justification is offered for this transfer, and to many people it would appear invalid and (thus) unjustifiable.

However, if we ignore interpretative difficulties for the moment and continue to argue along the lines of Example 8.1, we might make a natural demand for an *interval estimate* of μ. This will be obtained as an interval within which μ lies with a prescribed (fiducial) probability, and is called a **fiducial interval** for μ. As in classical or Bayesian inference we could consider one-sided or two-sided intervals. For the latter, however, the Bayesian approach offers the better parallel, in that there is a similar appeal in using the 'equal ordinates' interval. We have a direct (fiducial) probability interpretation for μ being included in the interval, so that it seems to make sense to exclude from the interval values of μ having lower (fiducial) probability density than any that are included. In Example 8.1 it is easily confirmed that the fiducial distribution of $[(\mu - \bar{x})\sqrt{n}]/s$ is Student's t-distribution with $(n-1)$ degrees of freedom. Thus if $t_{n-1}(\alpha)$ is the double-tailed α point of this t-distribution, the $100(1-\alpha)$ per cent *fiducial interval* for μ is (in view of the symmetry of the t-distribution) just

$$(\bar{x} - t_{n-1}(\alpha)s/\sqrt{n}, \bar{x} + t_{n-1}(\alpha)s/\sqrt{n}).$$

But this is simply the *central* $100(1-\alpha)$ *per cent confidence interval*. Although the confidence interval for μ does not echo the *uniqueness* of the fiducial interval even this distinction is not fundamental. The fiducial argument needs the additional 'equal ordinates' convention to produce a unique two-sided interval; the central confidence interval in this situation (and in other simple cases) is also singled out by additional criteria of 'shortest length' or 'greatest accuracy'. (See Chapter 5.) The essential distinction that does remain is one of interpretation!

The coincidence in the expressions for fiducial intervals and traditionally employed confidence intervals left a legacy of confusion. Early discussions of both concepts were informally expressed and illustrated for simple one-parameter problems where the two approaches lead to identical results. Furthermore, illustrations of the derivation of fiducial intervals often appeared to use identical methods. Example 8.1 is easily misinterpreted as leading to an interval estimate based on the acceptance criterion for a test of significance—the *confidence interval* criterion. As a result, it was tacitly believed for some time that the two approaches were identical in principle as well as expression. The terms 'confidence' and 'fiducial' were used indiscriminately, particularly the latter where in fact the former was appropriate. This was especially true of elementary presentations of inference, and the textbooks are still not entirely free of this misconception.

It was not until two-parameter problems were studied that the methods were really seen to produce different results, as well as to be using different principles. In the so-called Behrens–Fisher problem the construction of an interval estimate for the difference between two normal means (where the variances are unknown) played a notable part in crystallizing the distinction. It was, and still seems to be, somewhat uncertain how the fiducial argument should be extended to multi-parameter problems. However, using what to some seems an intuitively reasonable extension it was apparent that the confidence and fiducial arguments diverged in this problem. Not only were the solutions different in form but the fiducial interval was seen for the first time to lack the frequency interpretation of the confidence interval. (Much later work by Fraser[4], however, provided a modified frequency interpretation for the fiducial argument.) This apparent lack was regarded in confidence interval terms as an inherent failing in the fiducial argument—from the opposite standpoint it was seen to be irrelevant! The dispute, and its ramifications, have reverberated around the statistical world for a long time.

Kendall and Stuart (1979, Chapter 21) provide a detailed study of this problem, as well as outlining the basic ideas of fiducial inference and contrasting these with the other approaches to interval estimation. Their *Discussion* section is most informative, as indeed is the section on *Fiducial Probability* in Plackett[5], Edwards[6] and three recent detailed assessments of the fiducial argument (Pedersen[7], Wilkinson[8] and Seidenfeld, 1979) reviewed at the end of this section.

Let us return to the formal basis of the fiducial argument, in as far as it has been presented. Fisher[1] did in fact state that if $\tilde{\theta}(X)$ is a sufficient statistic for θ, with distribution function $F_\theta[\tilde{\theta}(x)]$, then *the fiducial distribution of θ has probability density function*

$$g(\theta;x) = -\frac{\partial}{\partial\theta}F_\theta[\tilde{\theta}(x)], \tag{8.1.2}$$

hinting that this depends upon $F_\theta[\tilde{\theta}(x)]$ having certain monotonicity properties as a function of θ, for any x. Later (Fisher, 1956) he re-iterates this form but again gives no detailed derivation. This has acted as the stimulus for discussions of the way in which the fiducial distribution should be derived.

Two procedures have been proposed.

Monotone Distribution Function

Suppose that we have a single parameter θ with parameter space the interval (θ_0, θ_1). It may happen that for any sample data x the distribution function $F_\theta[\tilde{\theta}(x)]$ of the sufficient statistic $\tilde{\theta}(X)$ is *monotone decreasing* in θ, varying from a value 1 at the lower limit θ_0 to a value 0 at the upper limit θ_1. Then

$$G(\theta;x) = 1 - F_\theta[\tilde{\theta}(x)]$$

has all the formal properties of a distribution function over (θ_0, θ_1). It is suggested that *we adopt $G(\theta; x)$ as a measure of the cumulative intensity of our belief in different values of θ*, engendered by the data x through the sufficient statistic $\tilde{\theta}(X)$. It is called the **fiducial distribution function** for θ. On this principle (which seems to have some intuitive, if ill-defined appeal) the fiducial distribution will then have the probability density function (8.1.2) proposed by Fisher. (The argument is easily modified if $F_\theta[\tilde{\theta}(x)]$ is monotone *increasing* in θ.)

Example 8.2. A random sample of size n from $N(\mu, \sigma_0^2)$, *with known variance* σ_0^2, *has mean* $\bar{x}$. $\bar{X}$ *is sufficient for μ and has distribution function* $\Phi\{[(\bar{x}-\mu)\sqrt{n}]/\sigma_0\}$, *where* $\Phi(z)$ *is the distribution function of the standardized distribution,* $N(0, 1)$. *Now μ varies over* $(-\infty, \infty)$ *and* $G = 1 - \Phi$ *satisfies the required conditions. Thus the fiducial distribution of μ, for given $\bar{x}$, has density*

$$g(\mu; x) = \sqrt{\left(\frac{n}{2\pi\sigma_0^2}\right)}\exp\left\{-\frac{n}{2\sigma_0^2}(\mu - \bar{x})^2\right\}.$$

That is to say, μ has a normal fiducial distribution $N(\bar{x}, \sigma_0^2/\sqrt{n})$, and the fiducial and central confidence interval are identical.

Note that the fiducial distribution for *this* example is equivalently obtained by attributing a probability interpretation to the likelihood function $p_\mu(x)$, *suitably normalized.* For

$$p_\mu(x) \propto g(\mu; x),$$

so that if we regard the likelihood function as measuring relative densities of credence in different values of θ we again obtain the same 'probability distribution'.

But such a simple re-interpretation of the likelihood function does not always suffice, as is illustrated by the following example.

Example 8.3. Suppose we have a random sample of size n from an exponential distribution with mean θ,

$$f_\theta(x) = \frac{1}{\theta} e^{-x/\theta} \ (x > 0).$$

Again $\bar{X}$ is sufficient and has distribution function

$$F_\theta(\bar{x}) = \int_0^{n\bar{x}/\theta} \frac{u^{n-1} e^{-u}}{(n-1)!} \, du.$$

$G(\theta; x) = 1 - F_\theta(\bar{x})$ *is monotone increasing from 0 to 1 over* $(0, \infty)$, *and we obtain the fiducial distribution of θ as*

$$g(\theta; x) = \frac{1}{\theta}\left(\frac{n\bar{x}}{\theta}\right)^n \frac{e^{-n\bar{x}/\theta}}{(n-1)!}.$$

But this does not come directly by considering the normalized likelihood function. We would need to introduce a further factor $1/\theta$.

The compensation which needs to be applied to the likelihood function in this example illustrates a modified principle that is widely adopted for the construction of fiducial distributions: namely that we employ the normalized likelihood function but multiply by a factor $1/\theta$ for any scale parameter θ (but not for location parameters). In this way certain multi-parameter fiducial distributions have been constructed.

But this principle suggests a further possibility. Adopting a Bayesian approach, and employing the Jeffreys' representation of prior ignorance about θ, we would obtain posterior distributions identical to the fiducial distributions in the Examples 8.2 and 8.3. It might be tempting, therefore, to think that the fiducial argument is just a re-expression of Bayesian inference with a conventional statement of prior ignorance about θ. However, this idea is readily rejected in the way described by Lindley[9] who shows that such an interpretation is valid only when we are sampling from a distribution in which the random variable and the parameter may be separately transformed so as to yield a new parameter which is a *location* parameter for the new random variable.

Thus to the classical statistician, as well as to the Bayesian, the fiducial approach will be unacceptable. It contains elements outside their respective frameworks; on the one hand a distribution over Ω and the possible lack of a frequency interpretation cannot be entertained, on the other the 'posterior' distribution will not always accord with any accepted expression of prior ignorance and takes no account of any prior information which may exist.

The second general method makes use of:

Monotone Pivotal Functions

Suppose it is possible to find some function $h[\theta, \tilde{\theta}(X)]$ of θ and $\tilde{\theta}(X)$, which is monotone increasing in θ for fixed $\tilde{\theta}$ (and in $\tilde{\theta}$ for fixed θ) and which has a distribution which does not depend on θ except through $h[\theta, \tilde{\theta}(X)]$. Then $h[\theta, \tilde{\theta}(X)]$ is called a **pivotal function**, and the fiducial distribution of θ is obtained merely by re-interpreting the probability distribution of $h[\theta, \tilde{\theta}(X)]$ over $\mathscr{X}$ as a probability distribution over Ω. The procedure I above is a special case of this, since $G(\theta; X)$ is just such a pivotal function, having a uniform distribution over $(0, 1)$, which is independent of θ. Thus Examples 8.2 and 8.3 also illustrate this more general procedure.

What of the current status of fiducial inference? Over the years a steady flow of research has been carried on: concerning the construction of fiducial distributions in a variety of different situations, the critical examination of the fiducial argument and its re-interpretation in alternative inference frame-works. The work by Fraser[4] and Lindley[9] are examples of this. Others are Dempster[10] and Godambe and Thompson[11]. A large group of papers on fiducial inference were presented, with discussion, in the 1964 issue of the *Bulletin of the International Statistics Institute* (volume 40). In more general surveys of statistical inference the comments on fiducial inference by Birnbaum[12], Cox[13] and Plackett[5] are interesting. Substantial recent contributions have been made by Pedersen[7], Wilkinson[8] and Seidenfeld (1979).

Pedersen[7], who claims that 'the fiducial argument has had a very limited success and is now essentially dead', nonetheless presents perhaps the most detailed review of this topic available to date. A central theme is that earlier reports have failed to give proper weight to what Fisher regarded as an essential requirement for the validity of the fiducial argument: namely, the existence of a 'relevant subset' of the reference set of all samples which may be used as the *collective* on which to define the fiducial probability concept. Pedersen attributes some of the stated inconsistencies to lack of recognition of the 'relevant subset' requirement, but feels that this notion is 'insufficiently explored' by Fisher (or others) for any final judgment of the fiducial method.

Wilkinson[8] claims that the 'inherently noncoherent' nature of inferential probabilistic assessment of observational data must be recognized and that in doing so we have a basis for reconciling the *confidence* and *fiducial* arguments. He offers a 'unified theory of confidence-based inferential probability'. The ideas are complicated ones and many commentators remain unconvinced that they provide an answer to the essential nature of fiducial probability or to the controversy surrounding this topic.

Seidenfeld (1979) subtitles his discussion of the philosophical problems of statistical inference 'Learning from R. A. Fisher'. He offers a detailed account of his views of Fisher's contributions to statistical inference, with particular reference to fiducial probability and claims to find (yet another) 'fatal flaw' in the argument. We have referred earlier to the interesting review of historical and conceptual aspects of (largely classical) inference provided in this book.

In spite of the continuing interest described above there still seems to be no generally accepted *definition* of fiducial probability, and it would probably be fair to say that there is little widespread acceptance of, or sympathy with, the fiducial argument.

The following comments represent various attitudes, the first attributed to Fisher himself.

I don't understand yet what fiducial probability does. We shall have to live with it for a long time before we know what it does for us. But it should not be ignored, just because we don't yet have a clear interpretation.

(Savage[14].)

... Fisher's ... concept of fiducial probability, often dismissed as too enigmatic for further consideration, was his attempt to delimit the class of problems which *could* yield probability statements about parameters without using Bayes' theorem and its controversial prior distribution.

(Edwards[6])

... if we do not examine the fiducial argument carefully, it seems almost inconceivable that Fisher should have made the error which he did in fact make. It is because (i) it seemed so unlikely that a man of his stature should *persist* in the error, and (ii) because, as he modestly says (... [1959], p. 54) his 1930 'explanation left a good deal to be desired', that so many people assumed for so long that the argument was correct. They lacked the *daring* to question it.

(Good[15])

Both currently and in the foreseeable future, fiducial probability seems likely to offer little or nothing towards the advance of statistical inference.

(Plackett[5])

... the fiducial argument makes a contribution to our exploration of the nature of uncertainty. ...

(Barnard[16])

8.2   LIKELIHOOD INFERENCE

The concept of *likelihood*, or the *likelihood function*, plays a crucial role in all three of the basic approaches to inference and decision-making. It acts as an expression of the information provided by sample data about unknown parameters (or more rudimentary features) of the probability model. In *classical inference* it is used as the basis for assessing the *sufficiency* of sample statistics and in the construction of estimators and tests of significance. The principles of *maximum likelihood* and of *likelihood ratio tests* occupy a central place in statistical methodology. Its role is even more fundamental to *Bayesian inference* (and *decision theory*). Rather than serving merely as an ingredient in the processing of sample data, *it is viewed as the sole and complete measure of the import of the data.* Thus in simple examples, the fact that the method of sampling does not effect the form of the likelihood function renders the sampling method itself irrelevant in the Bayesian approach. This is a direct implication of the **likelihood principle**. (See § 5.6)

We have already commented on the superficially appealing prospect of attributing some *probability* measure or *credibility* measure (call it what you will) to the likelihood itself, to measure the 'support' that the data provide for different probability models. The likelihood ratio criterion in classical statistics does this to some extent in representing the relative likelihoods of different hypotheses. But its application is via the probability structure on the sample space, $\mathscr{X}$, so that in the first place the likelihood is not necessarily the complete message that the data convey concerning the probability model and secondly no probability concept is transferred to the space of probability models itself (or equivalently in the parametric set-up, to the parameter space Ω). In the classical approach no statement of probabilities about a parameter θ can be entertained—the likelihood can serve no such purpose.

Not so in Bayesian inference, where the total inference is precisely a probability distribution over Ω. Furthermore, there is a special class of situations where the prior knowledge about θ is 'diffuse with respect to the sample data', in the sense that the prior distribution varies little over the region where the likelihood varies appreciably. This leads to a posterior distribution which is merely proportional to the likelihood function. This arises on certain conventional expressions of prior ignorance, or with vague prior knowledge through the operation of the 'principle of precise measurement'. (See § 6.5.) Here the likelihood function certainly does take on the role of measuring the probabilities (often degrees-of-belief) that should be attributed to different θ values. But there is no question of *transferring* the probability concept from $\mathscr{X}$ to Ω. In philosophical terms all that is happening is that an existing probability distribution over Ω is being up-dated by the data through the application of Bayes' theorem. Of course, where prior information is substantial even the utilitarian significance of the likelihood function as a measure of the credence to be attached to different θ-values is unacceptable from the Bayesian standpoint. The prior information is an additional crucial source of information about θ.

However, some statisticians are prepared to adopt a much more radical view of the likelihood function, and to regard it not only as the sole expression of the import of the data but (in certain circumstances, and in specific respects) as the only form of relevant information (in total). From this viewpoint the likelihood provides a meaningful relative numerical measure of 'propriety' or 'support' for different possible models, or for one θ-value compared with another. The concept is specifically a *relative* one. The term 'probability' is seldom used to describe the numerical concept embodied in the likelihood function. Undoubtedly this arises for emotive reasons, and because the *absolute* nature of the probability concept is regarded as inappropriate to the expression of 'beliefs' about different possible models. Instead we encounter measures of relative 'possibility', 'plausibility', 'credibility', 'support' and so on. Whatever the term used, the idea is that we can employ the data *through the likelihood function alone* to distinguish between different models, or possible parameter values. *No other information ingredient is utilized*, whether in the form of a prior distribution over Ω, or expressing the nature of the experimental sampling basis. There can be no Bayesian justification for such a **likelihood approach** to inference. Often the subjective basis (and expression) of prior information is quite unacceptable to its advocates. Commonly, associated probability calculations are specifically restricted to the *frequency* viewpoint. And yet, since the message of the data is seen to rest entirely in the likelihood function, sampling procedures and stopping rules become irrelevant and classical criteria and procedures such as *unbiasedness* or *tests of significance* appear utterly unjustifiable. (Since they have regard for what *might* have happened as well as for what *did* happen.)

Precedent for such a view concerning the special role of the likelihood function can be sought in isolated remarks of writers over a long period of time. However little detailed working out of its implications and applications is to be found until the last 30 years. Quite a vocal following has developed in this period, and a great deal of philosophical comment, methodology and practical investigation has been presented.

The basic ideas of the likelihood approach were established and examined by Barnard[17,18] and Birnbaum[12,19]. The book by Edwards (1972) entitled *Likelihood* offers a detailed study of 'the statistical concept of likelihood and its application to scientific inference'. With a refreshing lack of dogmatism, the author seeks to remedy his dissatisfaction with alternative approaches to statistics (and the conflicting demands they make for the adoption of a universal form for the probability concept) by granting 'likelihood an independent existence' as the only appropriate basis for statistical method. Apart from developing its thesis in

detail, this book provides some interesting comparative argument concerning the
classical, Bayesian (and *fiducial*) approaches, which are for various reasons found
unsatisfactory; likewise,

> the Method of Maximum Likelihood [which], *qua* estimation, I now think a red
> herring. (p. 212.)

The emphasis throughout is on the construction of a relative measure of 'support'
for rival hypotheses based on the likelihood function alone; and on its use in
developing statistical methods.

Some examples of the application of the likelihood approach to quite sophisticated problems are to be found in Sprott and Kalbfleisch[20], Sprott[21], Kalbfleisch
and Sprott[22], and Kalbfleisch[23], the last two being particularly concerned with
the handling of *nuisance parameters*. These last two papers also incorporate
fiducial distributions, the first of them prompting Edwards (in the discussion) to
welcome a 'bold and important paper', whilst remarking that 'the flash of insight
necessary to follow the fiducial argument has not yet visited me'. One of the few
detached commentaries on the likelihood approach is given by Plackett[5].

Let us now consider in simple terms and through simple examples some of the
basic tenets of the approach. It is a remark of R. A. Fisher[24], in a paper in 1934,
where he suggested that it is sometimes necessary to consider 'the entire course' of
the likelihood function, that is seen by some to herald its extended inferential
function as embodied in the *likelihood approach, per se*. That he had such an
extension in mind in quite clear from an example discussed in *Statistical Methods
and Scientific Inference* (Fisher, 1959), his last major work. He suggests that
likelihood functions and probability distributions 'supply complementary specifications of the same situation' but concludes that

> . . . . the values of the Mathematical Likelihood are better fitted to analyse,
> summarize, and communicate statistical evidence of types too weak to supply
> true probability statements. . . . (p. 72.)

This is illustrated by the suggestion that we should consider values of the
parameter where the likelihood has dropped to say 1/2, 1/5 and 1/15 of its
maximum value. He claims that such a statement of likelihoods, or corresponding
log-likelihoods, over a comprehensive range will 'convey all that is needed' about
a set of data; by considering such relative values of the likelihood we have a basis
for assessing 'what values of the parameter become implausible'. It is interesting
to note that in remarking that the areas under such relative likelihood curves are
irrelevant, and that only the ordinates matter, he is apparently ruling out a direct
probability interpretation.

The same is true of Barnard[25] who comments:

> If the ordinate of the likelihood at $\theta = \theta_1, \ldots$, is higher than the

ordinate. . . at $\theta = \theta_2$, we can say that the data point towards the value θ_1 rather than towards θ_2, or that on this evidence θ_1 is more plausible than θ_2. . . .

This attitude has met with some considerable support in different fields of application, notably physics and genetics. Its exponents are less hesitant in adopting a probability interpretation for relative likelihood; there is an obvious lay appeal in the proposed principle.

Barnard[17] seems to have been the first to attempt a formal approach to inference from such a likelihood standpoint. His proposals stem from the likelihood principle (see §5.6) which for present purposes may be restated as follows in two parts.

(i) *If the ratio of the likelihoods for two sets of data is constant for all values of a relevant parameter θ, then inferences about θ should be the same whether they are based on the first, or the second, set of data. This implies that the likelihood function conveys all the information provided by a set of data concerning the relative plausibility of different possible values of θ.*

(ii) *The ratio of the likelihoods, for a given set of data, at two different θ values is interpretable as a numerical measure of the strength of evidence in favour of the one value relative to the other.*

Among the properties of likelihood advanced by Barnard[25] in support of this approach are the ability to represent the power function of a test of significance as a weighted sum of likelihood functions, and more fundamentally its property of *minimal sufficiency*. Barnard's review article (Barnard[25]) makes fascinating reading both for its elementary exposition of the likelihood approach, and also for its critical comparison of the range of conflicting attitudes to inference. His tone here is a conciliatory one and many of his comments on the use of the likelihood function for the summarization of data (when there is little dispute concerning an appropriate parametric model) are important and largely un-exceptionable. He stresses the limited utility of hypothesis tests and point estimates, admits the relevance on occasions of prior probability assignments, but emphasizes most of all the 'data analysis' interests in modern statistical work: the use of data to suggest hypotheses and structure rather than formally to examine a limited range of possibilities. He sees the likelihood function as of pre-eminent importance in pointing the way—and with the facilities of modern computing aids suggests that this should, and can, go well beyond the mere two-point summary provided by the maximum likelihood principle. It is important to recognize that Barnard's approach to inference over the years has been a catholic one—stressing the fact that no single concept or attitude is sufficient to cover the range of different needs in statistical inference. His recent proposals for a unified approach are considered briefly below (§8.5) under the heading *Pivotal Inference*.

Edwards (1972) concentrates more on the **support** that a set of data provides for one hypothesis (or parameter value) compared with another, as measured by *the natural logarithm of the likelihood ratio*. (See also §8.4.) He commends the additive property of this measure over independent sources of data, and also its use for incorporating prior information, although he is at pains to distinguish this latter feature from the 'unacceptable' Bayesian approach which has only a 'superficial' similarity. **Support tests** of hypotheses are described as a replacement for the 'irrelevant' (or 'illogical') 'conventional tests of significance', and Fisher's **information** concept is widely embraced.

Birnbaum[12] adopted a rather more basic standpoint to justify the likelihood approach. He commenced by defining a concept of the *experimental* **evidence** $Ev(E, x)$ provided by data x, from a specified experiment E (involving Ω, $\mathscr{X}$ and $p_\theta(x)$]. He proposed what he regarded as two natural **principles** concerning this measure, that of **sufficiency** and of **conditionality**. The first of these says essentially that

$$Ev(E, x) = Ev[E', t(x)],$$

where $t(x)$ is a sufficient statistic and E' is the experiment which consists of replacing x by $t(x)$.

The second principle (which needs to be distinguished from that with a similar name described in §5.6) concerns experiments E made up of a probability mixture of several component experiments E_i, in the sense that observing x in E is equivalent to observing i according to the mixing rule and then observing x_i for that E_i. The *principle of conditionality* then declares that 'component experiments not actually performed are irrelevant' so that

$$Ev[E, (E_i, x_i)] = Ev(E_i, x_i).$$

Birnbaum then proceeded to show that these two principles *in conjunction* are equivalent to the first part of the *likelihood principle* as previously stated, which he paraphrased as the 'irrelevance of outcomes not actually observed'.† He regarded this equivalence as having 'immediate radical consequences for the everyday practice as well as the theory of informative inference', and concluded that 'in principle' statements of significance levels or interval estimates in practical problems should be replaced by detailed numerical statements of likelihood functions. Where he appeared to stop short (on mature reflection) of the Fisher–Edwards application of the likelihood approach is in the use of the likelihood ratio as a *numerical* measure of *weight of evidence*. He suggested that this has demonstrable justification only for a simple two-point parameter space, and is otherwise not amenable to any plausible interpretation and likely to produce misleading inferences with high probability. (See Birnbaum[19].) For a critical discussion of Birnbaum's views, and a later expression of them in the

---

† Birnbaum's reasoning is questioned by Durbin[26], and defended in Birnbaum[27].

context of alternative approaches, see Giere[28] and Birnbaum[29], respectively.

Grounds for criticism of the likelihood approach are self-evident. We have seen how it cannot appeal either to the convinced Bayesian, or confirmed classical statistician, basically because of its omission of what each, in their own terms, regard as essential information ingredients or interpretative yardsticks (prior information in a possibly subjective environment; the method of sampling and the long-run frequency interpretation of inferences). Even from a less entrenched standpoint, there is considerable dissatisfaction with what is seen by many to be an ill-defined, or uninterpretable, assignment of numerical measures of relative weight of evidence to alterative models,hypoheses or parameter values. On the other hand, genuine sympathy seems to exist for a fuller use of the likelihood function for data representation.

8.3 PLAUSIBILITY INFERENCE

The likelihood approach to inference inevitably places stress on the notion of *relative likelihood*: the ratio of the likelihood at some parameter value θ, $p_\theta(x)$, to its maximum possible value over Ω, $p_{\hat\theta}(x)$, for the observed data x. This arises as a measure of natural interest as a result of the likelihood principle, and has been used for determining interval estimators of θ; e.g. as those values of θ for which the relative likelihood exceeds some prescribed value.

Another form of standardization of the likelihood function has been proposed by Barndorff–Nielsen[30] as the basis of an approach to inference. Instead of standardizing with respect to $p_{\hat\theta}(x) = \sup_\Omega p_\theta(x)$, he proposes that we consider the ratio of $p_\theta(x)$ to its maximum value with respect to *different possible data sets* x that might have been encountered. Thus we consider

$$\Pi_\theta(x) = p_\theta(x)/\sup_{\mathscr{X}} p_\theta(x),$$

termed the *plausibility function*, as a measure of inferential import of the data x on the parameter θ. Parallel to the likelihood approach, it is suggested that we consider the **maximum plausibility estimator**

$$\check\theta = \{\theta: \Pi_\theta(x) = \sup_\Omega \Pi_\theta(x)\}$$

and **plausibility ratio tests** developed just as likelihood ratio tests, but using $\Pi_\theta(x)$ rather than $p_\theta(x)$.

In proposing this individual approach to inference, Barndorff–Nielsen does not claim it to be an improved alternative to the likelihood approach, but a complementary adjunct: plausibility providing an 'equally valid' representation of a 'different aspect—of the evidence in the data'. It is also presented as a development, an example, or an alternative representation, of earlier concepts. It illustrates the notion of an *ods function* (Barnard[17]), as a function of the data expressing relative 'credability' in different possible values of the parameter, θ.

Barndorff–Nielsen[30] (and 1978) also further develops his earlier ideas of *ancillarity* and *nonformation* (no information) through the plausibility concept. Central to such ideas is that of *universality* of a family, $\mathscr{P}$, of probability distributions, essentially requiring that every member p_θ of $\mathscr{P}$ should have a *mode point t* for some statistic T in the sense that

$$p_\theta(t) = \sup_T p_\theta(T).$$

Space does not permit any detailed study of the nature, or implications, of plausibility inference. But it is of interest to take one simple example to illustrate the distinctions between likelihood and plausibility. Barndorff–Nielsen sees discrete models as of prime importance in inference, and his illustrations reflect this. In his book (1978) he shows the (normed) likelihood function, and plausibility function, for a very simple situation—observing the outcome $x = 1$ of a binomial random variable $X \sim \mathbf{B}(n, \theta)$ for the two cases $n = 1$ and $n = 3$. The results are shown in Figure 8.1 and the principal difference is that using the likelihood an isolated point estimate is obtained for θ (1 and 1/3, respectively) whereas the maximum plausibility estimators are ranges of values, $[\frac{1}{2}, 1]$ and $[\frac{1}{4}, \frac{1}{2}]$, respectively.

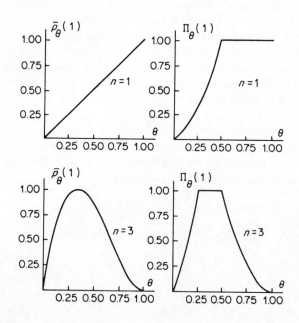

Figure 8.1 The (normed) likelihood $\bar{p}_\theta(1)$ and plausibility functions $\pi_\theta(1)$ corresponding to the observation $x = 1$ of a binomial variate with trial number $n = 1$ or 3. (Reproduced from Barndorff–Nielsen, 1978, by permission of John Wiley & Sons Ltd.)

A major criticism of the plausibility approach is that it violates the likelihood principle! It is not easy to resolve the distinction between the types of information provided by the likelihood, and plausibility functions. Barndorff–Nielsen[30] remarks:

> As an indication of the root of the difference between likelihood inference and plausibility inference it may be said that the former relates to how well the hypotheses explain the data whereas the latter pertains to the predictability of the data on the various hypotheses.

8.4 STRUCTURAL INFERENCE

Yet another identifiable approach to statistics is known as **structural inference**. It is a recent entry to the field, and again the creation of one man, D. A. S. Fraser, who has attempted to exhibit a whole new unified approach to inference in his book *The Structure of Inference* (Fraser, 1968). The ideas embodied in this approach are developed at great length in Fraser's book, expanding on earlier indications of basic principle (Fraser[31-33]). In spite of its claims to be an *introductory* text on mathematical statistics, this book is couched in a rigorous mathematical style which (after the introductory chapter) makes relatively little concession to the needs of the simple man for a simple explanation of basic principles and concepts. Indeed, there is hardly any discussion of the fundamental interpretative nature of the approach. In its short history *structural inference* has stimulated a deal of critical comment, much of which is also often expressed in sophisticated mathematical argument. The occasional down-to-earth criticism, attributing to the structural argument merely a re-interpretation of familiar *classical, Bayesian* or *fiducial* arguments (you name it!), makes one question whether anyone (apart possibly from Fraser himself) truly sees the conceptual wood for the mathematical trees! But in view of the interest and the guarded praise of some commentators, it is undoubtedly important to consider at least a simple illustrated description of the basic philosophy of this approach. However, we shall stop short of the detailed application (or extension) to progressively more complicated situations considered in Parts II and III of Fraser's book.

The structural approach is firmly rooted in the *classical* tradition in many respects. It is concerned with statistical *inference* defined as 'the theory that describes and prescribes the argument from observation and measurement to conclusions about the unknowns', and it identifies its origins as lying 'somewhere between' the Fisher and Neyman–Pearson schools of *classical* statistics. Its basis is again in the construction and study of a statistical model representing the fortuitous elements relating to repeated observations of a situation under assumed identical circumstances. There are, however, two absolutely crucial differences of attitude as compared with the *classical approach* described in Chapter 5. These concern the *formulation of the statistical model, and the manner in which inferences are expressed* about 'unknown quantities' ('physical constants, relationships'): broadly interpretable as what we have previously called *parameters*.

The Statistical Model

A distinction is drawn at the outset between the 'exterior' nature of the usual *classical* model, and the concern of the 'measurement model' (or its generalization as a 'structural model') of the *structural* approach with the *internal* mechanism governing the observable behaviour of the situation under study. In the classical model we typically observe some 'response variable' whose value, x, as a reflection of an unknown quantity θ, is governed by a probability distribution $p_\theta(x)$ on the sample space $\mathscr{X}$. Fraser refers to this as a 'black box' approach, which ignores the fact that we often know (directly or indirectly) much more about the internal relationship of x and θ than is represented through such a model. He argues that situations commonly contain *identifiable* sources of variation, such as errors of measurement, variations in the quality of products, effects of randomization in designed experiments, and so on. He calls these **error variables**, though in the wider applications of the method this term needs to be liberally interpreted. In many situations they truly determine the probabilistic structure of the situation, their variational behaviour is well understood, and they constitute the natural basis for expressing a statistical model.

These considerations provide the motivation for formulating the statistical model in a particular way. Instead of regarding the data x as being generated by the distribution $p_\theta(x)$, a quantity e (the *error variable*) is introduced which is expressed in terms of x and θ, and which is assumed to have a known distribution which does not depend on θ.

Thus the model consists of two parts: a statement of *the probability distribution of the error variable* (independent of θ), and a statement of the *relationship* between *the observational data x* (*which are known*) *and the unknown* θ, on the one hand, and *the unknown but realized value, e, of the error variable*, on the other.

Statistical Inference

The structural model tells us how x is related to θ and e and also the probability mechanism under which e was generated. This relationship frequently expresses the observed response (the data x) in terms of a simple kind of transformation of the error, e, governed by the value of θ. *The basis for statistical inference in the structural approach is essentially to reverse this relationship* and to interpret θ as being obtained as an associated transformation of e by the operation of the known data x. Thus θ is expressed in terms of x and e, the probability mechanism by which e is obtained is assumed known (in principle), so that through this inverse relationship we have (to quote Fraser) '*ipso facto* a probability statement concerning θ'.

Before considering the philosophy or interpretation of this principle let us consider a simple example of the type discussed by Fraser involving what he calls a **simple measurement model**.

Suppose we are interested in the value of some physical quantity θ, and have available an instrument which produces a measurement x of θ. Knowing something about the characteristics of the instrument we may be able to declare that x differs from θ by a measurement error e which has a probability density function $f(e)$ (independent of θ). The proposed statistical model for this situation then consists of two parts, a **structural equation** expressing x in terms of θ and e and an **error distribution** $f(e)$. In this model:

$$x = \theta + e, \quad f(e), \tag{8.4.1}$$

x is *known*, θ *unknown* and e has a *known distribution*. The general principle of structural inference then suggests that we reverse the structural equation. In the same way that x is regarded as being obtained from e merely by a translation of the realized error e by an (unknown) amount θ, so θ can be regarded as a translation of $-e$ by the (known) amount x:

$$\theta = x - e. \tag{8.4.2}$$

The argument then proceeds by remarking that general inferences about θ derive directly from the right-hand side of (8.4.2)—that is, since there is a probability distribution describing the variability of e, *and* we have

$$e = x - \theta$$

then this same distribution applies to $x - \theta$. As such it is called the **structural distribution**.

This primitive example concerns a single measurement, and a simple translation relationship. The model and principle are readily extended to multiple measurements $x_1, x_2, \ldots, x_n$ with corresponding realized (but unknown) errors $e_1, e_2, \ldots, e_n$. The *structural equation* relates the *known* values $x_1, x_2, \ldots, x_n$ to the *unknown* values $\theta, e_1, \ldots, e_n$. With independent errors the *error distribution* has density $\Pi_{i=1}^{n} f(e_i)$. Sometimes, however, it is convenient to work with a **reduced model**, which essentially takes the same form in relation to some summary measures of the data $x_1, \ldots, x_n$ and errors $e_1, \ldots, e_n$. Using conditional probability arguments, and exploiting the fact that the translation operation generates a *location group*, **reduced measurement models** are generated with the same *structural* form, and analogous inferential implications about θ. Then again, the extension to multi-parameter systems can also be developed.

For example, a somewhat more detailed situation employing the measurement model is one in which the measurements x may be expressed as depending on measurement error e through two unknowns μ, σ (location and scale parameters), in the form

$$x = \mu + \sigma e.$$

If the error variable has a probability density function $f(e)$ (independent of μ and σ) and we take independent observations $x_1, \ldots, x_n$ then the structural model is

$$x_i = \mu + \sigma e_i \ \ (i = 1, \ldots, n), \quad \prod_{i=1}^{n} f(e_i).$$

Fraser demonstrates (and justifies) the *reduction* of this model to the situation where the data are summarized by $\bar{x}$ and s_x^2, the sample mean and *classical* unbiased variance estimate, respectively. To illustrate this, if $\bar{e}$ and s_e^2 are the corresponding quantities for the (unknown) error terms $e_1, \ldots, e_n$ then he shows that we have a valid reduced model in the case of *standardized normal independent errors* which can be expressed as

$$\left. \begin{array}{l} \bar{x} = \mu + \sigma \bar{e} \\ s_x = \sigma s_e \end{array} \right] \quad \left[\begin{array}{l} \bar{e} = Z/\sqrt{n} \\ s_e = \sqrt{(W_{n-1}/(n-1))} \end{array} \right.$$

where Z is $N(0,1)$ and W_{n-1} is a χ^2 random variable with $n-1$ degrees of freedom, independent of Z. The implications of this model for inferences about μ and σ, *on the structural argument*, are that $(n-1)s_x^2/\sigma^2$ *has a χ^2 distribution with $n-1$ degrees of freedom*, and that $\sqrt{n}(\bar{x}-\mu)/s_x$ has (independently) Student's t-distribution with $n-1$ degrees of freedom.

The apparent similarity of these results to those obtained in other approaches is worth commenting on. In classical statistics the same distributional results hold but with quite different interpretation: the 'random variables' are $\bar{x}$ and s_x and the ditributions are *their* sampling distributions. On the current approach it is μ and σ which have the accredited distributions. Similar formal results to those given by the structural argument were obtained on the Bayesian approach using the Jeffreys' formulation for prior ignorance about μ and σ; likewise using a *fiducial* argument.

But this type of *structural* model is also readily employed for different inference purposes, such as the construction of *tests of significance*, without the need for overt use of the structural distribution of θ. Fraser illustrates this with simple numerical examples, which echo the form of the *classical* tests in their regard for 'tail area probabilities' as a criterion of accepting or rejecting an hypothesis. (In this respect the structural approach seems entirely at variance with *Bayesian* principles.)

As so far described, the structural approach is severely limited in the types of situation to which it can be applied (with measurement errors subject to only location or scale transformation in their promotion of observed measurements).

However, its scope is greatly extended in subsequent development. We can do no more than sketch this progress. The key to wider application is in the *group transformation* representation of the structural equation. This leads to a more general model, known as the **structural model**.

Recosidering the stuctural equation for the simple measurement model, shown in (8.4.1), we can rewrite this as

$$x = \theta e,$$

where θ is now regarded as a translation *operator*, transforming the realized error e to $\theta + e$ to yield the observation x. As θ varies over the real line these transformations constitute the *location group* on $\mathbf{R}^1$. The model is generalized by extending our consideration to *any one-to-one group* of transformations, G, with typical element θ, with the *further property that if $\theta_1 y = \theta_2 y$ for any y in the space operated on by G then $\theta_1 = \theta_2$.* (Fraser calls this a *unitary* group, but this does not appear to be entirely standard terminology.) In this way we obtain a structural model to describe a situation in which internal error exists with a component, the error variable E, which yields the potential response X by the operation of some transformation $\theta \in G$ which is itself interpretable as a basic unknown quantity of interest in the situation being studied. It is assumed that the space of values taken by both E and X is common, $\mathscr{X}$ say, that E has a constant probability distribution over $\mathscr{X}$ (independent of θ) and that this distribution is *absolutely continuous*. (This latter condition is obviously convenient from the mathematical standpoint, but seems greatly restricting.)

Such a model enables us to go well beyond the *affine group* of transformations representing the simple shifts of location and scale described above. But whilst opening up a much wider field, it still implies a severe limitation on the range of problems that can be studied.

Thus the general **structural model** is expressed again by two components: the *structural equation* and the *error distribution* (represented by the error variable),

$$X = \theta E; E. \tag{8.4.3}$$

The method of making inferences about θ in this general model may be summarized as follows. If the set of values into which any X is transformed by G is the whole space $\mathscr{X}$, then $\theta^{-1} X$ *has the same probability distribution as E*: the so-called *structural distribution*; if this set of values is merely a subspace $\mathscr{X}_G \subset \mathscr{X}$ then it is necessary to condition the distribution on $\mathscr{X}_G$.

In the further development of his *structural approach*, Fraser (1968) considers its application to linear models and to problems involving 'additional quantities . . . not in direct correspondence with the simple kind of transformation', and extends the argument to situations in which the error variable is not immediately identifiable. The principles of **conditionality** and **marginal likeli-**

hood play a role in this development. The use of likelihood for general inference (cf. §8.2), and concepts of **precision** and **information**, are advanced (cf. §8.6).

In conclusion we must attempt to summarize the advantages claimed for this approach, and the criticisms directed against it.

Fraser claims essentially two principal advantages over the *classical* approach. Firstly, that it sets the emphasis in model construction in the proper place: in considering the *internal* error structure rather than its *external* manifestation through the observed responses. Secondly, that it yields a general principle of statistical inference producing 'unique solutions . . . in terms of classical frequency-based probability' without the need of 'additional principles and techniques' or specific assumptions about the 'error form'. (For example, least squares as a principle relies on an assumed *normal* error structure for justification through the method of maximum likelihood, or the criterion of minimizing the variance of linear unbiased estimates; see §5.3.3.)

What response do these claims encounter? We have seen that the group representation of the structural model implies a restriction of the types of situation which may be analysed. This causes dissatisfaction—that what purports to be a *general* method of inference can be applied only in special circumstances. Lindley[34] is

> . . . suspicious of any argument, . . ., that only works in some situations, for inference is surely a whole and the Poisson distribution [is] not basically different in character from . . . the normal.

The reference to the Poisson distribution presumably concerns the restriction to *continuous* distributions.

Fraser[35] refers to modifications by other writers to the nature of the group of transformations to be employed, but these do not appear greatly to extend the range of application.

A major problem is in the interpretation of the *probability concept* in the structural distribution of θ. Although there is little mention of the word probability in Fraser (1968) he does claim there, and elsewhere (Fraser[35]) that it is the central concept and in the classical frequency-based tradition. In spite of this claim the transference of a probability distribution to θ in the manner described surely needs some interpretation and justification. The fact that Fraser offers none is a wide source of dissatisfaction among commentators. In his explanation of the propriety of attributing probabilities to 'unknown constants', Fraser (1968) merely comments that this is no different in kind to the entirely accepted practice of discussing the probability of a realized (but unobserved) outcome of an experiment. If we deal a card face down from a pack its designation is (he claims) an '*unknown constant*', but we do not hesitate to attribute probabilities to the different possible designations! This analogy is seen by some to be naive, if not somewhat disingenuous. In the strictly frequency sense, what is to be the '*collective*' induced by the error structure as a basis for carrying probability statements about θ?

It is tempting to re-interpret or re-express the structural argument in terms of alternative concepts and principles. The idea of invariance is crucial; the central consideration of the *pivotal quantity* $\theta^{-1} X$ (with a constant distribution, used to transfer probability from E to θ) is strongly reminiscent of Fisher's fiducial approach. Fraser's[31] earlier concern with 'the fiducial method and invariance' adds further stimulus to such an attribution. Stone[36] talks of 'making transplants on ... the nearly expired corpus of fiducial theory', whilst Lindley[34] sees 'Fraser's argument [as] an improvement upon and an extension of Fisher's in the special case where the group structure is present'. Yet Fraser himself suggests no such association; on the contrary he seems quite clearly to reject the fiducial approach (Fraser, 1968, p. 42).

As is to be expected we also encounter a deal of re-interpretation in terms of the principles of Bayesian inference, and a corresponding evaluation from this standpoint. After all, the Bayesian method is specifically concerned with making probability statements about θ, and it is natural to seek parallels (or contrasts) between the effects of the two approaches in this respect. Fraser himself seems to support no general correspondence in principle between the Bayesian and structural arguments.

Essential reading on the debate surrounding structural inference is Fraser's[35] reply to the review of his book by Lindley[34], and the ensuing discussion. Two points that he makes, among many, stress that the structural model *should* be used but only where justified by the nature of the problem (otherwise it has an exploratory function in relation to the classical model), and that the *structural distribution* of θ is only a '*small portion of the development*' of his argument (the 'major portion ... [being] concerned with tests of hypotheses, with distributions of test statistics, with distributions for concealed error values, [and so on]"). This latter point is important; the part of the development which does not lean on a transferred probability measure for θ presents us with an interesting new concept in statistical modelling. Explicit regard for the error structure, where this is appropriate, may prove to be a most powerful basis for drawing inferences. Nonetheless, it seems that contraversy will inevitably continue to centre on the *interpretation* of the structural distribution.

> Professor Fraser has written a book that, despite the obscurity of its restrictions, will stimulate interesting research. I hope, however, that it will not inspire any slavish, routine application; this is jungle territory and there will be snakes in the grass for a long time.
>
> (Stone[36])

> One thing is certain: this is *not* a trivial contribution to our understanding of the inference process.
>
> (Lindley[34])

8.5 PIVOTAL INFERENCE

Another important individual initiative in the development of general approaches to inference is that of G. A. Barnard and entitled **pivotal inference**. At the present time it would seem that there is no detailed (formally) published account of the general principles, or resulting methodology, of this approach and it is necessary to extract the basic ideas from fairly brief comment in published articles (e.g. Barnard[37]) or from more discussive informal reports and preprints. It is to be hoped that a more detailed description of this approach may be published in the near future.

Whilst admitting the importance of various different approaches to inference, each relevant in appropriate circumstances, Barnard presents a more general *pivotal model* for a range of inference problems. By appropriate specification of the components of the model it is possible to incorporate any available prior information about some or all of the parameters, or with equal facility to avoid the need to include prior information if none is available. This does not seem to overcome the basic controversy about whether or not it is appropriate to make probabilistic statements concerning parameters. But it is undoubtedly attractive to be able to examine within a single system (and without conventional assumptions concerning the representation of prior ignorance) the implications of incorporating, or excluding, prior information about parameters. The approach is also particularly concerned with the problems of handling nuisance parameters.

The central concept is one that has already been encountered in fiducial inference and structural inference: namely, that of a **pivotal quantity**, or a **pivotal**. This is a function $p(x, \theta)$ of the observations x and the parameters θ, whose distribution is independent of θ. *The basic form of pivotal inference consists essentially of a statement of the distribution of some appropriately chosen pivotal, possibly conditional on the observed values of some statistics together with a statement of the observed values of the conditioning statistics.* A critical consideration is what constitutes an 'appropriate' pivotal, and we must give some attention to this matter.

The general pivotal model has five components $\{\mathscr{X}, \Omega, P, p, D\}$ where $\mathscr{X}$ and Ω are the sample space and parameter space, respectively. P is the pivotal space on which the basic pivotal function p: $\mathscr{X} \times \Omega \to P$ takes values and D is a family of distributions for the pivotal function. The statement of a model for a particular situation amounts to declaring the nature of D, often as a *set* of fully specified distributions which are (in some sense) not too different one from another, rather than as just a *single* distribution. This flexibility aids the use of the approach in robustness studies.

The basic pivotal is required to satisfy certain fundamental conditions (invertibility, and 'robustness' with respect to the distributions in D). With appropriate restrictions on the form of the family of distributions, D, it turns out that for any pivotal $p(x, \theta)$ we can define a (robust) **maximal ancillary** $a(x) = g[p(x, \theta)]$ which is *unique* up to functional equivalence. The ancillary $a(x)$ is *maximal* in the sense that any function of $p(x, \theta)$ which is constant over the parameter space (the basic requirement of an ancillary) must be a function of $a(x)$. We note in passing that $a(x)$ is itself a pivotal—of a special kind in the respect that it does not functionally involve θ.

We now proceed by appropriate transformations to partition the basic pivotal $p(x, \theta)$ into two components $a(x)$ and $q(x, \theta)$, say, and adopt the pair $[a(x), q(x, \theta)]$ as the basic pivotal in place of $p(x, \theta)$. We must, of course, correspondingly transform the family D of distributions for p, to one for (a, q).

Pivotal inference consists of observing data x and employing the distribution of $q(x, \theta)$ conditional on the observed value of a (x).

We have so far implicitly assumed that the only available information consists of sample data x.

Suppose, however, that we have some prior information about θ or some component of θ, in the form of a prior distribution. Since the distribution of θ (or its component) is fully specified we can introduce θ [or some convenient function, $b(\theta)$] as an element in the basic pivotal, and include its distributional behaviour in the specification of D.

Thus in its most general form the pivotal model consists of a pivotal with three components

$$\begin{bmatrix} a(x) \\ q(x, \theta) \\ b(\theta) \end{bmatrix},$$

where $a(x)$ is the (robust) *maximal ancillary* (constant on Ω), $b(x)$ is the *Bayesian pivotal* (again maximal, and constant on $\mathscr{X}$) and $q(x, \theta)$ varies over both Ω and $\mathscr{X}$. Inferences take the form of statements of the joint distribution of (q, b) conditional on a.

Let us consider a simple example. Suppose the data consist of a random sample $x_1, x_2, \ldots x_n$ of n observations of a random variable X having a distribution with location and scale parameters μ and σ, respectively, and with distribution function of the form $F[(x - \mu)/\sigma]$. We can take

$$q(x, \theta) = [(\bar{x} - \mu)/\sigma, s_x/\sigma]$$

and

$$a(x) = [a_1(x), \ldots, a_{n-1}(x)],$$

where $\bar{x}$ and s_x^2 are the sample mean and variance, respectively, and $a_i(x) = (x_i - \bar{x})/s_x$. The primary inference consists of stating, for any member of D, the joint distribution of $(\bar{x} - \mu)/\sigma$ and s_x/σ (conditional on the observed value of $a(x)$) and the values of $\bar{x}$ and s_x. If we had available a prior distribution for σ, we would augment $p(x, \theta)$ by a Bayesian pivotal component, σ, and reduce $q(x, \theta)$ to one element, $\bar{x} - \mu$. The primary inference would now consist, for any member of D, of a statement of the joint distribution of $(\bar{x} - \mu)$ and σ, conditional on observed values of $x_1 - \bar{x}, \ldots, x_{n-1} - \bar{x}$, and of the value of $\bar{x}$. If we can go further, in the respect of having available a proper joint prior distribution for (μ, σ), then the maximal ancillary is just the whole data set and the pivotal inference becomes a statement of the usual Bayesian posterior distribution.

There is undoubtedly an attraction in the synthesis of sample space and parameter space considerations provided by this approach. We must await with interest and anticipation the fuller formulation of principle and method. It is clear that many of the basic criticisms will be refocused on the pivotal approach. In particular, the approach seems not to conform to the strong likelihood principle. But perhaps the most fundamental issue concerns the probabilistic interpretation of any inference. Probability statements relate to the pivotal quantity which involves not only the data x but also the parameter θ; indeed, it might just involve θ alone. Again it seems crucial to ask what mechanism allows transference of the probability concept from sample space to parameter space and what interpretation can legitimately be placed on any probabilistically expressed inferences.

8.6   INFORMATION

The word 'information' figured in the earliest pages of this book, as a general description of the raw material to be processed in an inferential or decision-making enquiry. It was subsequently given more specific form in this respect, as a label for the three components: *prior knowledge, sample data* and *consequential costs*. No formal meaning was attached to the term 'information', its significance was global and intuitive. On the other hand, there certainly have been attempts made to develop a more formal technical definition of the **information** concept.

One example has already been mentioned (§5.3.2). Fisher, as early as 1925, introduced such a definition based on the idea of measuring the importance of sample data, or some statistic, in terms of the likelihood function. This definition has been widely applied in classical statistics, particularly in relation to the theory of estimation. But this is not the only technical form that has been advanced to describe the concept of information.

It seems most apt that in the closing pages we should return to this topic, and discuss briefly some of the efforts that have been made to set up useful mathematical expressions to represent this appealing idea of the 'information' obtained in a statistical enquiry.

We start with Fisher's concept of the information contained in a set of data, briefly introduced in §5.3.2. Consider the familiar parametric model where data x arise from a probability distribution $p_\theta(x)$ over the sample space $\mathscr{X}$, where the parameter θ takes some value in the parameter space Ω. The import of x for drawing inferences about θ is represented by $p_\theta(x)$, thought of as a function of θ

for fixed x and called the *likelihood function*. As we have seen, Fisher defined the
information in the sample (that is, in the data x) as the quantity

$$I_x(\theta) = E_{\mathscr{X}}\{-\partial^2[\log_e p_\theta(x)]/\partial\theta^2\} = E\{-\partial^2 L/\partial\theta^2\}. \qquad (8.6.1)$$

If $\tilde{\theta}(x)$ is some statistic, with sampling distribution $g_\theta(\tilde{\theta})$, then analogously the
information in the statistic, $\tilde{\theta}$, is defined as

$$I_\theta(\theta) = E\{-\partial^2[\log_e g_\theta(\tilde{\theta})]/\partial\theta^2\}. \qquad (8.6.2)$$

Several properties of these functions support their definition as measures of
'information' about θ in the terms of classical statistics. We should expect that
there can be no more 'information' in the statistic $\tilde{\theta}$ than there is in the total data x.
We readily confirm that

$$I_{\tilde{\theta}}(\theta) \leqslant I_x(\theta)$$

with equality (as we should hope) only if $\tilde{\theta}$ is *sufficient* for θ. Then again we should
intuitively expect the 'information' to be non-negative, to provide a means of
assessing the 'accuracy' of inferences about θ and to increase with the 'extent' of
the data. These demands are also met by $I_x(\theta)$ under appropriate conditions, and
in particular respects. $I_x(\theta)$ is certainly non-negative. Under the *regularity
conditions*, the Cramér-Rao lower bound to the variance of unbiased estimators
of θ is $[I_x(\theta)]^{-1}$. In this respect the *information* $I_x(\theta)$ represents the accuracy of
the best possible estimator (the MVB estimator) that could be obtained. In any
situation where the MVB estimator does not exist, $I_x(\theta)$ is less compelling as a
measure of accuracy, although some comfort is derived from the fact that the
maximum likelihood estimator is fully efficient (asymptotically) and also has
variance $[I_x(\theta)]^{-1}$. Finally, if data x is augmented with further data y then

$$I_{x+y}(\theta) \geqslant I_x(\theta).$$

What of the role of $I_x(\theta)$ outside the *classical* approach? In its intrinsic basis in
sample space considerations, it cannot appeal as a general principle to adherents
of the *Bayesian* approach. This is not to say that it does not appear in the Bayesian
literature. It is, for example, used in a particular method of constructing prior
distributions to describe prior ignorance. (As we saw in §6.4.)

Other mathematical forms for the information concept have been suggested,
and applied to statistical problems. The most well known finds its formal origins
in attempts to measure disorder in statistical mechanics and thermodynamics.
The indigenous probablistic concepts of *entropy* and *information* have been
applied to problems of communications, leading to *communication theory* as a
system for representing and analysing the transmission of information in the
presence of random disturbances. Within this theory information is given a
logarithmic definition in terms of the prevailing probability structure. A simple
treatment of the mathematical foundations of this topic is given by Khinchin
(1957). It is quite a recent branch of investigation, having been developed only

over the last 35 years. The independent pioneering efforts of Shannon[38] and Wiener (1948) sparked off an energetic interest and activity. Kullback (1959) attributes to the remark by Wiener (1948), that the logarithmic measure of information might usefully replace Fisher's concept in *statistical* investigations, the transference of the logarithmic information concept from a purely *probabilistic* (model representation) function to a truly *statistical* (inference) one. Much work has gone into developing this idea (see, for example, Savage, 1954) and using it for constructing statistical procedures (though not without criticism). A lengthy detailed discussion of 'measures of logarithmic information and their application to the testing of statistical hypothesis' is given by Kullback (1959).

The basic ideas in Kullback's treatment are easily summarized, but space does not permit any discussion of his extensive consideration of the properties of the information concept or their application to a large range of different problems. In our current notation if θ_1 and θ_2 are two possible values of the parameter θ, *then the* **information** *in data x for discrimination between θ_1 and θ_2 is given by the logarithm (to some base exceeding unity) of the likelihood ratio*:

$$\log \{p_{\theta_1}(x)/p_{\theta_2}(x)\}. \tag{8.6.3}$$

On the assumption that our data arise from the distribution with $\theta = \theta_1$, the **mean information** provided by such data in favour of θ_1 against θ_2 is given by the expected value of (8.6.3)

$$I(1:2) = \int_{\mathscr{X}} p_{\theta_1}(x) \log \{p_{\theta_1}(x)/p_{\theta_2}(x)\}. \tag{8.6.4}$$

If we were to have prior probabilities $\pi(\theta_1)$ and $\pi(\theta_2)$ for $\theta \in (\theta_1, \theta_2) = \Omega$, then (8.6.4) can be rewritten as

$$I(1:2) = \int_{\mathscr{X}} p_{\theta_1}(x) \log \{\pi(\theta_1|x)/\pi(\theta_2|x)\} - \log \{\pi(\theta_1)/\pi(\theta_2)\}$$

and may be interpreted as the difference between what are called the posterior, and prior, mean **log-odds** in favour of θ_1 against θ_2. (See Good, 1965, Edwards, 1972, and earlier references.) Kullback considers the extent to which this idea of *information* satisfies intuitively desirable conditions. He demonstrates its *additivity* for independent sets of data (true also of Fisher's concept), *convexity* and *invariance* properties, and establishes a tie-up with the earlier Fisher concept. He considers also the effects of the existence of sufficient statistics (no processing of data can lead to more information; sufficient statistics contain *all* the information present in the data). Jeffreys (1961) has considered such a measure of information as a basis for justifying the choice of prior distributions expressing ignorance about θ. Good[39] has presented a fully axiomatic justification of the logarithmic form of *information*, and has considered (earlier) associated ideas of **corroboration** and **weight of evidence**. Related concepts of **support**, and **support tests**, are described by Edwards (1972), see §8.2.

Lindley[40] (summarized in Lindley, 1971*b*) describes an extended Bayesian application of the logarithmic form of information. Motivated by the communications theory concept of information, due to Shannon, the quantity

$$\int_\Omega \pi(\theta|x)\log\pi(\theta|x) - \int_\Omega \pi(\theta)\log\pi(\theta) \qquad (8.6.5)$$

is proposed to measure the difference in the information about θ that we possess before and after obtaining the data x. Apart from a change of sign reflecting the different objectives in the statistical and communications theory applications, (8.6.5) is just the difference in the Shannon measures of information for the posterior and prior distributions of θ.

Further averaging over $\mathcal{X}$ with respect to possible variations in the data x leads to the *expected* (increase in) *information* from collecting data in the prevailing situation. This can be shown to have the symmetric form

$$\int_{\mathcal{X}} \int_\Omega p(x,\theta)\log\left\{\frac{p(x,\theta)}{p(x)\pi(\theta)}\right\}, \qquad (8.6.6)$$

where $p(x,\theta)$ is the joint probability (density) of x and θ, and $p(x)$ the marginal probability (density) of x. (8.6.6) provides a measure of the import of the experiment we are performing to obtain our data. As such, it is used as the basis for constructing a Bayesian approach to the *design of experiments*.

This concept of *information* has also been employed in the Bayesian approach with reference to the idea of 'equivalent sample size' in *conjugate* prior distributions. (See §6.6.1.)

8.7 THE 'MATHEMATICS OF PHILOSOPHY'

In the study of any situation involving uncertainty (or non-deterministic elements) the aim will be to delimit, measure and employ any relevant material (be it diameters of rivet heads, or opinions on the suitability of different medical treatments) to provide a clearer insight, or facilitate some choice of action. It is natural, by the very nature of this process, that we should consider whether everyday intuitive concepts of content, benefit or ameliorization might be usefully represented in mathematical form to this end. Of course, we would like to measure 'weight of evidence', 'amount, or relevance, of information', and so on. This desire is undoubtedly the stimulus for using such terms as *likelihood* and *admissibility* in formal inference or decision-making procedures.

We have already encountered many other examples of the use of *mathematical* constructs designed, and labelled, to represent concepts and principles with an obvious appeal in terms of everyday meaning. A notable example is *information*, discussed in the previous section; others are *consistency, power, bias, sufficiency, coherence, dominance, risk, evidence, utility, support*, and so on. These have occurred usually as *internal* components of some formal approach to statistics or probability theory. But they also figure on occasions as the external stimulus for some basic system—in Birnbaum's approach to inference, for example, the concept of *evidence* is advanced as an important one; it is formalized, desirable features are expressed mathematically and a system of inference is developed which embraces the *likelihood principle*. (See §8.2.)

Obvious precautions must be taken in transferring the desire to represent some property, into the use of a *mathematical expression* (or logical system) labelled in the same way. Mere labelling achieves nothing unless the mathematical expression can be shown independently to possess properties that seem appropriate or desirable. This point has already been made in relation to *likelihood*, and *admissibility*. As another example, suppose I choose to define a formal measure of **stupidity** for a generalized statistical procedure, and to then effect a choice of procedure on the **principle of minimum stupidity**. It sounds appealing, but achieves nothing if my concept of stupidity is *stupid*. But due caution does not justify nihilism. Few press this point to the extent of Bross[41] in his attack on 'quibbling' in the naming of concepts or principles, and in his concern for the possible deception in the use of 'highly specialized languages' employing everyday words endowed with an extended formal meaning.

Undoubtedly the past and current fever of interest in what Good[42] calls the 'mathematics of philosophy' will remain with us. In its application to non-deterministic situations it is very much part of the statistical scene, and deserves our attention. Existing statistical practice and *mores* owe much to philosophical argument; new ideas will develop with a corresponding indebtedness. It is unbelievable that any single attitude to inference and decision-making will ever be demonstrated to be *the correct one*, or ever find universal acceptance. But in exploring rival claims a proper mixture of philosophical argument and practical application will continue to provide an appropriate yardstick.

The 'mathematics of philosophy', even in relation to *statistical* problems, is a vast area of study, largely untouched in the pages of this book. Merely to sketch its development, range and structural form would be an enormous task. In terms of the breadth of his ideas, and the prolific extent of his publications, I. J. Good has undoubtedly made a major contribution. His extensions of *subjective probability* arguments and *Bayesian inference* deserve attention. See, for example, Good (1950, 1965). At a more fundamental level are his multifarious contributions to 'the application of mathematics in the philosophy of science'. His quite individual style, and complex argument, make interesting reading. He is much criticized, and may be much misunderstood. As an *entrée* to Good's work in this area, and related references, see his paper (Good[42]) 'The probabilistic explication of information, evidence, surprise, causality, explanation and utility' which contains as an appendix what he calls '27 priggish principles of rationality'.

The 200 pages or so of highly formal, symbolic, meta-linguistic development in the second half of Kyburg (1974) illustrates the lengths to which some authors feel that they must go to discuss in a 'rich enough' language the logical foundations of inference.

Needless to say, the 'mathematics of philosophy' is by no means restricted to the study of non-deterministic structures: in their representation through the *probability* concept and allied factors, or their processing by means of *statistical* ideas. At its heart is the very form and function of *mathematics* in terms of its logical structure and any facility it provides for understanding or representing the world we live in, in all its manifestations. This is a fascinating area of study, for the mathematician, statistician and philosopher, alike.

References

1. Fisher, R. A. (1930). 'Inverse probablity', *Proc. Camb. Phil. Soc.*, **26**, 528–535.
2. Fisher, R. A. (1933). 'The concepts of inverse probability and fiducial probability referring to unknown parameters', *Proc. Roy. Soc. A*, **139**, 343–348.
3. Fisher, R. A. (1935). 'The fiducial argument in statistical inference', *Ann. Eugenics*, **VI**, 391–398.
† 4. Fraser, D. A. S. (1964). 'On the definition of fiducial probability', *Bull. Int. Statist. Inst.*, **40**, 842–856.
† 5. Plackett, R. L. (1966). 'Current trends in statistical inference', *J. Roy. Statist. Soc. A*, **129**, 249–267.
† 6. Edwards, A. W. F. (1976). 'Fiducial probability', *The Statistician*, **25**, 15–35.
† 7. Pedersen, J. G. (1978). 'Fiducial inference', *Int. Statist. Rev.*, **46**, 147–170.
8. Wilkinson, G. N. (1977). 'On resolving the controversy in statistical inference' (with Discussion), *J. Roy. Statist. Soc. B*, **39**, 119–171.
† 9. Lindley, D. V. (1958). 'Fiducial distributions and Bayes' Theorem', *J. Roy. Statist. Soc. B*, **20**, 102–107.
10. Dempster, A. P. (1963). 'Further examples of inconsistencies in the fiducial argument', *Ann. Math. Statist.*, **34**, 884–891.
11. Godambe, V. P., and Thompson, M. E. (1971). 'Bayes, fiducial and frequency aspects of statistical inference in regression analysis in survey sampling' (with Discussion), *J. Roy. Statist. Soc. B*, **33**, 361–390.
† 12. Birnbaum, A. (1962). 'On the foundations of statistical inference' (with Discussion), *J. Amer. Statist. Assn.*, **57**, 269–326.
† 13. Cox, D. R. (1958). 'Some problems connected with statistical inference', *Ann. Math. Statist.*, **29**, 357–372.
14. Savage, L. J. (1964). 'Discussion of session on fiducial probability'. *Bull. Int. Statist. Inst.*, **40**, 925–927.
† 15. Good, I. J. (1971). In reply to comments on his paper 'The probabilistic explication of information, evidence, surprise, causality, explanation and utility', in Godambe and Sprott (1971).
† 16. Barnard, G. A. (1964). 'Logical aspects of the fiducial argument', *Bull. Int. Statist. Inst.*, **40**, 870–883.
† 17. Barnard, G. A. (1949). 'Statistical inference' (with Discussion), *J. Roy. Statist. Soc. B*, **11**, 115–149.
18. Barnard, G. A., Jenkins, G. M. and Winsten, C. B. (1962). 'Likelihood inference and time series' (with Discussion), *J. Roy. Statist. Soc. A*, **125**, 351–352.
19. Birnbaum, A. (1968). 'Likelihood', in *International Encyclopedia of the Social Sciences*, Vol. 9. New York: Macmillan and The Free Press.

20. Sprott, D. A. and Kalbfleisch, J. D. (1969). 'Examples of likelihoods and comparisons with point estimates and large sample approximations', *J. Amer. Statist. Assn.*, **64**, 468–484.
21. Sprott, D. A. (1973). 'Normal likelihoods and their relation to large sample theory of estimation', *Biometrika*, **60**, 457–465.
22. Kalbfleisch, J. D. and Sprott, D. A. (1970). 'Application of likelihood methods to problems involving large numbers of parameters' (with Discussion), *J. Roy. Statist. Soc. B*, **32**, 175–209.
23. Kalbfleish, J. D. (1971). 'Likelihood methods of prediction', in Godambe and Sprott (1971).
24. Fisher, R. A. (1934). 'Two new properties of mathematical likelihood', *Proc. Roy. Soc. A*, **144**, 285–307.
25. Barnard, G. A. (1967). 'The Bayesian controversy in statistical inference', *J. Inst. Actuaries*, **93**, 229–269. (Paper presented to the Institute of Actuaries on 27 February 1967.)
26. Durbin, J. (1970). 'On Birnbaum's theorem on the relation between sufficiency, conditionality and likelihood', *J. Amer. Statist. Assn.*, **65**, 395–398.
27. Birnbaum, A. (1970). 'On Durbin's modified principle of conditionality', *J. Amer. Statist. Assn.*, **65**, 402–403.
† 28. Giere, R. N. (1977). 'Allan Birnbaum's conception of statistical evidence', *Synthese*, **36**, 5–13.
† 29. Birnbaum, A. (1977). 'The Neyman–Pearson theory as decision theory, and as inference theory; with a criticism of the Lindley–Savage argument for Bayesian theory', *Synthese*, **36**, 19–49.
† 30. Barndorff-Nielsen, O. (1976). 'Plausibility inference' (with Discussion), *J. Roy. Statist. Soc. B*, **38**, 103–131.
31. Fraser, D. A. S. (1961). 'The fiducial method and invariance', *Biometrika*, **48**, 261–280.
32. Fraser, D. A. S. (1966). 'Structural probability and a generalization', *Biometrika*, **53**, 1–9.
† 33. Fraser, D. A. S. (1968). 'A black box or a comprehensive model', *Technometrics*, **10**, 219–229.
† 34. Lindley, D. V. (1969). Review of Fraser (1968), *Biometrika*, **56**, 453–456.
† 35. Fraser, D. A. S. (1971). 'Events, information processing, and the structured model' in Godambe and Sprott (1971).
36. Stone, M. (1969). Review of Fraser (1968). *J. Roy. Statist. Soc. A*, **132**, 447–449.
† 37. Barnard, G. A. (1980). 'Pivotal inference and the Bayesian controversy', in Bernardo, De Groot, Lindley and Smith (1980).
38. Shannon, C. E. (1948). 'A mathematical theory of communication', *Bell System Tech. J.*, **27**, 379–423 and 623–656.
39. Good, I. J. (1966). 'A derivation of the probabilistic explication of information, *J. Roy. Statist. Soc. B*, **28**, 578–581.
† 40. Lindley, D. V. (1956). 'On a measure of the information provided by an experiment', *Ann. Math. Statist.*, **27**, 986–1005.
41. Bross, I. D. J. (1971). Comment on Good[42].
† 42. Good, I. J. (1971). 'The probabilistic explication of information, evidence, surprise, causality, explanation and utility', in Godambe and Sprott (1971).

CHAPTER 9

Perspective

It seems appropriate in these concluding pages to attempt briefly to place in perspective what we have seen of the range and variety of attitudes to statistical enquiry: the controversial and constructive arguments they generate and their implications for the practical studies of situations involving uncertainty. We can do this by posing the following question: *What purpose is served by a comparative discussion of alternative approaches to inference and decision-making*?

Our response needs to distinguish between different levels of interest and activity.

(i) *Basic Principles*. In essence the aim of all approaches is common; to harness what we know about some situation involving elements of uncertainty. But having admitted this as a common aim it is apparent that there is wide scope in a choice of appropriate means of pursuing it. In the first place we have seen how attitudes may differ, from one situation to another or one individual to another, on what constitutes legitimate raw material in a statistical enquiry—sample data, prior information, consequential gains and losses all make a claim for consideration. Then again, we need to declare the function of the enquiry—inferential or decision-making. Finally, some attitude must be adopted to what is an appropriate means of representing uncertainty—the philosophical and inter- pretative nature of the probability concept. Interactions are inevitable in these considerations.

Philosophical and practical arguments constrain our choice. In the pursuit of 'objectivity' we may reject certain types of 'raw material' (prior information, or consequential costs), and reject all but the frequency view of probability. On grounds of the 'intangibility' of the raw material in certain circumstances we may rule out the decision-making function of a statistical enquiry, since this cannot constitute a 'universally viable' aim. At the opposite extreme, philosophical views on what constitutes an appropriate model for 'rational behaviour' in the face of uncertainty may compel the decision-making function: render 'inevitable' the use of prior information and consequential costs within a subjective expression of probabilities.

We have seen how the different approaches have developed in response to such considerations. Much has been achieved if we can now appreciate how the different approaches have been thus constrained, particularly if we can appreciate the arbitrary (and personal) nature of the different stimuli. This should surely place us in a better position to stand aside, consider the different views on their merits, understand the meaning of the factional divisions and labels and follow the general prescription offered by Plackett (§1.6) of 'patiently' exploring 'differences of opinion' in the spirit that whatever our 'assumptions' we might

. . . reach essentially the same conclusions on given evidence, if possible.

(ii) *Methodology*. Alternative basic principles must lead to the development of different internal concepts. How we choose to represent formally (or mathematically) the import of the raw material, how we express the practical aims, and what criteria we use to assess the extent to which we achieve these aims, will all be correspondingly affected. This aspect has been considered in some detail. We have traced the internal machinery of the different approaches, and have compared the approaches at this level. Implied distinctions between inference and decision-making techniques, and between the assessment of initial, and final, precision, have been explored. Methods of estimation, hypothesis testing, and the construction of rules for action have been developed. Assessment of the performance of such methods has been discussed in terms of criteria prescribed in the different approaches. The extent to which such concepts as likelihood, sufficiency, admissibility, bias (and so on) have universal (or merely local) relevance has been considered. This was seen to depend on (or to constrain) the way in which we interpret the basic idea of probability, and how we view the function of a statistical investigation.

An understanding of such basic distinctions is also important in the day-to-day application of statistical methods. In applying a particular technique in a practical problem it is vital to understand the philosophical and conceptional attitudes from which it derives if we are to be able to interpret (and appreciate the limitations of) any conclusions we draw. In reverse, if the practical problem presents us with certain types of raw material, or seems to demand a particular emphasis in the expression of probabilities or in the function of the conclusions we wish to draw, we should be in a better position to choose an appropriate methodological framework within which to act.

(iii) *Statistical Practice*. When all is said and done, we must not lose sight of the fact that the ultimate (and sole) aim of statistics is to provide the wherewithal for drawing valid conclusions in the practical problems of the world we live in. As the discussion in the earlier chapters has emphasized, any amount of philosophical argument, or mathematical ingenuity, is of little value (*qua* statistics) if it does not serve this end.

It is possible to understand the disquiet that statisticians feel at times for the expression of the subject in some books and professional journals. Often the treatment of a topic gives the appearance of having prime concern for the mathematical elegance of its development, and correspondingly little concern for the reality of the model or interest in the detailed application of the results to real problems. Even when problems are motivated by real-life interest and the treatment extended to the derivation of explicit formulae or computational algorithms, it frequently turns out that the model either misrepresents the real situation (perhaps through ignorance of the fine detail of the practical discipline from which it derives) or else so oversimplifies the real situation (to render the solution tractable) that it has little relevance or applicability. This represents a real dilemma in professional statistical practice—the conflicts between the genuine virtue in obtaining an armoury of *explicit* results for *potential* use, the vast complexity of factors and interrelationships in *real-life* problems, and the dual demands on the statistician to be both proficient in his own highly sophisticated area and knowledgeable (or at least conversant) in a variety of applied disciplines. Is it reasonable to expect the statistician to be philosopher, logician, mathematician, calculator, agriculturalist, manager and zoologist?

Various attempts have been made to describe the role of the statistician. See, for example, Barnett[1]. One solution to the ubiquitous demands on the statistician is to encourage multidisciplinary team-work—a principle becoming more and more commonly adopted in industrial (operational research), medical and agricultural applications. See Healy[2]. Wider use of this principle is discussed by Sprent[3] in relation to 'statistical consultancy'. Individual needs of particular disciplines have also been considered; for example, Rosenbaum[4] (social sciences), Benjamin[5] (management), Skellam[6] (biology), Moser[7] (Government Statistical Service). Other attitudes to the role of the statistician are advanced by Welch[8] ('a vocational or a cultural study?'), Kendall[9] (the 'future of Statistics'), Bartholomew[10] ('need for . . . exchange of experience') and contributors to Watts (1968).

Without a doubt the enormous expansion in the power and availability of computers has had great influence on statistical practice. New scope in calculations has been opened up. Statistical packages have helped (if judiciously employed) both the statistician and the non-statistical applied researcher to take advantage of detailed analyses with a minimum of organizational effort. But the computer has also redirected to some extent the basic emphasis in statistical methodology. That we need no longer shy away from large-scale problems due to their computational intractability has led to a new wave of statistical activity—what some distinguish by the term 'data analysis'. (See Tukey[11], Tukey (1977).) This re-emphasis, described by Yates[12] as the 'second revolution in Statistics' is bearing fruit in some exciting developments in the fields of (for example) multivariate analysis and classification; such as in cluster analysis and multidimensional scaling. See for example, Barnett (1981). A distinguishing feature in this work is its playing down of the modelling or inferential aspects of statistics, and prime concern with the desire to express economically and usefully the *descriptive* features of large sets of data (often drawn from areas of application new to statistical analysis). At this stage, the philosophical and formal methodological aspects of statistics have little influence; techniques have a largely *ad hoc*, intuitive, basis. With increasing confidence and expertise in large-scale data handling it seems reasonable to believe that such issues will reassert themselves. It is likely that when they do so the basic comparative aspects of statistical theory will be seen to be equally, if not even more, important than they are in the traditional areas of statistical practice. We have already begun to recognize how anomalies and inconsistencies can become aggravated in multidimensional situations. (See Dawid, Stone and Zidek[13] for example.)

The new wave of descriptive, data-analytic, methods has (for reasons just outlined) had no natural place in our study of comparative inference. But we must not minimize the present, and ever-growing, importance of such an attitude to statistical analysis. The principle of taking a long hard look at the data and describing succinctly but forcefully just what they have to say can be regarded almost as a distinct further general approach to statistical inference.

Another marked trend which must deserve mention in these closing pages is the increasing interest in statistical methods where *sample data* are allowed to play an *interventionalist role* in constructing models, determining tests or estimators, protecting against anomolous eventualities, etc. This body of work includes **cross-validation**, which in its simplest form consists of dividing a set of data into two parts, using one to formulate a model or construct an estimator and the other to assess the validity of the model or properties of the estimator (Stone[14]). Then there is the development of so-called **adaptive procedures** for inference where not only the value taken by an estimator or test statistic, *but its very form*, is allowed to depend on the set of data to hand (Hogg[15]). Concern for actual outcome rather than data-generating mechanism (model) is reflected in the interest in methods of **prediction**. [Bayesian prediction was discussed in §6.3; see also Cox and Hinkley (1974, Chapter 7).] Another aspect of a redirection of emphasis from model to data is found in the recent interest in **robustness**: the construction of statistical techniques which when employed on a data set will lead to valid conclusions (and perhaps not widely differing precision) over quite a wide range of possible models for the generation of the data. This is illustrated by the use of order statistics in estimating a location parameter; for example, by the *median*, or *tri-mean*, or a *trimmed* or *Winsorized mean* (See Huber[16]; Andrews, *et al*, 1972.) Study of **outliers** also involves allowing an intervensionalist role for the data, both in development of method and analysis of data (Barnett and Lewis, 1978). In most situations the methods just described are employed in a *classical* framework—processing just sample data using a frequency-interpretable probability concept. But there is a crucial shift of emphasis. Rather than developing and justifying a method and *then* applying it to a set of data; the processes of development, justification and application are not clearly separated in time sequence and the very data we wish to analyse may play some role in conducting their own cross-examination.

Finally, let us return from the statistical method to the statistical practitioner. Whatever the role of the statistician or the nature of his craft, *he must be trained.* Continuing and increasing interest is being directed to needs in this area, including consultancy, the influences of 'data analysis' and the effects of more powerful and more accessible computing aids. (See references 17–24; Volume 25(2) of *The Statistician,* and Barnett, 1981).

What seems most important is that we make every effort to maintain a proper balance of emphasis in teaching statistics. On the one hand the vast strength of computers for digesting masses of data is something that the student statistician must be taught to understand and exploit. On the other hand he needs to recognize the computer for what it is—a sophisticated tool, not a substitute for thought. We need to develop an intuitive appreciation, a 'feel' for the import of data. It is debatable to what extent this 'green fingers' aspect of the subject can be formally taught. It is unlikely to come from a total preoccupation with computer-based analyses—wide experience of real-data handling (however inefficient in time and effort this might appear) seems to continue to hold out the best hope. But apart from questions of instruction in data-processing and data-handling, one thing seems incontrovertible. The teaching of statistics must continue to place major emphasis on basic principles and concepts, and on their implications in the form of practical statistical techniques. Exposure to the *range* of philosophical and conceptional attitudes to statistical theory and practice must be an essential ingredient.

References

† 1. Barnett, V. D. (1976). 'The statistician: Jack of all trades, master of one?', *The Statistician,* **25,** 261–279.

2. Healy, M. J. R. (1973) 'The varieties of statistician', *J. Roy. Statist. Soc. A,* **136,** 71–74.

† 3. Sprent, P. (1970). 'Some problems with statistical consultancy' (with Discussion), *J. Roy. Statist. Soc. A,* **133,** 139–164.

4. Rosenbaum, S. (1971) 'A report on the use of statistics in Social Science research' (with Discussion), *J. Roy. Statist. Soc. A,* **134,** 534–610.

5. Benjamin, B. (1971). 'The statistician and the manager', *J. Roy. Statist. Soc. A,* **134,** 1–13.

6. Skellam, G. J. (1964). 'Models, inference and strategy', *Biometrics,* **25,** 457–475.

7. Moser, C. A. (1973). 'Staffing in the Government Statistical Service', *J. Roy. Statist. Soc. A,* **136,** 75–88.

† 8. Welch, B. L. (1970). 'Statistics—a vocational or a cultural study?' (with Discussion), *J. Roy. Statist. Soc. A,* **133,** 531–554.

† 9. Kendall, M. G. (1968). 'On the future of Statistics—a second look' (with Discussion), *J. Roy. Statist. Soc. A,* **131,** 182–204.

10. Bartholomew, D. J. (1973). 'Post-experience training for statisticians', *J. Roy. Statist. Soc. A,* **136,** 65–70.

† 11. Tukey, J. W. (1962). 'The future of data analysis', *Ann. Math. Statist.,* **33,** 1–67.

† 12. Yates, F. (1966). 'Computers: the second revolution in statistics', *Biometrics*, **22**, 233–251.

† 13. Dawid, A. P., Stone, M. and Zidek, J. (1973). 'Marginalization paradoxes in Bayesian and structural inference' (with Discussion), *J. Roy. Statist. Soc. B*, **35**, 000–000.

14. Stone, M. (1978). 'Cross-validation: a review', *Math. Operationsforsch. Statistik*, **9**, 127–140.

15. Hogg, R. V. (1974). 'Adaptive robust procedures: a partial review and some suggestions for future applications and theory' (with comments), *J. Amer. Statist. Assn*, **69**, 909–927.

16. Huber, P. J. (1972). 'Robust statistics: a review' (The 1972 Wald Lecture), *Ann. Math. Statist.*, **43**, 1041–1067.

† 17. Yates, F. and Healy, M. J. R. (1964). 'How should we reform the teaching of statistics?', *J. Roy. Statist. Soc. A*, **127**, 199–210.

18. Bishop, H. E. (1964). 'The training of government statisticians', *J. Roy. Statist. Soc. A*, **127**, 211–215.

19. Deane, Marjorie (1964). 'The training of statisticians for economics and business', *J. Roy. Statist. Soc. A*, **127**, 216–218.

20. Spicer, C. C. (1964). 'The training of medical statisticians', *J. Roy. Statist. Soc. A*, **127**, 219–221.

21. Cox, C. P. (1968). 'Some observations on the teaching of statistical consultancy', *Biometrics*, **24**, 789–802.

(References 17–21, with Discussion, constituted a Symposium on the Teaching of Statistics.)

22. Finney, D. J. (1968). 'Teaching biometry in the University', *Biometrics*, **24**, 1–12.

23. Evans, D. A. (1973). 'The influence of computers on the teaching of statistics', *J. Roy. Statist. Soc. A*, **136**, 153–190.

24. Mead, R. and Stern, R. D. (1973). 'The use of the computer in the teaching of statistics', *J. Roy. Statist. Soc. A*, **136**, 191–205.

(References 23 and 24, introduced a Discussion on Computers in the Teaching of Statistics.)

Bibliography

Andrews, D. F., Bickel, P. J., Hampel, F. R., Huber, P. J., Rogers, W. H. and Tukey, J. W. (1972). *Robust Estimates of Location: Survey and Advances*. Princeton, N.J.: Princeton University Press.

Aitchison, J. (1970). *Choice against Chance: An Introduction to Statistical Decision Theory*. Reading, Mass.: Addison-Wesley.

Aitchison, J. and Dunsmore, I. R. (1975). *Statistical Prediction Analysis*. Cambridge: C.U.P.

Barlow, R. E. and Proschan, F. (1965). *Mathematical Theory of Reliability*. New York: Wiley.

Barndorff-Nielsen, O. (1978). *Information and Exponential Families in Statistical Theory*. Chichester: Wiley.

Barnett, V. (ed.) (1981). *Interpreting Multivariate Data. Proceedings of the Conference Entitled 'Looking at Multivariate Data'*. Chichester: Wiley.

Barnett, V. and Lewis, T. (1978). *Outliers in Statistical Data*. Chichester: Wiley.

Bartlett, M. S. (1962). *Essays on Probability and Statistics*. London: Methuen.

Bellman, R. E. (1957). *Dynamic Programming*. Princeton: Princeton University Press.

Bennett, J. H. (ed.) (1971–74). *Collected Papers of R. A. Fisher*, Vols 1–5. Adelaide: The University of Adelaide Press.

Berger, J. O. (1980). *Statistical Decision Theory, Foundations, Concepts and Methods*. New York: Springer-Verlag.

Bernardo, J. M., De Groot, M. H., Lindley, D. V. and Smith, A. F. M. (eds) (1980). *Bayesian Statistics. Proceedings of the First International Meeting, Valencia (Spain) 1979*. Valencia: Valencia University Press.

Bernoulli, J. (1713). *Ars Conjectandi*. Basel.

Borel, E. (1950). *Elements of the Theory of Probability*. (English translation by Freund of French edition, 1950.) Englewood Cliffs, N. J.: Prentice-Hall.

Box, G. E. P. and Tiao, G. C. (1973). *Bayesian Inference in Statistical Analysis*. Reading, Mass.: Addison-Wesley.

Box, J. F. (1978). *R. A. Fisher. The Life of a Scientist*. New York: Wiley.

Carnap, R. (1962). *Logical Foundations of Probability*, 2nd edn. Chicago: University of Chicago Press.

Carnap, R. and Jeffrey, R. C. (1971). *Studies in Inductive Logic and Probability*, Vol. 1. Los Angeles: University of California Press.

Chernoff, H. and Moses, L. E. (1959). *Elementary Decision Theory*. New York: Wiley.

Cohen, J. (1960). *Chance, Skill and Luck*. Harmondsworth: Penguin Books.

Cox, D. R. and Hinkley, D. V. (1974). *Theoretical Statistics*. London: Chapman and Hall.

Cramér, H. (1946). *Mathematical Methods of Statistics*. Princeton: Princeton University Press.

311

David, F. N. (1962). *Games, Gods and Gambling: A History of Probability and Statistical Ideas*. London: Griffin.

David, H. A. (1981). *Order Statistics*, 2nd edn. New York: Wiley.

Davidson, D., Suppes, P. and Siegel, S. (1957). *Decision Making: An Experimental Approach*. Stanford, California: Stanford University Press.

De Groot, M. H. (1970). *Optimal Statistical Decisions*. New York: McGraw-Hill.

De Morgan, A. (1847). *Formal Logic*. London: Taylor and Walton.

Dubins, L. E. and Savage, L. J. (1965). *How to Gamble if You Must. Inequalities for Stochastic Processes*. New York: McGraw-Hill.

Edwards, A. W. F. (1972). *Likelihood*. Cambridge: C.U.P.

Ferguson, T. S. (1967). *Mathematical Statistics: A Decision Theoretic Approach*. New York: Academic Press.

Fine, T. L. (1973). *Theories of Probability – An Examination of Foundations*. New York: Academic Press.

de Finetti, B. (1974, 1975). *Theory of Probability*, Vols 1 and 2. Chichester: Wiley.

Fishburn, P. C. (1970). *Utility Theory for Decision Making*. New York: Wiley.

Fisher, R. A. (1950). *Contributions to Mathematical Statistics*. New York: Wiley.

Fisher, R. A. (1959). *Statistical Methods and Scientific Inference*, 2nd edn. Edinburgh: Oliver and Boyd.

Fisher, R. A. (1966). *The Design of Experiments*, 8th edn. Edinburgh: Oliver and Boyd.

Fisher, R. A. (1970). *Statistical Methods for Research Workers*, 14th edn. Edinburgh: Oliver and Boyd.

Fraser, D. A. S. (1968). *The Structure of Inference*. New York: Wiley.

Fraser, D. A. S. (1976). *Probability and Statistics: Theory and Applications*. Massachusetts: Duxberry Press.

Gnedenko, B. V. (1968). *The Theory of Probability*. New York: Chelsea. (English translation of augmented 2nd edition of Russian text.)

Godambe, V. P. and Sprott, D. A. (eds) (1971). *Foundations of Statistical Inference*. Toronto: Holt, Rinehart and Winston of Canada.

Good, I. J. (1950). *Probability and the Weighing of Evidence*. London: Griffin.

Good, I. J. (1965). *The Estimation of Probabilities: An Essay on Modern Bayesian Methods*. Cambridge, Mass.: M.I.T. Press.

Gupta, S. S. and Moore, D. S. (eds) (1977). *Statistical Decision Theory and Related Topics II*. New York: Academic Press.

Gupta, S. S. and Yackel, J. (eds) (1971). *Statistical Decision Theory and Related Topics*. New York: Academic Press.

Guttmann, I. (1970). *Statistical Tolerance Regions: Classical and Bayesian*. London: Griffin.

Hacking, I. (1965). *Logic of Statistical Inference*. Cambridge: C.U.P.

Hoel, P. G. (1962). *Introduction to Mathematical Statistics*, 3rd edn. New York: Wiley.

Hogben, L. (1957). *Statistical Theory*. London: George Allen and Unwin.

Jeffreys, H. (1973). *Scientific Inference*, 3rd edn (1st edn 1931). Cambridge: C.U.P.

Jeffreys, H. (1961). *Theory of Probability*, 3rd edn. Oxford: Clarendon Press.

Kendall, M. G. and Plackett, R. L. (eds) (1977). *Studies in the History of Statistics and Probability*, Vol. 2. London: Griffin.

Kendall, M. G. and Stuart, A. (1977). *The Advanced Theory of Statistics*, Vol. I, 4th edn. London: Griffin.

Kendall, M. G. and Stuart, A. (1979). *The Advanced Theory of Statistics*, Vol. II, 4th edn. London: Griffin.

Kendall, M. G. and Stuart, A. (1976). *The Advanced Theory of Statistics*, Vol. III, 3rd edn. London, Griffin.

Keynes, J. M. (1921). *A Treatise on Probability*. London: Macmillan.

Khinchin, A. I. (1957). *Mathematical Foundations of Information Theory*. New York: Dover.

Kolmogorov, A. N. (1956). *Foundations of the Theory of Probability*. New York: Chelsea.

Kullback, S. (1959). *Information Theory and Statistics*. New York: Wiley.

Kyburg, Jr., H. E. and Smokler, H. E. (eds.) (1964). *Studies in Subjective Probability*. New York: Wiley.

Kyburg, H. E. (1974). *The Logical Foundations of Statistical Inference*. Dordrecht: Reidel.

Laplace, P. S. de (1951). *A Philosophical Essay on Probabilities* (an English translation by Truscott and Emory of 1820 edition). New York: Dover.

Lehmann, E. L. (1959). *Testing Statistical Hypotheses*. New York: Wiley.

Lindgren, B. W. (1971). *Elements of Decision Theory*. New York: Macmillan.

Lindley, D. V. (1965*a*). *Introduction to Probability and Statistics from a Bayesian Viewpoint*. Part 1, *Probability*. Cambridge: C.U.P.

Lindley, D. V. (1965*b*). *Introduction to Probability and Statistics from a Bayesian Viewpoint*. Part 2, *Inference*. Cambridge: C.U.P.

Lindley, D. V. (1971*a*). *Making Decisions*. London: Wiley.

Lindley, D. V. (1971*b*). *Bayesian Statistics Review*. Philadelphia: Society for Industrial and Applied Mathematics.

Luce, R. D. (1959). *Individual Choice Behaviour*. New York: Wiley.

Luce, R. D. and Suppes, P. (1965). *Preference, Utility and Subjective Probability*, (pp. 249–410 of Vol. 3, *Handbook of Mathematical Psychology*; edited by Luce, Bush and Galanter). New York: Wiley.

Maĭstrov, L. E. (1974). *Probability Theory: A Historical Sketch*. New York: Academic Press.

Maritz, J. S. (1970). *Empirical Bayes Methods*. London: Methuen.

von Mises, R. (1957). *Probability, Statistics and Truth*, 2nd English edn. London: George Allen & Unwin.

von Mises, R. (1964). *Mathematical Theory of Probability and Statistics*. New York: Academic Press.

Mood, A. M., Graybill, F. A. and Boes, D. C. (1973). *Introduction to the Theory of Statistics*, 3rd edn. Kogokusha: McGraw-Hill.

Mosteller, F. and Wallace, D. L. (1964). *Inference and Disputed Authorship: The Federalist*. Reading, Mass.: Addison-Wesley.

Neter, J., Wasserman, W. and Whitmore, G. A. (1978). *Applied Statistics*. Boston: Allyn and Bacon.

von Neumann, J. and Morgenstern, O. (1953). *Theory of Games and Economic Behaviour*, 3rd edn. Princeton: Princeton University Press.

Neyman, J. (1952). *Lectures and Conferences on Mathematical Statistics and Probability*, 2nd edn. Washington: U.S. Department of Agriculture.

Neyman, J. (1967). *A Selection of Early Statistical Papers*. Berkeley: University of California Press.

Neyman, J. and Pearson, E. S. (1967). *Joint Statistical Papers*. Cambridge: C.U.P.

Pearson, E. S. (1966). *The Selected Papers of E. S. Pearson*. Cambridge: C.U.P.

Pearson, E. S. and Kendall, M. G. (eds) (1970). *Studies in the History of Statistics and Probability*. London: Griffin.

Raiffa, H. (1968). *Decision Analysis: Introductory Lectures on Choices under Uncertainty*. Reading: Mass.: Addison-Wesley.

Raiffa, H. and Schlaifer, R. (1961). *Applied Statistical Decision Theory*. Boston: Division of Research, Graduate School of Business Administration, Harvard University.

Reichenbach, H. (1949). *The Theory of Probability*. Berkeley: University of California Press.

Rényi, A. (1972). *Letters on Probability*. Detroit: Wayne State University Press.

Salmon, W. C. (1966). *The Foundations of Scientific Inference*. Pittsburgh: University of Pittsburgh Press.

Savage, L. J. (1954). *The Foundations of Statistics*. New York: Wiley.

Savage, L. J., *et al.* (1962). *The Foundations of Statistical Inference*. London: Methuen.

Schlaifer, R. (1959). *Probability and Statistics for Business Decisions*. New York: McGraw-Hill.

Seidenfeld, T. (1979). *Philosophical Problems of Statistical Inference*. Dordrecht: Reidel.

Todhunter, I. (1949). *A History of the Mathematical Theory of Probability*. New York: Chelsea. (First published 1865.)

Venn, J. (1962). *The Logic of Chance*. New York: Chelsea. (Reprint of 1888 edition published by Macmillan, London.)

Wald, A. (1950). *Statistical Decision Functions*. New York: Wiley.

Watts, D. G. (ed.) (1968). *The Future of Statistics*. New York: Academic Press.

Wetherill, G. B. (1969). *Sampling Inspection and Quality Control*. London: Methuen.

Wiener, N. (1948). *Cybernetics*. New York: Wiley.

Wilks, S. S. (1962). *Mathematical Statistics.* New York: Wiley.

Winkler, R. L. (1972). *Introduction to Bayesian Inference and Decision*. New York: Holt, Rinehart and Winston.

Zachs, S. (1971). *The Theory of Statistical Inference*. New York: Wiley.

Zellner, A. (1971). *An Introduction to Bayesian Inference in Econometrics*. New York: Wiley.

Subject and Author Index